T0305574

# Statistics for
# Compensation

# Statistics for Compensation

## A Practical Guide to Compensation Analysis

JOHN H. DAVIS

WILEY

A JOHN WILEY & SONS, INC., PUBLICATION

Published by John Wiley & Sons, Inc., Hoboken, New Jersey
Published simultaneously in Canada

For general information on our other products and services or for technical support, please contact our Customer Care Department within the United States at (800) 762-2974, outside the United States at (317) 572-3993 or fax (317) 572-4002.

Wiley also publishes its books in a variety of electronic formats. Some content that appears in print may not be available in electronic formats. For more information about Wiley products, visit our web site at www.wiley.com.

*Library of Congress Cataloging-in-Publication Data:*

Davis, John H., Ph.D.
Statistics for compensation : a practical guide to compensation analysis / John H. Davis.
        p. cm.
Includes bibliographical references and index.
ISBN 978-0-470-94334-2 (cloth)
1. Compensation management–Statistical methods. 2. Executives–Salaries, etc.
I. Title.
HF5549.5.C67D38 2011
658.3'2015195–dc22

2010033577

10 9 8 7 6 5 4 3 2

# Contents

# Preface

The objective of this book is to provide statistical and analytical techniques and guidance that will help compensation professionals and human resources professionals responsible for compensation make better decisions as they go about their day-to-day analyses.

This book presents the most needed and most useful statistical techniques in compensation analysis that I have found through my experience as a practitioner, consultant, and teacher. The book is designed to be a thorough reference for the practitioner. It may also serve as a textbook for a college-level course for those specializing in compensation or human resources. In all the chapters following the Introduction, there are practice problems to help reinforce the material presented. Answers to them are given at the end of the book.

The reader will gain the most benefit by working the examples given in the case studies and the practice problems. The data for the examples and practice problems are all contained in the book, and should be entered in a spreadsheet program or statistical package. In addition, more lengthy data sets are available for downloading from the Wiley website mentioned in the Introduction.

The content of this book has evolved over the years from analyzing internal and external compensation data for many clients, statistics courses taught in college classrooms, courses developed and taught as part of the WorldatWork (formerly, American Compensation Association) certification program, workshops given by Davis Consulting and later Davis & Neusch, in-house courses given to organizations, workshops given to various compensation associations, and in particular as part of the professional development program of the North Texas Compensation Association.

During that time, clients and students have provided examples to share and have given feedback on improving the usefulness and presentation of this material. To all of them I owe a great debt of gratitude. In particular, I want to thank my colleagues Sarah Hutchinson, Becky Wood, Janet Koechel, and Billie Day, who so graciously read the entire text of earlier versions of this book and gave me honest and most valuable feedback on improving it. I could not have done it without them.

I sincerely hope that the reader will find this book highly useful and informative. As any teacher or consultant will tell you, our greatest pleasure comes from knowing that the material presented has been grasped by the student or client and that it has been found to be valuable in helping to make sound decisions.

JOHN H. DAVIS

# CHAPTER 1

# Introduction

The purpose of this book is to provide statistical tools and guides to compensation professionals and human resources professionals responsible for compensation to enable them to conduct sound statistical analyses, focusing on the descriptive statistics that are most used and needed in compensation. This in turn will help their organizations make sound decisions to better attract, retain, motivate, and align the kinds and numbers of people the organizations need.

Compensation is the branch of human resources dealing with the elements of pay provided by an employer to its employees for work performed. Elements of pay include base pay, variable pay, and stock. Human resources is the function of an organization dealing with the management of people employed by the organization.

In a broad sense, compensation helps decide how much jobs are worth and how to pay employees fairly for the work they do. Compensation professionals get involved with analyzing both internal and external pay and organizational data, developing salary structures (the range of pay for jobs), recommending salary increase budgets, creating guidelines for individual salary increases, designing incentive programs, and developing performance management systems. They do all this in the context of an organization's mission, operational and financial considerations, and compensation philosophy, the latter of which they may have helped developed, and integrate all the compensation programs with the other branches of human resources.

The thrust of most professional jobs involves a great deal of decision making. Indeed, on a fundamental basis, making decisions *is* the job of a compensation (and human resources) professional—decisions that will help the organization

*Statistics for Compensation: A Practical Guide to Compensation Analysis,* By John H. Davis
Copyright © 2011 John Wiley & Sons Inc.

achieve its goals and the employees achieve their job-related goals. In the analytical realm, decisions are made in deciding what questions to ask, what related issues are important contextually, what data to use, how to analyze the data, what to recommend, how to present it, and how to act upon the recommendations.

Many times we need to decide and act in the face of uncertainty, as facts are always limited. Furthermore, we often work under great time pressure. All this occurs in the context that the answer to most questions is, "It depends."

Most of the time we have to make a business case for a recommendation, and the executives to whom we are making a presentation tell us, "Prove it to us using data." Behind the scenes we have to perform a number of critical thinking steps.[1]

- Identify the question behind the question, and identify implications and impacts on business plans, budgets, employee engagement, legal/regulatory restrictions, and so on.
- Translate the question into analyses and quickly assess how the analyses are best conducted and presented.
- Get the right data from internal and external data sources, and ensure they are accurate and appropriate.
- Organize and conduct the analyses and draw conclusions, which is usually an iterative process. The initial analyses may raise more questions than provide answers.
- Identify the underlying assumptions (i.e., what the answer "depends" on) and implications of the conclusions.
- Prepare executive-ready analyses and conclusions.

Although this book is directed mainly toward compensation, we will sometimes illustrate techniques with other human resources issues, as compensation professionals, with their statistical and analytical skills, often get involved in broader human resources projects.

## 1.1 WHY DO STATISTICAL ANALYSIS?

A compensation professional encounters many issues, such as

- What is our market position?
- What should our salary increase budget be this year?
- What should the new accounting supervisor be paid?
- What should our strategy be in balancing pay and benefits?
- Do we have any pay inequities?

---

[1] I am indebted to Sarah Hutchinson for articulating this approach.

and might be involved in helping analyzing other human resources issues, such as

- How do we decide who gets released when we have a reduction in force?
- Can we justify the training program?
- How effective are our employee communications programs?
- Should our policies for time off for two plants be the same?
- How do the benefits costs of our company compare with those of our competitors?
- What is the company day care usage?

These examples illustrate three types of issues addressed.

- Specific issues, such as what should the new accounting supervisor be paid.
- Policy issues, such as what should be our pay policy with respect to competition.
- Broad strategy issues, such as how should we balance pay and benefits to attract, retain, and motivate the kinds and numbers of employees we need to achieve the company goals.

In addressing these issues, we use numbers as one of our starting points—all kinds of numbers that represent all kinds of things such as number of employees, average salary, benefits premiums, performance level, time to vesting, turnover rate, percent using day care facilities, training utilization, productivity, and so on.

But numbers alone will not help us. We have to do something with them to help lead to sound decisions.

## Example Analysis

Throughout this book, we will be using a fictitious manufacturing company, BPD, to illustrate the various techniques.

Suppose you are the compensation manager of BPD and the vice president of finance asks you to conduct a quick competitive analysis on accounting supervisors to help determine if a subordinate supervisor is paid competitively. The supervisor in question is paid $68,580.

You gather the data in Table 1.1 on what accounting supervisors are paid in the market.

Can you make sense out of these raw data? Would you present these as the final result of your analysis? Hopefully, the answer to both questions is "no." Data in its raw form are often of little use.

| TABLE 1.1 | RAW DATA MARKET PAY ACCOUNTING SUPERVISOR | | | |
|---|---|---|---|---|
| 70,200 | 59,800 | 57,200 | 75,200 | 59,800 |
| 55,000 | 60,400 | 61,600 | 60,200 | 63,000 |
| 59,800 | 65,000 | 66,200 | 59,800 | 65,200 |
| 76,800 | 65,200 | 61,400 | 59,400 | 56,400 |
| 56,400 | 53,800 | 63,600 | 63,000 | 62,400 |
| 55,200 | 68,400 | 66,600 | 60,200 | 59,000 |

So you decide to organize, analyze, and describe the data, and provide a tabular and graphical summary as shown in Table 1.2 and Figure 1.1.

The table lists various statistics that are a summary of the raw data. In Chapters 3, 4 and 5, we will define and calculate all the terms (and many more), and discuss how to interpret them, as well as how to construct the graph. The graph is a picture of the distribution of the data. On the horizontal axis are salary

| TABLE 1.2 | SUMMARY MARKET PAY ACCOUNTING SUPERVISOR |
|---|---|
| No. of Incumbents | 30 |
| Average | 62,207 |
| Low | 53,800 |
| P10 | 56,280 |
| P25 | 59,500 |
| P50 | 60,900 |
| P75 | 65,150 |
| P90 | 68,580 |
| High | 76,800 |
| Standard Deviation | 5,465 |
| CV | 8.8% |
| P90/P10 | 1.22 |

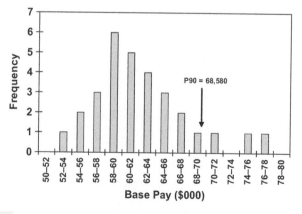

FIGURE 1.1    MARKET PAY ACCOUNTING SUPERVISOR

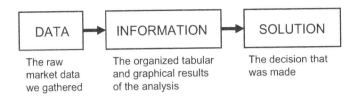

**FIGURE 1.2    DECISION MODEL**

categories in $2,000 buckets. The vertical axis is a scale indicating how many data points are in each category. For example, there are six salaries that are between $58,000 and $60,000.

Shown on the chart is the location of what the BPD accounting supervisor is paid. Her salary is $68,580, which is at the 90th percentile, meaning that 90% of the market salaries are below or equal to her salary. She is very well paid with respect to the market.

Now the vice president has the information needed to make a sound decision. The subordinate supervisor is paid at the 90th percentile. The vice president's decision is to leave the pay as is.

The decision model in Figure 1.2 describes the process we just completed.

We started with the statement of the problem, namely to determine if the accounting supervisor was being paid competitively. The pending decision was that if the pay level was too low, an adjustment would be made; otherwise the pay would remain unchanged.

The bottom line is that we gather, summarize, describe, analyze, interpret, and present information to help make better decisions—decisions on your part as to what to do or recommend and decisions on the organization's part as to what to do. In other words, we use statistics to help make better decisions.

> The purpose of this book is to give basic statistical and analytical tools to help you, as a compensation professional, and the organization make *smart* decisions.

## 1.2 STATISTICS

Just what is statistics? Following are three definitions.

Statistics text [1] . . .

> A body of methods for obtaining, organizing, summarizing, presenting, interpreting, analyzing, and acting upon numerical facts related to an activity of interest.

Another statistics text [11] . . .

> A branch of applied mathematics that constitutes the science of decision making in the face of uncertainty.

A book on the history of statistics [7] . . .

> A logic and methodology for the measurement of uncertainty and for an examination of the consequences of that uncertainty in the planning and interpretation of experimentation and observation.

Although statistics is all of these, we will use the following definition.

> Statistics is the branch of mathematics concerned with the measurement of uncertainty. Mathematics is the science of measurement.

Statistics is divided into two broad categories: descriptive statistics and inferential statistics. Descriptive statistics is a branch of statistics that describes variables and the strength and nature of relationships between them. Inferential statistics is a branch of statistics in which conclusions or generalizations are made about a population based on the results from a properly constructed sample from that population.

Most of the statistics used and needed in compensation are descriptive statistics. We seldom are taking random samples from a population. Most of the time we *have* the population, such as compensation, benefits, employment, and training data on our employees, or sales, expense, operational, marketing, and facilities data on our organization or industry group, and are simply trying to make sense out of the data.

Rarely do we get involved in true random sampling where we can apply inferential statistics, and then, it is important to have someone with statistical expertise as our partner so that both the sampling procedure and the analysis are appropriate.

Hence, as mentioned at the beginning of this chapter, the focus of this book is on the descriptive statistics that are most used and needed in compensation. The topic of inferential statistics is briefly discussed in the Appendix.

## 1.3 NUMBERS RAISE ISSUES

In statistical analyses, whether analyzing a salary survey, modeling employee engagement, identifying causes of turnover, or assessing managerial effectiveness, the numbers themselves often do not answer our questions. They just raise issues and challenge our assumptions and policies. Here are some examples from the author's experience.

- A market analysis indicates that all the competition has a base–bonus mix of 50–50 for its sales force. An analysis of the company's internal sales force shows a base–bonus mix of 80–20. Should the company structure its program to be similar to the competition? Is the company satisfied with the sales results it now gets? If it does change the mix, what are the implications on employee engagement? What do the employees think about the current mix? How will the company transition pay to be more heavily weighted on variable pay?

- A company thinks it needs to pay at the 75th percentile to attract employees with certain skills. An analysis of the market and of employee pay indicates that the cost would be prohibitive, given the current market position. This raises several issues. Does the company go ahead and spend the money for these employees, risking a short-term financial loss but with a hope of a long-term gain? Does the company forego hiring these employees and contract the work out? Does the company seek new technology or a different way of doing things to accomplish the same results?

- An analysis of turnover indicates that pay is not the issue. Rather there were indications that ineffective supervision and lack of training opportunities greatly influence employees' decisions to leave the organization. How should these two factors be dealt with?

- An analysis of benefits costs indicates that the company must "hold the line" on total costs. Further analysis shows that some costly programs are not very popular with the rank and file, but are with the executives. What should the company do?

- The company's policy for research scientists states that over time, there is no pay difference between those whose highest degree is a BS and those whose highest degree is an MS. An analysis of the actual practice shows that there is a definitely higher level of pay for MS employees than for the BS employees throughout their careers. How should the company resolve the difference between its policy and its practice?

- The company has certain benefits for its employees, such as time off for family illnesses and time off for community service. However, an analysis shows that those who use these benefits do not receive as many promotional opportunities as those who don't. There is an apparent cultural stigma against those who use them. What should the company do to resolve this apparent anomaly?

All these situations raise issues on priorities and address various strategies and policies. Ultimately, the organization seeks to determine the appropriate balance that will help attract, retain, motivate, and align the employees in accordance with both the short- and long-term business strategies.

Furthermore, in almost all cases, there isn't a single right answer. There is a myriad of solutions to various problems, and often subsequent statistical analyses can help identify implications of these solutions.

## 1.4 BEHIND EVERY DATA POINT, THERE IS A STORY

Every data point and every number represents some fact of reality along with the context as to how it came to be what it is. The number may indicate the number of employees in a division, and the number is low because the division just had a reduction in force. The number may indicate a large bonus that is large because the sales representative brought in a huge account. To discover what that fact is often takes some digging, but with enough effort, it can be done.

Figure 1.3 is a simple example of pay and experience for scientists in BPD. The figure is a scatter plot, or chart, that shows the relationship between pay and experience. The horizontal axis (x-axis) shows the scale for the years of experience. The vertical axis (y-axis) shows the scale for the pay. A dot on the chart represents the pay and experience for an individual scientist.

Notice that there seems to be a "nice" and positive relationship—in general, the more the experience, the higher the pay, except for one data point at 8 years of experience.

Before including this scatter plot, or chart, as part of a report or a presentation to management, you need to know "who is that point," because that will be the first question asked. You need to look behind the scenes to understand it. In this case, it might have been a former supervisor who is now an individual contributor and the pay was not cut when the demotion occurred. Or, it might be the president's nephew or niece. Whatever the story, there *is* a story.

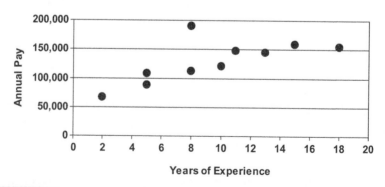

**FIGURE 1.3    PAY AND EXPERIENCE FOR SCIENTISTS**

## 1.5  AGGRESSIVE INQUISITIVENESS

In general, as in the above example, what should you do if you see something that seems out of line, or does not fit, or is an outlier, or does not meet your preconceived notions or expectations? The answer, though simple, has a bit of philosophy behind it. The answer is that you should try to explain the anomaly or the situation. You should look behind the scenes and try to discover why it is like it is.

The *reason* you should do this is the philosophical part. If you want to be successful in compensation, or for that matter, any field of endeavor (or life), you must be *aggressively inquisitive*. You must have an inner drive, a natural curiosity, and a persistent inclination to figure out why things are like they are. Why are they different? Why are they same? Why …? You must want to be able to understand things and explain things. The drive to do this is the human mind constantly asking, "What is it?" and "Why?" about all aspects of reality. The mind's main role is to discover the facts of reality. This aggressive inquisitiveness is a definite and distinct human mode of survival.

## 1.6  MODEL BUILDING FRAMEWORK

The approach used to analyze problems and find solutions is through the use of models. This topic is discussed in more detail in Chapter 6, *Model Building*, but it is important to know the highlights of the general framework as we begin.

When you are addressing an issue or solving a problem, it is instructive to draw a picture of a model, such as the general model in Figure 1.4.[2]

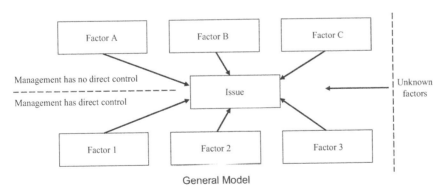

General Model

**FIGURE 1.4    GENERAL MODEL**

---

[2] I am indebted to Marc J. Wallace for this view of models.

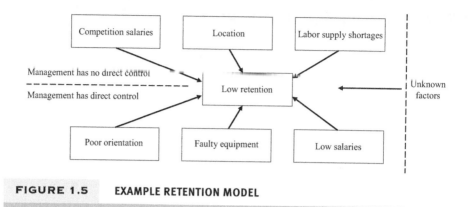

**FIGURE 1.5     EXAMPLE RETENTION MODEL**

The issue is the problem you are working on, the problem for which you are trying to find a solution. With subject matter experts, you identify factors that may influence the issue. Some of these are under direct control of management, and some are not. It is helpful to segregate them into these two categories. In addition, there are usually unknown factors that you are unable to identify, as we are not omniscient.

### Example Model

You have been assigned to address turnover in the BPD call center that is believed to be too high. With recruiting costs rising, the turnover is affecting the bottom line. This is an important business issue. You meet with the manager and a small focus group of call center employees to brainstorm factors that could influence turnover. Figure 1.5 shows the resulting model.

This model identifies fruitful areas to investigate. As is typical, a problem is not one-dimensional. There is work for compensation, training, and management, and these factors are all interrelated.

We will use this general approach throughout this book.

## 1.7  DATA SETS

Throughout this book, we will use examples that draw on data from Data Sets located on the website ftp://ftp.wiley.com/public/sci_tech_med/statistics_compensation. The reader should download the data into a spreadsheet program or a statistical software program and follow along. In those cases, the data are also presented in tables in this book. For smaller sets of data, the data appear only in this book.

## 1.8 PREREQUISITES

This book is written for compensation professionals or human resources professionals who have responsibility for compensation. A course in college algebra and/or statistics (e.g., business statistics) will help the reader understand and use the material presented. However, one does not need to be a mathematician or statistician to comprehend and use these methods.

## RELATED TOPICS IN THE APPENDIX

A.1   Value Exchange Theory
A.2   Factors Determining a Person's Pay

# CHAPTER 2

# Basic Notions

Many problems can be analyzed with the simple notion of a percent. We will describe a situation in which the use of percents was key in developing a solution.

---

**CASE STUDY 1, PART 1 OF 2**

BPD has a fleet of almost 500 trucks, and recruits, trains, and obtains certification for all the drivers. In the past, the hiring process was decentralized, with the six regions doing their own hiring and training. Management was concerned with the low retention rate, which was causing high recruiting costs. The task force that was addressing the issue developed the following model shown in Figure 2.1.

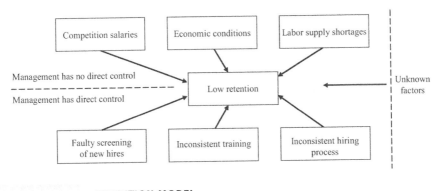

**FIGURE 2.1    RETENTION MODEL**

---

*Statistics for Compensation: A Practical Guide to Compensation Analysis,* By John H. Davis
Copyright © 2011 John Wiley & Sons Inc.

It was thought that centralizing the hiring process would help, so the entire process was centralized for a year, starting July 1, 2011. It is now 1 year later. You have been asked to answer the question "Is the driver hiring process better centralized or decentralized?"

The cost of the hiring process, which includes recruiting, training, and certification, is estimated to be an average of $6,500 per driver. You realize that this is an estimate but it is the best number you can come up with.

You also realize there may be other factors affecting turnover, such as changes in the business climate, the economy, the unemployment rate, the company's reputation, and changes in management during the past year.

You gather the data from each region for the year before centralization and for the first year of centralization. You decide to use data for 12-month periods to eliminate any seasonal fluctuations.

The raw data are shown in Table 2.1. For each region, the table shows how many drivers were hired during the 12-month period, what the turnover was, and how many were retained. Since you are concerned about the cost of hiring, the turnover includes both voluntary and involuntary terminations.

**TABLE 2.1**   **RAW DATA RETENTION**

| Region | Decentralized Hiring (7/1/10–6/30/11) | | | Centralized Hiring (7/1/11–6/30/12) | | |
|---|---|---|---|---|---|---|
| | Hired | Turnover | Retained | Hired | Turnover | Retained |
| Texas | 73 | 49 | 24 | 54 | 19 | 35 |
| California | 57 | 30 | 27 | 36 | 7 | 29 |
| Florida | 62 | 32 | 30 | 76 | 30 | 46 |
| Central | 70 | 50 | 20 | 86 | 42 | 44 |
| West | 39 | 18 | 21 | 65 | 27 | 38 |
| East | 241 | 153 | 88 | 272 | 79 | 193 |
| Total | 542 | 332 | 210 | 589 | 204 | 385 |

You decide that since the absolute numbers vary between decentralized hiring and centralized hiring, the most meaningful comparison would be to use percentages and examine the percent of drivers retained. This will allow a comparison on a relative basis.

Before we continue with the analysis, we will discuss in detail the notion of percent.

## 2.1 PERCENT

A percent, or percentage, is a representation of the amount of a particular quantity *relative* to a reference quantity, expressed as a proportion or ratio that is multiplied by 100. Often for an item, it is the proportion of the whole of which the item is a part, multiplied by 100. Here are two examples.

**Example 2.1**
In the BPD Texas region under decentralized hiring, 73 drivers were hired and only 24 were retained. The reference quantity is 73 drivers hired. The particular quantity is 24 drivers retained.

$$\text{Proportion of drivers retained} = 24/73 = 0.33 \text{ (rounded)}$$
$$\text{Percent of drivers retained} = (0.33)(100) = 33\%$$

**Example 2.2**
There are 380 employees at BPD's Denver factory location. The number using the day care facilities for their children is 35. The reference quantity is 380 employees. The particular quantity is 35 employees using day care.

$$\text{Proportion of employees using day care facilities} = 35/380 = 0.092 \text{ (rounded)}$$
$$\text{Percent of employees using day care facilities} = (0.092)(100) = 9.2\%$$

Sometimes the proportion is larger than the whole. Here are two examples.

**Example 2.3**
The sales target for a BPD sales representative was $300,000. The sales representative sold $390,000. The reference quantity is the target of $300,000. The particular quantity is $390,000 sales achieved.

$$\text{Proportion of sales target achieved} = 390,000/300,000 = 1.30$$
$$\text{Percent of sales target achieved} = (1.30)(100) = 130\%$$

**Example 2.4**
The BPD Denver office had a charitable campaign to collect 150 toys for Christmas. A total of 192 toys were collected. The reference quantity is 150 toys as the goal. The particular quantity is 192 toys collected.

$$\text{Proportion of goal collected} = 192/150 = 1.28$$
$$\text{Percent of goal collected} = (1.28)(100) = 128\%$$

Sometimes we want to go in the opposite direction and calculate the absolute number from a percent. To convert from a percent of a quantity to the absolute amount, first divide the percent by 100 to get a proportion, and then multiply the proportion by the total of which the quantity is a part. Here is an example.

**Example 2.5**
Thirty percent of the 300 employees in the BPD Ohio factory are exempt (i.e., exempt from legally mandated overtime payments) and we want to calculate

the number of exempts. The reference quantity is 300 employees. We want to calculate the particular quantity.

$$\text{Proportion of exempts} = 30/100 = 0.30$$
$$\text{Number of exempts} = (0.30)(300) = 90$$

Sometimes we are given the percentage and the particular quantity and want to calculate the reference quantity.

### Example 2.6

As part of a sales promotion from one of its suppliers, BPD received 15 all expense-paid trips worth $4,000 each. You want to combine them with a certain number of cash prizes, each worth $4,000, and all to be given away to employees in a company-wide random drawing to celebrate the 25th anniversary of the company. You want the number of trips to be 20% of the total number of prizes, and have to calculate that total.

The particular quantity is 15, which is 20% of the reference quantity. We want to calculate the reference quantity. We have to use some algebra here. Let $x =$ the reference quantity.

$$15 = 20\% \text{ of } x$$
$$15 = 20/100 \text{ of } x$$
$$15 = 0.20 \text{ of } x$$
$$15 = 0.20x$$

Dividing both sides by 0.20, we get

$$15/0.20 = x$$
$$75 = x$$

Total number of prizes $= 75$

With this information, you will need $75 - 15 = 60$ cash prizes of $4,000 each, for a total of $240,000. You investigate if this will fit in with the budget for the anniversary celebration.

## Graphical Displays of Percents

When the tabular presentation of percents is supplemented with a picture, you often get a better understanding of the analysis. There are several graphical methods for showing percents. Two common ones are pie charts and bar graphs.

**Pie Charts of Percents**   Pie charts are useful to show a total picture of something with a breakdown of the major components, which are shown as pieces of the pie, with the areas of the pieces indicating the relative amounts of the components. Figure 2.2 is an example for the BPD human resources budget.

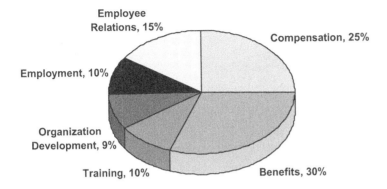

BPD Human Resources Budget
Percent for Each Department

Employee Relations, 15%

Compensation, 25%

Employment, 10%

Organization Development, 9%

Training, 10%

Benefits, 30%

Note: Percents do not add to 100% due to rounding.

**FIGURE 2.2** BPD HUMAN RESOURCES BUDGET—WHOLE NUMBERS

This shows the relative budget amounts for each department in human resources. Keep in mind that it does not show the actual amounts, but just the relative amounts.

**Rounding Issue** This also illustrates a major issue with respect to showing percents that should add up to 100%. These don't. They add up to 99%. How can that be? Table 2.2 shows the raw data.

Figure 2.2 shows the percents rounded to the nearest whole number, and they add up to 99. From Table 2.2, it can be seen that five departments had one to four tenths of a percent that were "lost" in the rounding, so the total was an entire percent less than 100.

If you had shown the chart with the percents rounded to the nearest tenth, then they would add up to 100, as shown in Figure 2.3.

**TABLE 2.2** BPD HUMAN RESOURCES BUDGET

| Department | Budget | % Rounded to Tenth | % Rounded to Whole No. |
|---|---|---|---|
| Compensation | 583,000 | 25.4 | 25 |
| Benefits | 695,000 | 30.3 | 30 |
| Training | 225,000 | 9.8 | 10 |
| Organization Development | 210,000 | 9.2 | 9 |
| Employment | 232,000 | 10.1 | 10 |
| Employee Relations | 349,000 | 15.2 | 15 |
| **Total** | **2,294,000** | **100.0** | **99** |

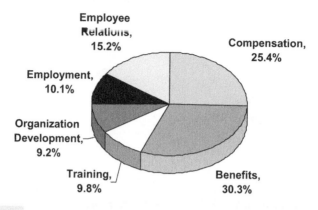

**FIGURE 2.3    BPD HUMAN RESOURCES BUDGET—NEAREST TENTH**

So, as a word of caution, when presenting pie charts or some other display with percents that should add up to 100%, make sure they do, or if they don't, understand why and note it in your presentation.

*Technical Note*: Some software will force the total to equal 100 when you tell it that you are showing percents, causing at least one component to show more or less than what is really there. You need to make sure this isn't happening.

**Bar Charts of Percents**    Figure 2.4 is another graphical display of percents that compares the budget for this year to that for next year, each as a percent of the total human resources budget.

As compensation increased from 25% of the budget this year to 30% of the budget next year, does this mean that there is more money in next year's budget for compensation? The answer is that we can't tell. There may be more, there

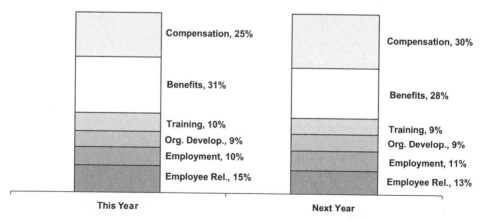

**FIGURE 2.4    BPD HUMAN RESOURCES BUDGET—YEARLY COMPARISON**

**TABLE 2.3    COMPENSATION DOLLARS—THREE POSSIBILITIES**

|  | This Year | Next Year |
|---|---|---|
| *Possibility 1: Compensation Has More* | | |
| HR Budget | 3,000,000 | 2,900,000 |
| % for compensation | 25% | 30% |
| $for compensation | 750,000 | 870,000 |
| *Possibility 2: Compensation Has Less* | | |
| HR Budget | 3,000,000 | 2,100,000 |
| % for compensation | 25% | 30% |
| $for compensation | 750,000 | 630,000 |
| *Possibility 3: Compensation Has the Same* | | |
| HR Budget | 3,000,000 | 2,500,000 |
| % for compensation | 25% | 30% |
| $for compensation | 750,000 | 750,000 |

may be less, or there may be the same dollars for next year. It all depends on the reference amount for each year, which is the total human resources budget. All that percents show is the *relative* amount for each year.

The three possibilities for compensation are shown in Table 2.3, each with a different decreased human resources budget for next year.

The first possibility shows compensation with more. The second possibility shows compensation with less. The third possibility shows compensation with the same.

Of course, if the total human resources budget for next year is equal to or greater than the budget for this year, then compensation will get more money.

**CASE STUDY 1, PART 2 OF 2**

Returning to our original problem, here are the results of your analysis in Table 2.4, using percentages.

**TABLE 2.4    RETENTION ANALYSIS PERCENTS**

| Region | Decentralized Hiring (7/1/10–6/30/11) | | | | Centralized Hiring (7/1/11–6/30/12) | | | |
|---|---|---|---|---|---|---|---|---|
|  | Hired | Turnover | Retained | % Retained | Hired | Turnover | Retained | % Retained |
| Texas | 73 | 49 | 24 | 33 | 54 | 19 | 35 | 65 |
| California | 57 | 30 | 27 | 47 | 36 | 7 | 29 | 81 |
| Florida | 62 | 32 | 30 | 48 | 76 | 30 | 46 | 61 |
| Central | 70 | 50 | 20 | 29 | 86 | 42 | 44 | 51 |
| West | 39 | 18 | 21 | 54 | 65 | 27 | 38 | 58 |
| East | 241 | 153 | 88 | 37 | 272 | 79 | 193 | 71 |
| **Total** | **542** | **332** | **210** | **39** | **589** | **204** | **385** | **65** |

For the Texas region, with decentralized hiring, 73 drivers were hired, and after a turnover of 49, 24 were retained. The percent retained was $(24/73)(100) - 33\%$. Similar calculations were done for the other entries. In each region, there was a higher percentage of drivers retained with centralized hiring than with decentralized hiring. Overall, the retention rate increased from 39% to 65%.

Even though there are other factors involved in retention, the dramatic increase in percentage retention for centralized hiring presents a convincing case to continue the centralization. To finalize the business case, you calculate the estimated savings in the hiring process by comparing the additional drivers retained using centralized hiring over what it would have been if the historical retention rates from decentralized hiring had held. This is shown in Table 2.5.

**TABLE 2.5     ESTIMATED SAVINGS FROM CENTRALIZED HIRING**

| Actual Hires (7/1/11–6/30/12) | | If Hiring Had Been Decentralized | | Actual Hiring Was Centralized | | Estimated Savings from Centralization | |
|---|---|---|---|---|---|---|---|
| Region | Hired | % Retained | No. Retained | % Retained | No. Retained | Additional Retained | @ $6,500 |
| Texas | 54 | 33 | 18 | 65 | 35 | 17 | 110,500 |
| California | 36 | 47 | 17 | 81 | 29 | 12 | 78,000 |
| Florida | 76 | 48 | 37 | 61 | 46 | 9 | 58,500 |
| Central | 86 | 29 | 25 | 51 | 44 | 19 | 123,500 |
| West | 65 | 54 | 35 | 58 | 38 | 3 | 19,500 |
| East | 272 | 37 | 99 | 71 | 193 | 94 | 611,000 |
| **Total** | **589** | | **230** | | **385** | **155** | **1,001,000** |

For the Texas region, for the 12-month period under centralized hiring, 54 drivers were hired. If the historical retention rate under decentralized hiring of 33% had held, there would have been only 18 drivers retained. In actuality under centralized hiring, the retention rate was 65%, resulting in 35 drivers retained. The increase in retained drivers is 35 minus 18, or 17 additional drivers. At an estimated recruiting cost of $6,500 per driver, the estimated savings from centralization is $110,500. Similar calculations were done for the other entries.

This is the final table in your presentation and recommendation to continue centralized hiring. If the actual 589 hires from 7/1/11 to 6/30/12 had been decentralized with the historical decentralized retention rate, there would have been an estimated 230 drivers retained. In fact, with centralized hiring, there were 385 drivers retained. At an estimated hiring cost of $6,500 per driver, decentralization would have cost an estimated $1 million more than centralization to hire the additional 155 drivers.

## 2.2  PERCENT DIFFERENCE

Percent difference on the surface may seem to be a relatively simple concept, but, as the following example shows, there are two ways to describe differences. It is important that you know there are two ways and what the difference is, so that you will be able to use the correct method in any given situation.

### Example 2.7

Suppose you as a BPD hiring manager made an offer of $100,000 to a senior accountant, and she said she had another offer of $120,000. There are two ways we can describe the situation in terms of percent difference.

1. The other offer is 20.0% more than your offer.
2. Your offer is 16.7% less than the other offer.

We have two correct statements with different percentages based on the same data. Which one is preferable? The correct answer is "It depends." It depends on what you intend to do with the information.

First, let us look at how we arrived at the two figures.

*Alternative 1*

Difference of the *other offer from your offer* $= 120,000 - 100,000 = 20,000$.

Percent difference of *other offer from your offer* $= (20,000/100,000)(100) = 20.0\%$ above your offer. The reference is 100,000.

*Alternative 2*

Difference of *your offer from the other offer* $= 100,000 - 120,000 = -20,000$.

Percent difference of *your offer from the other offer* $= (-20,000/120,000)(100) = -16.7\%$, or 16.7% below the other offer. The reference is 120,000.

Each of these two cases has a different reference, or different base. In the first case, the reference is your offer. In the second case, the reference is the other offer.

Now, which percent figure from these two correct statements will you use? If you intend to match the other offer, then you must raise your offer by 20.0%. 100,000 is the figure you will be taking action on.

$$(100,000)(20.0\%) = 20,000$$

$$100,000 + 20,000 = 120,000$$

So the preferable method is to have the data you will be taking action on as the reference point, and hence as the value in the denominator of the calculation.

$$\text{Percent difference} = \frac{(\text{other data}) - (\text{data you will be taking action on})}{\text{data you will be taking action on}} \times 100$$

or

$$\text{Percent difference} = \frac{\text{data} - \text{reference}}{\text{reference}} \times 100$$

Sometimes you might report both percentages and say "Our offer for this senior accountant is 16.7% below the other offer, and we need to raise our offer by 20.0%." Of course, you would have to explain the two figures.

A way around this is to say that "Our offer for this senior accountant is $20,000 below the other offer, and we need to raise our offer by $20,000." But here you have avoided the notion of percent difference altogether.

### Example 2.8

A common situation where percent difference occurs is when stating your overall market position.

You have just completed a market analysis for BPD and will present the results as input to a salary budget increase decision. You obtain the following summary figures in Table 2.6.

You calculate the percent difference both ways, because your executives want to know where the company is with respect to the competition as well as what kind of salary increase budget will be needed.

*Alternative 1*

Reference is market average. Data is BPD total pay.

$$\% \text{ from market} = \frac{318{,}750{,}000 - 337{,}875{,}000}{337{,}875{,}000} \times 100$$

$$\% \text{ from market} = \frac{-19{,}125{,}000}{337{,}875{,}000} \times 100$$

$$\% \text{ from market} = -5.7\%$$

**TABLE 2.6**    **DOLLAR COMPARISON TO MARKET**

| No. of Employees | Projected at Start of Plan Year | |
| --- | --- | --- |
| | BPD Total Pay | Market Average |
| 4,250 | 318,750,000 | 337,875,000 |

TABLE 2.7    PERCENT COMPARISON TO MARKET

| No. of Employees | Projected at Start of Plan Year | | Market Position (% from Market) | Market-Based Salary Increase Budget (% to Match Market) |
|---|---|---|---|---|
| | BPD Total Pay | Market Average | | |
| 4,250 | 318,750,000 | 337,875,000 | −5.7 | 6.0 |

*Alternative 2*

Reference is BPD total pay. Data is market average.

$$\% \text{ to match market} = \frac{337{,}875{,}000 - 318{,}750{,}000}{318{,}750{,}000} \times 100$$

$$\% \text{ to match market} = \frac{19{,}125{,}000}{318{,}750{,}000} \times 100$$

$$\% \text{ to match market} = 6.0\%$$

These results are summarized in Table 2.7.

Your total pay is 5.7% below the market average. It would take a 6.0% salary increase budget to match the market. The numerator is the same (except for the sign) but the denominators, or reference points, are different.

Of course, there are other factors involved in the final salary budget increase decision, but this gives the perspective of the market position.

## 2.3  COMPOUND INTEREST

---

**CASE STUDY 2, PART 1 OF 3**

This case study is about compound interest, which extends the notion of percent.

The president of BPD wants to know two things. The first item concerns the future base salary payroll costs. The current total base salary is $318,750,000. If it increases by 5% a year, what will the total base salary be in 4 years? The second item concerns a promised stay bonus. One of the vice presidents was promised a stay bonus of $50,000 at the end of 3 years. If the company can earn 8%, compounded quarterly, on its money, how much should be set aside now in order to have the $50,000 in 3 years?

---

Before we continue with the calculations, we will discuss the details of the notion of compound interest.

The notion of compound interest is concerned with the time value of money, although the concept may be applied to nonmonetary situations as well.

To introduce the concept of compound interest, consider a savings account. When you deposit an original amount of money, called the principal, and it earns interest and you do not withdraw the interest but leave it in the account, then that unwithdrawn interest also earns interest along with the original amount deposited. The interest earned on the principal and on the accumulated unwithdrawn interest earned previously is the compound interest. This process is called compounding.

This notion underlies the entire concept of compound interest, and underlies many subsequent complex financial concepts. We will discuss two of these concepts: future value and present value.

The future value is how much a given amount of money now is worth in the future.

The present value is how much a given amount of money in the future is worth now (the present).

## Future Value

To illustrate the concept of determining the future value of a present amount of money, we will use an interest-bearing credit union account. Suppose you deposited $1,000 in your credit union account, and it paid 6% interest annually. At the end of the first year, you would earn $60 in interest.

$$\text{Interest} = (6/100)(1,000) = (0.06)(1,000) = 60$$

At the end of the first year, you have a total of $1,000 + $60 = $1,060.

Now suppose you leave the interest in the account and let it also earn interest the second year. Then the second year, still at 6%, you earn $63.60 in interest.

$$\text{Interest} = (6/100)(1,060) = (0.06)(1,060) = 63.60$$

At the end of the second year, you have a total of $1,060 + $63.60 = $1,123.60.

And suppose you just left it in for a third year. Still at 6%, the interest on the total amount would be $67.42.

$$\text{Interest} = (6/100)(1,123.60) = (0.06)(1,123.60) = 67.42$$

At the end of the third year, you have a total of $1,123.60 + $67.42 = $1,191.02.

We could continue for year 4, year 5, and so on, but will stop here.

For this example, the future value of $1,000 after 3 years at 6% annual interest is $1,191.02.

The notion of calculating the future value using compound interest has been put into a formula, as shown below, to enable a direct calculation rather than doing it stepwise.

Let

PV = present value, or principal

FV = future value

$i$ = interest rate per period, expressed as the decimal equivalent of the percent

$n$ = number of periods

Then,

$$FV = PV(1 + i)^n$$

In this formula, $n$ is an exponent, which means to raise the quantity $(1 + i)$ to the $n$th power. For example, $(1 + 0.06)^3 = (1.06)^3 = (1.06)(1.06)(1.06)$.

This formula allows us to calculate the future value in one step, rather than doing it year by year. In the above example,

PV = $1,000

$i$ = 0.06 per year

$n$ = 3 years

$FV = (1,000)(1 + 0.06)^3 = (1,000)(1.06)^3 = (1,000)(1.19102)$ (rounded)

= $1,191.02

In the formula, it is important to note that $i$ is the interest rate *per period* and $n$ is the number of *periods*.

Suppose in the first example the credit union compounded quarterly at an annual rate of 6%. In this case, the period would be 3 months in length, so in 3 years there would be 12 periods (3 years × 4 periods per year). The interest rate per period would be 6% × 3/12, or 1.5% per period. Hence, the amount we would have at the end of 3 years is

PV = $1,000

$i$ = 0.015 per period

$n$ = 12 periods

$FV = (1,000)(1 + 0.015)^{12} = (1,000)(1.015)^{12} = (1,000)(1.19562)$ (rounded)

= $1,195.62

Here, we end up with $1,195.62 with quarterly compounding, compared to $1,191.02 with just annual compounding. This example shows that for a given annual interest rate, you get more money in the future if the compounding takes place more frequently.

---

**CASE STUDY 2, PART 2 OF 3**

Returning to Case Study 2, where the BPD president wants to know what the total base salary for the company will be in 4 years, if it increases by 5% a year, and where the current total base salary is $318,750,000, we apply the future value formula.

Future base salary payroll costs

PV = current total base salary, 318,750,000

FV = calculated future total base salary

$i$ = annual increase rate, 0.05

$n$ = number of years, 4

$FV = (318,750,000)(1.05)^4 = (318,750,000)(1.2155) = 387,441,000$

Assuming that the employee population stays stable over the next 4 years, assuming that the annual increase rate is 5%, and assuming nothing else changes, the future payroll costs will increase to $387,441,000.

---

## Present Value

Now that we have calculated the future value of a given present amount of money, let us do the reverse, and calculate the present value of a future amount of money. Here, we use the concept of compound interest to determine how much money would have to be invested today (the present value) at a certain interest rate to produce at a certain time in the future a certain amount (the future value).

Many compensation items today represent deferred payments, such as pensions or retention bonuses. The use of their future values for comparison purposes against today's dollars would be misleading. What has to be done is to convert the future values to present values.

Using the same symbols as before, and rearranging the previous formula, we can calculate the present value.

$$PV = FV/(1 + i)^n$$

This formula tells us how much we would have to set aside today, PV, at a given interest rate per period, $i$, for a given number of periods, $n$, to have a certain amount in the future, FV.

---

**CASE STUDY 2, PART 3 OF 3**

We return again to Case Study 2, where one of the vice presidents was promised a stay bonus of $50,000 at the end of 3 years. If the company can earn 8%, compounded quarterly, on its money, how much should be set aside now in order to have the $50,000 in 3 years?

Amount set aside for retention bonus

$PV$ = calculated amount to set aside now

$FV$ = promised bonus, 50,000

$i$ = quarterly increase = 8%/year divided by 4 quarters/year = 0.08/4 = 0.02

$n$ = number of quarters = 3 years × 4 quarters/year = 12

$PV = 50,000/(1.02)^{12} = 50,000/1.2682 = 39,426$

The company would have to invest $39,426 at 8% interest compounded quarterly to have $50,000 in 5 years. This assumes that the company is able to get 8% annual interest compounded quarterly on its money over the next 3 years.

---

Additional formulas using the notion of compound interest are given in the Appendix.

## Translating

Sometimes you have to translate, or convert, some of the units used in the formulas to those appropriate for the problem at hand.

*Example 2.9*

The BPD employee population has grown at an annual rate of 4%. If that trend continues, what will be the employee population in 5 years?

$PV$ = current number of employees, 4,250

$FV$ = calculated future number of employees

$i$ = annual growth rate, 0.04

$n$ = number of years, 5

$FV = (4,250)(1.04)^5 = (4,250)(1.2167) = 5,171$

*Example 2.10*

You want to estimate what the average individual health insurance premium for BPD employees will be in 4 years at an annual increase rate of 7%.

$PV$ = current premium amount, 15,150

$FV$ = calculated future premium amount

$i =$ annual increase rate, 0.07

$n =$ number of years, 4

$FV = (15,150)(1.07)^4 = (15,150)(1.3108) = 19,859$

### Example 2.11

The number of hits on the BPD HR website has grown 1% a month. If that trend continues, what will be the number of hits in 2 years?

$PV =$ current number of hits, 3,400

$FV =$ calculated future number of hits

$i =$ monthly increase rate, 0.01

$n =$ number of months, 24

$FV = (3,400)(1.01)^{24} = (3,400)(1.2697) = 4,317$

## RELATED TOPICS IN THE APPENDIX

A.3 Types of Numbers

A.4 Significant Figures

A.5 Scientific Notation

A.6 Accuracy and Precision

A.7 Compound Interest—Additional

A.8 Rule of 72

## PRACTICE PROBLEMS

**2.1** You have 660 skilled and unskilled labor personnel, 420 office and clerical personnel, and 320 professional and managerial personnel in a manufacturing facility. What is the percent of office and clerical personnel in the facility?

**2.2** 12% of your workforce of 9,500 live in Texas. How many live in Texas?

**2.3** The profit goal for the Ohio division was $15,000,000. The division had a profit of $18,000,000. What percent of goal was achieved?

**2.4** A survey provider said that the 32 companies participating in a survey represented 40% of the companies in the industry. How many companies are in the industry?

**2.5** On January 1 of last year your company had 1,000 employees. During that year your company lost money and on December 31 there was a 20% reduction in force. By July of this year business picked up and you then increased your workforce by 20%. Do you now have 1,000 employees again? Why or why not? If not, how many do you have?

**2.6** You are adjusting your salary midpoint to match the survey mean.

| | |
|---|---|
| Your midpoint | 72,000 |
| Survey mean | 78,000 |

Calculate the two ways of percent difference between the two numbers and make a statement for each. Why are they different? Which one is preferable and why?

**2.7**   In the BPD London office, the vice president makes 200,000 pounds, and his administrative assistant makes 20,000 pounds. Each gets a 2,000 pound raise. To whom do you think the raise might be more meaningful in a positive way? Why? Suppose instead they each receive a 10% raise. What might they think now?

**2.8**   If your salary of $70,000 increases with annual raises of 6% a year, what will it be in 3 years? If instead you get semiannual raises of 3%, what will your salary be in 3 years?

**2.9**   You want to have $50,000 in 4 years for a cruise for you and your family. You are able to invest in a high-yielding investment that returns 9% compounding monthly. How much do you need to invest now?

**2.10**   BPD currently has 15 offices in cities in Europe. The business plans call for an annual increase of 10% in the number of offices over the next 5 years. How many offices is BPD projected to have in 5 years?

# CHAPTER 3

# Frequency Distributions and Histograms

Data in its raw form are often of little use in helping to make sound decisions. Data have to be summarized, described, and presented in a manner appropriate to the particular analyses and decisions. One such method of summarizing and presenting raw data is the frequency distribution and its graphical representation the histogram, or bar chart.

The adage, "A picture is worth a thousand words," is just as true in statistics as it is elsewhere. A key to understanding quantitative data is to display it in a manner that lends itself to easy understanding. The frequency distribution and histogram is one such method.

We will describe a situation in which frequency distributions and histograms were instrumental in identifying the cause of some major problems.

---

**CASE STUDY 3**

BPD has two manufacturing plants in Texas. You have been assigned to a team with Finance and Operations to investigate the profitability of the Dallas plant, as compared to that of the Greenville plant. It is suspected that there may be compensation issues as well as some employee relations and management issues involved. Your role is to explore and evaluate just the compensation issues, which we will address here.

In addition to the different levels of profitability, there has been a continuing high number of requests for transfers from the Greenville plant to the Dallas plant.

The plants at Dallas and Greenville are "identical" manufacturing facilities—the same products produced, the same physical layout, the same

---

machinery, and the same capacity, but different people as well as different locations (although only 50 miles apart).

Dallas has a profit margin of 4.9% and is struggling to meet its profit goals of 10%. However, there is high employee morale according to the latest employee opinion survey. The annual turnover is 18%.

Greenville has a profit margin of 13.8% and its management expects to receive a healthy bonus by well exceeding its profit goals of 10%. However, there are many employee complaints along with very low morale according to the latest employee opinion survey. The annual turnover is 35%.

Your team examined these initial data, completed for the most recent fiscal year, shown in Table 3.1.

**TABLE 3.1    MANUFACTURING PLANTS INITIAL DATA**

|  | Dallas | Greenville |
|---|---|---|
| No. of employees | 400 | 395 |
| Low salary | 35,000 | 31,000 |
| High salary | 71,000 | 61,000 |
| Average salary | 52,300 | 43,200 |
| Total salary | 20,920,000 | 17,064,000 |
| Salary burden | 52% | 52% |
| Burdened labor cost | 31,798,400 | 25,937,280 |
| Nonlabor cost | 22,847,000 | 23,350,000 |
| Revenue | 57,479,000 | 57,172,000 |
| Profit | 2,833,600 | 7,884,720 |
| Profit margin | 4.9% | 13.8% |
| Turnover | 18% | 35% |

Issues are rarely one-dimensional, and there were four main concerns that may have related underlying causes. These are shown in Figure 3.1 in the model, which the team adopted for the initial approach. It was strongly suspected that since salaries affect all four issues, they should be investigated first.

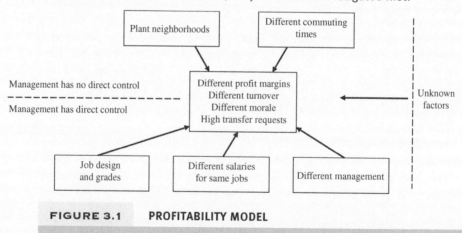

**FIGURE 3.1    PROFITABILITY MODEL**

You gather the salaries and grades for the nonmanagement employees in both plants, and decide that organizing them into frequency distributions and histograms might shed some light on the situation.

The data are in Table 3.2 and are also in Data Set 1.

**TABLE 3.2    SALARIES AND GRADES FOR MANUFACTURING PLANTS**

| Dallas Employee No. | Dallas Grade | Dallas Salary | Greenville Employee No. | Greenville Grade | Greenville Salary |
|---|---|---|---|---|---|
| 1 | 23 | 35,000 | 401 | 21 | 31,000 |
| 2 | 23 | 35,000 | 402 | 21 | 31,000 |
| 3 | 23 | 35,000 | 403 | 21 | 31,000 |
| 4 | 23 | 35,000 | 404 | 21 | 31,000 |
| 5 | 23 | 35,000 | 405 | 21 | 31,000 |
| 6 | 23 | 35,000 | 406 | 21 | 31,000 |
| 7 | 23 | 35,000 | 407 | 21 | 31,000 |
| 8 | 23 | 35,000 | 408 | 21 | 31,000 |
| 9 | 23 | 35,000 | 409 | 21 | 31,000 |
| 10 | 23 | 35,000 | 410 | 21 | 31,000 |
| 11 | 23 | 36,000 | 411 | 21 | 31,000 |
| 12 | 23 | 36,000 | 412 | 21 | 31,000 |
| 13 | 23 | 36,000 | 413 | 21 | 31,000 |
| 14 | 23 | 37,000 | 414 | 21 | 32,000 |
| 15 | 23 | 37,000 | 415 | 21 | 32,000 |
| 16 | 23 | 37,000 | 416 | 21 | 32,000 |
| 17 | 23 | 37,000 | 417 | 21 | 33,000 |
| 18 | 23 | 37,000 | 418 | 22 | 31,000 |
| 19 | 23 | 37,000 | 419 | 22 | 31,000 |
| 20 | 23 | 37,000 | 420 | 22 | 31,000 |
| 21 | 23 | 37,000 | 421 | 22 | 32,000 |
| 22 | 23 | 38,000 | 422 | 22 | 32,000 |
| 23 | 23 | 38,000 | 423 | 22 | 32,000 |
| 24 | 24 | 36,000 | 424 | 22 | 32,000 |
| 25 | 24 | 36,000 | 425 | 22 | 32,000 |
| 26 | 24 | 37,000 | 426 | 22 | 32,000 |
| 27 | 24 | 37,000 | 427 | 22 | 32,000 |
| 28 | 24 | 38,000 | 428 | 22 | 32,000 |
| 29 | 24 | 38,000 | 429 | 22 | 32,000 |
| 30 | 24 | 38,000 | 430 | 22 | 32,000 |
| 31 | 24 | 38,000 | 431 | 22 | 32,000 |
| 32 | 24 | 38,000 | 432 | 22 | 32,000 |
| 33 | 24 | 38,000 | 433 | 22 | 33,000 |
| 34 | 24 | 38,000 | 434 | 22 | 33,000 |
| 35 | 24 | 39,000 | 435 | 22 | 33,000 |
| 36 | 24 | 39,000 | 436 | 22 | 33,000 |
| 37 | 24 | 39,000 | 437 | 22 | 33,000 |
| 38 | 24 | 39,000 | 438 | 22 | 33,000 |
| 39 | 24 | 40,000 | 439 | 22 | 33,000 |
| 40 | 24 | 40,000 | 440 | 22 | 33,000 |

(*continued*)

**TABLE 3.2**   *(CONTINUED)*

| Dallas Employee No. | Dallas Grade | Dallas Salary | Greenville Employee No. | Greenville Grade | Greenville Salary |
|---|---|---|---|---|---|
| 41 | 24 | 40,000 | 441 | 22 | 33,000 |
| 42 | 24 | 40,000 | 442 | 22 | 33,000 |
| 43 | 24 | 40,000 | 443 | 22 | 33,000 |
| 44 | 24 | 40,000 | 444 | 22 | 33,000 |
| 45 | 24 | 40,000 | 445 | 22 | 33,000 |
| 46 | 24 | 40,000 | 446 | 22 | 33,000 |
| 47 | 24 | 41,000 | 447 | 22 | 33,000 |
| 48 | 24 | 41,000 | 448 | 22 | 33,000 |
| 49 | 24 | 41,000 | 449 | 22 | 33,000 |
| 50 | 24 | 42,000 | 450 | 22 | 33,000 |
| 51 | 24 | 42,000 | 451 | 22 | 33,000 |
| 52 | 24 | 42,000 | 452 | 22 | 34,000 |
| 53 | 24 | 42,000 | 453 | 22 | 34,000 |
| 54 | 24 | 42,000 | 454 | 22 | 34,000 |
| 55 | 24 | 42,000 | 455 | 22 | 34,000 |
| 56 | 24 | 42,000 | 456 | 22 | 34,000 |
| 57 | 24 | 42,000 | 457 | 22 | 34,000 |
| 58 | 24 | 42,000 | 458 | 22 | 34,000 |
| 59 | 24 | 42,000 | 459 | 22 | 34,000 |
| 60 | 24 | 43,000 | 460 | 22 | 34,000 |
| 61 | 24 | 43,000 | 461 | 22 | 34,000 |
| 62 | 24 | 43,000 | 462 | 22 | 34,000 |
| 63 | 24 | 43,000 | 463 | 22 | 35,000 |
| 64 | 24 | 43,000 | 464 | 22 | 35,000 |
| 65 | 24 | 43,000 | 465 | 22 | 35,000 |
| 66 | 24 | 43,000 | 466 | 22 | 35,000 |
| 67 | 24 | 43,000 | 467 | 22 | 35,000 |
| 68 | 24 | 43,000 | 468 | 22 | 35,000 |
| 69 | 24 | 43,000 | 469 | 22 | 35,000 |
| 70 | 24 | 43,000 | 470 | 22 | 35,000 |
| 71 | 24 | 44,000 | 471 | 22 | 35,000 |
| 72 | 24 | 44,000 | 472 | 22 | 36,000 |
| 73 | 24 | 44,000 | 473 | 23 | 34,000 |
| 74 | 24 | 44,000 | 474 | 23 | 35,000 |
| 75 | 24 | 44,000 | 475 | 23 | 35,000 |
| 76 | 24 | 45,000 | 476 | 23 | 35,000 |
| 77 | 24 | 45,000 | 477 | 23 | 35,000 |
| 78 | 24 | 46,000 | 478 | 23 | 35,000 |
| 79 | 24 | 47,000 | 479 | 23 | 35,000 |
| 80 | 25 | 44,000 | 480 | 23 | 36,000 |
| 81 | 25 | 44,000 | 481 | 23 | 36,000 |
| 82 | 25 | 44,000 | 482 | 23 | 36,000 |
| 83 | 25 | 45,000 | 483 | 23 | 36,000 |
| 84 | 25 | 45,000 | 484 | 23 | 36,000 |
| 85 | 25 | 45,000 | 485 | 23 | 36,000 |
| 86 | 25 | 45,000 | 486 | 23 | 36,000 |
| 87 | 25 | 45,000 | 487 | 23 | 36,000 |

**TABLE 3.2**     *(CONTINUED)*

| Dallas Employee No. | Dallas Grade | Dallas Salary | Greenville Employee No. | Greenville Grade | Greenville Salary |
|---|---|---|---|---|---|
| 88 | 25 | 45,000 | 488 | 23 | 36,000 |
| 89 | 25 | 45,000 | 489 | 23 | 36,000 |
| 90 | 25 | 45,000 | 490 | 23 | 36,000 |
| 91 | 25 | 45,000 | 491 | 23 | 36,000 |
| 92 | 25 | 45,000 | 492 | 23 | 36,000 |
| 93 | 25 | 46,000 | 493 | 23 | 36,000 |
| 94 | 25 | 46,000 | 494 | 23 | 36,000 |
| 95 | 25 | 46,000 | 495 | 23 | 37,000 |
| 96 | 25 | 46,000 | 496 | 23 | 37,000 |
| 97 | 25 | 46,000 | 497 | 23 | 37,000 |
| 98 | 25 | 46,000 | 498 | 23 | 37,000 |
| 99 | 25 | 46,000 | 499 | 23 | 37,000 |
| 100 | 25 | 46,000 | 500 | 23 | 37,000 |
| 101 | 25 | 46,000 | 501 | 23 | 37,000 |
| 102 | 25 | 46,000 | 502 | 23 | 37,000 |
| 103 | 25 | 46,000 | 503 | 23 | 37,000 |
| 104 | 25 | 46,000 | 504 | 23 | 37,000 |
| 105 | 25 | 46,000 | 505 | 23 | 37,000 |
| 106 | 25 | 46,000 | 506 | 23 | 37,000 |
| 107 | 25 | 46,000 | 507 | 23 | 37,000 |
| 108 | 25 | 46,000 | 508 | 23 | 37,000 |
| 109 | 25 | 47,000 | 509 | 23 | 37,000 |
| 110 | 25 | 47,000 | 510 | 23 | 37,000 |
| 111 | 25 | 47,000 | 511 | 23 | 37,000 |
| 112 | 25 | 47,000 | 512 | 23 | 37,000 |
| 113 | 25 | 47,000 | 513 | 23 | 37,000 |
| 114 | 25 | 47,000 | 514 | 23 | 37,000 |
| 115 | 25 | 47,000 | 515 | 23 | 37,000 |
| 116 | 25 | 47,000 | 516 | 23 | 38,000 |
| 117 | 25 | 47,000 | 517 | 23 | 38,000 |
| 118 | 25 | 47,000 | 518 | 23 | 38,000 |
| 119 | 25 | 47,000 | 519 | 23 | 38,000 |
| 120 | 25 | 47,000 | 520 | 23 | 38,000 |
| 121 | 25 | 47,000 | 521 | 23 | 38,000 |
| 122 | 25 | 47,000 | 522 | 23 | 38,000 |
| 123 | 25 | 47,000 | 523 | 23 | 38,000 |
| 124 | 25 | 47,000 | 524 | 23 | 38,000 |
| 125 | 25 | 47,000 | 525 | 23 | 38,000 |
| 126 | 25 | 47,000 | 526 | 23 | 38,000 |
| 127 | 25 | 47,000 | 527 | 23 | 38,000 |
| 128 | 25 | 47,000 | 528 | 23 | 38,000 |
| 129 | 25 | 47,000 | 529 | 23 | 38,000 |
| 130 | 25 | 47,000 | 530 | 23 | 38,000 |
| 131 | 25 | 47,000 | 531 | 23 | 38,000 |
| 132 | 25 | 47,000 | 532 | 23 | 38,000 |

*(continued)*

**TABLE 3.2**   (*CONTINUED*)

| Dallas Employee No. | Dallas Grade | Dallas Salary | Greenville Employee No. | Greenville Grade | Greenville Salary |
|---|---|---|---|---|---|
| 133 | 25 | 47,000 | 533 | 23 | 38,000 |
| 134 | 25 | 47,000 | 534 | 23 | 38,000 |
| 135 | 25 | 48,000 | 535 | 23 | 38,000 |
| 136 | 25 | 48,000 | 536 | 23 | 38,000 |
| 137 | 25 | 48,000 | 537 | 23 | 38,000 |
| 138 | 25 | 48,000 | 538 | 23 | 39,000 |
| 139 | 25 | 48,000 | 539 | 23 | 39,000 |
| 140 | 25 | 48,000 | 540 | 23 | 39,000 |
| 141 | 25 | 48,000 | 541 | 23 | 39,000 |
| 142 | 25 | 48,000 | 542 | 23 | 39,000 |
| 143 | 25 | 48,000 | 543 | 23 | 39,000 |
| 144 | 25 | 48,000 | 544 | 23 | 39,000 |
| 145 | 25 | 48,000 | 545 | 23 | 39,000 |
| 146 | 25 | 48,000 | 546 | 23 | 39,000 |
| 147 | 25 | 48,000 | 547 | 23 | 39,000 |
| 148 | 25 | 48,000 | 548 | 23 | 39,000 |
| 149 | 25 | 48,000 | 549 | 23 | 39,000 |
| 150 | 25 | 48,000 | 550 | 23 | 39,000 |
| 151 | 25 | 48,000 | 551 | 23 | 39,000 |
| 152 | 25 | 48,000 | 552 | 23 | 39,000 |
| 153 | 25 | 48,000 | 553 | 23 | 39,000 |
| 154 | 25 | 48,000 | 554 | 23 | 39,000 |
| 155 | 25 | 48,000 | 555 | 23 | 40,000 |
| 156 | 25 | 48,000 | 556 | 23 | 40,000 |
| 157 | 25 | 48,000 | 557 | 23 | 40,000 |
| 158 | 25 | 48,000 | 558 | 23 | 40,000 |
| 159 | 25 | 48,000 | 559 | 23 | 40,000 |
| 160 | 25 | 48,000 | 560 | 23 | 41,000 |
| 161 | 25 | 48,000 | 561 | 24 | 39,000 |
| 162 | 25 | 48,000 | 562 | 24 | 39,000 |
| 163 | 25 | 49,000 | 563 | 24 | 40,000 |
| 164 | 25 | 49,000 | 564 | 24 | 40,000 |
| 165 | 25 | 49,000 | 565 | 24 | 40,000 |
| 166 | 25 | 49,000 | 566 | 24 | 40,000 |
| 167 | 25 | 49,000 | 567 | 24 | 40,000 |
| 168 | 25 | 49,000 | 568 | 24 | 40,000 |
| 169 | 25 | 49,000 | 569 | 24 | 40,000 |
| 170 | 25 | 49,000 | 570 | 24 | 40,000 |
| 171 | 25 | 50,000 | 571 | 24 | 40,000 |
| 172 | 25 | 50,000 | 572 | 24 | 40,000 |
| 173 | 25 | 50,000 | 573 | 24 | 41,000 |
| 174 | 26 | 49,000 | 574 | 24 | 41,000 |
| 175 | 26 | 49,000 | 575 | 24 | 41,000 |
| 176 | 26 | 49,000 | 576 | 24 | 41,000 |
| 177 | 26 | 49,000 | 577 | 24 | 41,000 |
| 178 | 26 | 49,000 | 578 | 24 | 41,000 |
| 179 | 26 | 49,000 | 579 | 24 | 41,000 |

**TABLE 3.2**    *(CONTINUED)*

| Dallas Employee No. | Dallas Grade | Dallas Salary | Greenville Employee No. | Greenville Grade | Greenville Salary |
|---|---|---|---|---|---|
| 180 | 26 | 49,000 | 580 | 24 | 41,000 |
| 181 | 26 | 49,000 | 581 | 24 | 41,000 |
| 182 | 26 | 50,000 | 582 | 24 | 41,000 |
| 183 | 26 | 50,000 | 583 | 24 | 42,000 |
| 184 | 26 | 50,000 | 584 | 24 | 42,000 |
| 185 | 26 | 50,000 | 585 | 24 | 42,000 |
| 186 | 26 | 50,000 | 586 | 24 | 42,000 |
| 187 | 26 | 50,000 | 587 | 24 | 42,000 |
| 188 | 26 | 50,000 | 588 | 24 | 42,000 |
| 189 | 26 | 50,000 | 589 | 24 | 42,000 |
| 190 | 26 | 50,000 | 590 | 24 | 42,000 |
| 191 | 26 | 50,000 | 591 | 24 | 42,000 |
| 192 | 26 | 50,000 | 592 | 24 | 42,000 |
| 193 | 26 | 50,000 | 593 | 24 | 42,000 |
| 194 | 26 | 50,000 | 594 | 24 | 42,000 |
| 195 | 26 | 50,000 | 595 | 24 | 42,000 |
| 196 | 26 | 50,000 | 596 | 24 | 42,000 |
| 197 | 26 | 50,000 | 597 | 24 | 42,000 |
| 198 | 26 | 50,000 | 598 | 24 | 42,000 |
| 199 | 26 | 50,000 | 599 | 24 | 42,000 |
| 200 | 26 | 51,000 | 600 | 24 | 42,000 |
| 201 | 26 | 51,000 | 601 | 24 | 42,000 |
| 202 | 26 | 51,000 | 602 | 24 | 42,000 |
| 203 | 26 | 51,000 | 603 | 24 | 42,000 |
| 204 | 26 | 51,000 | 604 | 24 | 42,000 |
| 205 | 26 | 51,000 | 605 | 24 | 42,000 |
| 206 | 26 | 51,000 | 606 | 24 | 43,000 |
| 207 | 26 | 51,000 | 607 | 24 | 43,000 |
| 208 | 26 | 51,000 | 608 | 24 | 43,000 |
| 209 | 26 | 52,000 | 609 | 24 | 43,000 |
| 210 | 26 | 52,000 | 610 | 24 | 43,000 |
| 211 | 26 | 52,000 | 611 | 24 | 43,000 |
| 212 | 26 | 52,000 | 612 | 24 | 43,000 |
| 213 | 26 | 52,000 | 613 | 24 | 43,000 |
| 214 | 26 | 52,000 | 614 | 24 | 43,000 |
| 215 | 26 | 52,000 | 615 | 24 | 43,000 |
| 216 | 26 | 52,000 | 616 | 24 | 43,000 |
| 217 | 26 | 52,000 | 617 | 24 | 43,000 |
| 218 | 26 | 52,000 | 618 | 24 | 43,000 |
| 219 | 26 | 52,000 | 619 | 24 | 43,000 |
| 220 | 26 | 52,000 | 620 | 24 | 43,000 |
| 221 | 26 | 52,000 | 621 | 24 | 43,000 |
| 222 | 26 | 52,000 | 622 | 24 | 43,000 |
| 223 | 26 | 52,000 | 623 | 24 | 43,000 |
| 224 | 26 | 52,000 | 624 | 24 | 44,000 |
| 225 | 26 | 53,000 | 625 | 24 | 44,000 |

*(continued)*

**TABLE 3.2**    (*CONTINUED*)

| Dallas Employee No. | Dallas Grade | Dallas Salary | Greenville Employee No. | Greenville Grade | Greenville Salary |
|---|---|---|---|---|---|
| 226 | 26 | 53,000 | 626 | 24 | 44,000 |
| 227 | 26 | 53,000 | 627 | 24 | 44,000 |
| 228 | 26 | 53,000 | 628 | 24 | 44,000 |
| 229 | 26 | 53,000 | 629 | 24 | 44,000 |
| 230 | 26 | 54,000 | 630 | 24 | 44,000 |
| 231 | 27 | 52,000 | 631 | 24 | 44,000 |
| 232 | 27 | 53,000 | 632 | 24 | 44,000 |
| 233 | 27 | 53,000 | 633 | 24 | 44,000 |
| 234 | 27 | 53,000 | 634 | 24 | 44,000 |
| 235 | 27 | 53,000 | 635 | 24 | 44,000 |
| 236 | 27 | 53,000 | 636 | 24 | 45,000 |
| 237 | 27 | 53,000 | 637 | 24 | 45,000 |
| 238 | 27 | 53,000 | 638 | 25 | 43,000 |
| 239 | 27 | 53,000 | 639 | 25 | 43,000 |
| 240 | 27 | 53,000 | 640 | 25 | 44,000 |
| 241 | 27 | 53,000 | 641 | 25 | 44,000 |
| 242 | 27 | 54,000 | 642 | 25 | 44,000 |
| 243 | 27 | 54,000 | 643 | 25 | 44,000 |
| 244 | 27 | 54,000 | 644 | 25 | 44,000 |
| 245 | 27 | 54,000 | 645 | 25 | 45,000 |
| 246 | 27 | 54,000 | 646 | 25 | 45,000 |
| 247 | 27 | 54,000 | 647 | 25 | 45,000 |
| 248 | 27 | 54,000 | 648 | 25 | 45,000 |
| 249 | 27 | 54,000 | 649 | 25 | 45,000 |
| 250 | 27 | 54,000 | 650 | 25 | 45,000 |
| 251 | 27 | 55,000 | 651 | 25 | 45,000 |
| 252 | 27 | 55,000 | 652 | 25 | 45,000 |
| 253 | 27 | 55,000 | 653 | 25 | 45,000 |
| 254 | 27 | 55,000 | 654 | 25 | 45,000 |
| 255 | 27 | 55,000 | 655 | 25 | 46,000 |
| 256 | 27 | 55,000 | 656 | 25 | 46,000 |
| 257 | 27 | 55,000 | 657 | 25 | 46,000 |
| 258 | 27 | 55,000 | 658 | 25 | 46,000 |
| 259 | 27 | 55,000 | 659 | 25 | 46,000 |
| 260 | 27 | 55,000 | 660 | 25 | 46,000 |
| 261 | 27 | 55,000 | 661 | 25 | 46,000 |
| 262 | 27 | 55,000 | 662 | 25 | 46,000 |
| 263 | 27 | 55,000 | 663 | 25 | 46,000 |
| 264 | 27 | 55,000 | 664 | 25 | 47,000 |
| 265 | 27 | 56,000 | 665 | 25 | 47,000 |
| 266 | 27 | 56,000 | 666 | 25 | 47,000 |
| 267 | 27 | 56,000 | 667 | 25 | 47,000 |
| 268 | 27 | 56,000 | 668 | 25 | 47,000 |
| 269 | 27 | 56,000 | 669 | 25 | 47,000 |
| 270 | 27 | 56,000 | 670 | 25 | 47,000 |
| 271 | 27 | 56,000 | 671 | 25 | 47,000 |
| 272 | 27 | 56,000 | 672 | 25 | 47,000 |

**TABLE 3.2**     *(CONTINUED)*

| Dallas Employee No. | Dallas Grade | Dallas Salary | Greenville Employee No. | Greenville Grade | Greenville Salary |
|---|---|---|---|---|---|
| 273 | 27 | 56,000 | 673 | 25 | 47,000 |
| 274 | 27 | 56,000 | 674 | 25 | 47,000 |
| 275 | 27 | 57,000 | 675 | 25 | 47,000 |
| 276 | 27 | 57,000 | 676 | 25 | 47,000 |
| 277 | 27 | 57,000 | 677 | 25 | 47,000 |
| 278 | 27 | 57,000 | 678 | 25 | 47,000 |
| 279 | 27 | 57,000 | 679 | 25 | 47,000 |
| 280 | 27 | 57,000 | 680 | 25 | 47,000 |
| 281 | 27 | 57,000 | 681 | 25 | 47,000 |
| 282 | 27 | 57,000 | 682 | 25 | 48,000 |
| 283 | 27 | 58,000 | 683 | 25 | 48,000 |
| 284 | 27 | 58,000 | 684 | 25 | 48,000 |
| 285 | 27 | 58,000 | 685 | 25 | 48,000 |
| 286 | 27 | 58,000 | 686 | 25 | 48,000 |
| 287 | 27 | 58,000 | 687 | 25 | 48,000 |
| 288 | 27 | 58,000 | 688 | 25 | 48,000 |
| 289 | 27 | 58,000 | 689 | 25 | 48,000 |
| 290 | 27 | 58,000 | 690 | 25 | 48,000 |
| 291 | 27 | 58,000 | 691 | 25 | 48,000 |
| 292 | 27 | 58,000 | 692 | 25 | 48,000 |
| 293 | 27 | 58,000 | 693 | 25 | 48,000 |
| 294 | 27 | 58,000 | 694 | 25 | 48,000 |
| 295 | 27 | 58,000 | 695 | 25 | 48,000 |
| 296 | 27 | 58,000 | 696 | 25 | 48,000 |
| 297 | 27 | 58,000 | 697 | 25 | 48,000 |
| 298 | 27 | 58,000 | 698 | 25 | 49,000 |
| 299 | 27 | 58,000 | 699 | 25 | 49,000 |
| 300 | 27 | 59,000 | 700 | 25 | 49,000 |
| 301 | 27 | 59,000 | 701 | 25 | 49,000 |
| 302 | 27 | 59,000 | 702 | 25 | 49,000 |
| 303 | 27 | 59,000 | 703 | 25 | 50,000 |
| 304 | 27 | 59,000 | 704 | 25 | 50,000 |
| 305 | 27 | 60,000 | 705 | 25 | 51,000 |
| 306 | 28 | 59,000 | 706 | 26 | 48,000 |
| 307 | 28 | 59,000 | 707 | 26 | 48,000 |
| 308 | 28 | 59,000 | 708 | 26 | 49,000 |
| 309 | 28 | 60,000 | 709 | 26 | 49,000 |
| 310 | 28 | 60,000 | 710 | 26 | 49,000 |
| 311 | 28 | 60,000 | 711 | 26 | 49,000 |
| 312 | 28 | 60,000 | 712 | 26 | 50,000 |
| 313 | 28 | 60,000 | 713 | 26 | 50,000 |
| 314 | 28 | 60,000 | 714 | 26 | 50,000 |
| 315 | 28 | 61,000 | 715 | 26 | 50,000 |
| 316 | 28 | 61,000 | 716 | 26 | 50,000 |
| 317 | 28 | 61,000 | 717 | 26 | 50,000 |
| 318 | 28 | 62,000 | 718 | 26 | 50,000 |

*(continued)*

**TABLE 3.2** (*CONTINUED*)

| Dallas Employee No. | Dallas Grade | Dallas Salary | Greenville Employee No. | Greenville Grade | Greenville Salary |
|---|---|---|---|---|---|
| 319 | 28 | 62,000 | 719 | 26 | 51,000 |
| 320 | 28 | 62,000 | 720 | 26 | 51,000 |
| 321 | 28 | 62,000 | 721 | 26 | 51,000 |
| 322 | 28 | 62,000 | 722 | 26 | 51,000 |
| 323 | 28 | 62,000 | 723 | 26 | 51,000 |
| 324 | 28 | 62,000 | 724 | 26 | 52,000 |
| 325 | 28 | 62,000 | 725 | 26 | 52,000 |
| 326 | 28 | 63,000 | 726 | 26 | 52,000 |
| 327 | 28 | 63,000 | 727 | 26 | 52,000 |
| 328 | 28 | 63,000 | 728 | 26 | 52,000 |
| 329 | 28 | 63,000 | 729 | 26 | 52,000 |
| 330 | 28 | 63,000 | 730 | 26 | 52,000 |
| 331 | 28 | 63,000 | 731 | 26 | 52,000 |
| 332 | 28 | 63,000 | 732 | 26 | 53,000 |
| 333 | 28 | 63,000 | 733 | 26 | 53,000 |
| 334 | 28 | 63,000 | 734 | 26 | 53,000 |
| 335 | 28 | 63,000 | 735 | 26 | 53,000 |
| 336 | 28 | 64,000 | 736 | 26 | 53,000 |
| 337 | 28 | 64,000 | 737 | 26 | 53,000 |
| 338 | 28 | 64,000 | 738 | 26 | 53,000 |
| 339 | 28 | 64,000 | 739 | 26 | 54,000 |
| 340 | 28 | 64,000 | 740 | 26 | 54,000 |
| 341 | 28 | 64,000 | 741 | 26 | 54,000 |
| 342 | 28 | 64,000 | 742 | 26 | 54,000 |
| 343 | 28 | 65,000 | 743 | 26 | 55,000 |
| 344 | 28 | 65,000 | 744 | 26 | 55,000 |
| 345 | 28 | 65,000 | 745 | 26 | 55,000 |
| 346 | 28 | 66,000 | 746 | 26 | 55,000 |
| 347 | 28 | 66,000 | 747 | 26 | 55,000 |
| 348 | 28 | 67,000 | 748 | 26 | 56,000 |
| 349 | 29 | 65,000 | 749 | 26 | 56,000 |
| 350 | 29 | 65,000 | 750 | 26 | 57,000 |
| 351 | 29 | 66,000 | 751 | 27 | 54,000 |
| 352 | 29 | 66,000 | 752 | 27 | 55,000 |
| 353 | 29 | 66,000 | 753 | 27 | 55,000 |
| 354 | 29 | 67,000 | 754 | 27 | 56,000 |
| 355 | 29 | 67,000 | 755 | 27 | 56,000 |
| 356 | 29 | 67,000 | 756 | 27 | 56,000 |
| 357 | 29 | 67,000 | 757 | 27 | 57,000 |
| 358 | 29 | 67,000 | 758 | 27 | 57,000 |
| 359 | 29 | 67,000 | 759 | 27 | 57,000 |
| 360 | 29 | 67,000 | 760 | 27 | 57,000 |
| 361 | 29 | 67,000 | 761 | 27 | 57,000 |
| 362 | 29 | 67,000 | 762 | 27 | 57,000 |
| 363 | 29 | 68,000 | 763 | 27 | 57,000 |
| 364 | 29 | 68,000 | 764 | 27 | 57,000 |
| 365 | 29 | 68,000 | 765 | 27 | 57,000 |

**TABLE 3.2    (CONTINUED)**

| Dallas Employee No. | Dallas Grade | Dallas Salary | Greenville Employee No. | Greenville Grade | Greenville Salary |
|---|---|---|---|---|---|
| 366 | 29 | 68,000 | 766 | 27 | 58,000 |
| 367 | 29 | 68,000 | 767 | 27 | 58,000 |
| 368 | 29 | 68,000 | 768 | 27 | 58,000 |
| 369 | 29 | 68,000 | 769 | 27 | 58,000 |
| 370 | 29 | 68,000 | 770 | 27 | 58,000 |
| 371 | 29 | 68,000 | 771 | 27 | 58,000 |
| 372 | 29 | 68,000 | 772 | 27 | 58,000 |
| 373 | 29 | 69,000 | 773 | 27 | 58,000 |
| 374 | 29 | 69,000 | 774 | 27 | 59,000 |
| 375 | 29 | 69,000 | 775 | 27 | 59,000 |
| 376 | 29 | 69,000 | 776 | 27 | 59,000 |
| 377 | 29 | 69,000 | 777 | 27 | 59,000 |
| 378 | 29 | 69,000 | 778 | 27 | 59,000 |
| 379 | 29 | 69,000 | 779 | 27 | 60,000 |
| 380 | 29 | 69,000 | 780 | 27 | 60,000 |
| 381 | 29 | 69,000 | 781 | 27 | 60,000 |
| 382 | 29 | 70,000 | 782 | 27 | 60,000 |
| 383 | 29 | 70,000 | 783 | 27 | 61,000 |
| 384 | 29 | 70,000 | 784 | 27 | 61,000 |
| 385 | 29 | 70,000 | 785 | 27 | 61,000 |
| 386 | 29 | 70,000 | 786 | 28 | 59,000 |
| 387 | 29 | 70,000 | 787 | 28 | 59,000 |
| 388 | 29 | 70,000 | 788 | 28 | 60,000 |
| 389 | 29 | 70,000 | 789 | 28 | 61,000 |
| 390 | 29 | 70,000 | 790 | 28 | 61,000 |
| 391 | 29 | 71,000 | 791 | 28 | 61,000 |
| 392 | 29 | 71,000 | 792 | 28 | 61,000 |
| 393 | 29 | 71,000 | 793 | 28 | 61,000 |
| 394 | 29 | 71,000 | 794 | 28 | 61,000 |
| 395 | 29 | 71,000 | 795 | 28 | 61,000 |
| 396 | 29 | 71,000 | | | |
| 397 | 29 | 71,000 | | | |
| 398 | 29 | 71,000 | | | |
| 399 | 29 | 71,000 | | | |
| 400 | 29 | 71,000 | | | |

## 3.1  DEFINITIONS AND CONSTRUCTION

A frequency distribution is a classification of the data into certain categories, with subsequent counting (absolute frequency) and sometimes percentage calculations (relative frequency) that indicate the amount in each category. An example of an absolute frequency distribution is shown in Table 3.3. For example, there are

| TABLE 3.3 | FREQUENCY DISTRIBUTION OF SALARIES IN DALLAS |
| --- | --- |

| Frequency Distribution | |
| --- | --- |
| Salary ($000) | Dallas No. of Employees |
| ≤30 | 0 |
| >30 to ≤35 | 10 |
| >35 to ≤40 | 36 |
| >40 to ≤45 | 44 |
| >45 to ≤50 | 109 |
| >50 to ≤55 | 65 |
| >55 to ≤60 | 50 |
| >60 to ≤65 | 33 |
| >65 to ≤70 | 43 |
| >70 to ≤75 | 10 |
| >75 | 0 |
| **Total** | **400** |

10 salaries that are greater than $30,000 and less than or equal to $35,000. A histogram is a bar chart graphical display of the frequency distribution, as shown in Figure 3.2. The height of the bar indicates the frequency (or percentage) of data points in that category. For example, the tallest bar with height 109 indicates there are 109 employees with salaries in the category designated 45,000–50,000.

A frequency distribution and a histogram allow us to see in tabular and graphical forms, respectively, the distribution of a set of data. They indicate where the parts and amounts of data are with respect to each other. Often we are able to observe characteristics about the distribution—such as its shape, or

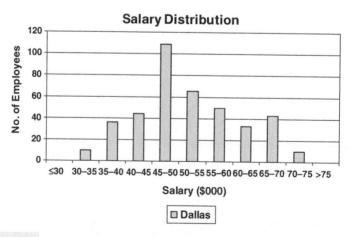

FIGURE 3.2   HISTOGRAM OF SALARY DISTRIBUTION IN DALLAS

comparison to another distribution—that lead to conclusions and decisions based on the data.

We will now go through the process of creating this frequency distribution and histogram.

In this case study, after some initial screening of the data for the two manufacturing plants you conclude that the salaries and grades are valid. Now you have to decide on the categories.

## Rules for Categories

There are four rules for categories to follow in constructing a frequency distribution and its histogram.

1. *Decide on the Categories.* Selecting how many categories to have is sometimes a trial-and-error process. Usually somewhere from 5 to 15 categories is satisfactory for analysis. Sometimes you have more; sometimes you have less. These numbers suggested here are a balance between having too many categories, in which case you are back to the raw data, and too few categories, in which case the distribution of the data is concealed. Sometimes you try different category definitions to see which makes more sense.

   An exception is that if you start with only two or three categories, such as eligible for bonus and not eligible for bonus, or Chicago, Houston, and San Antonio, then that is what you use. Sometimes you use more than 15 categories, such as if you have 18 grades.

2. *Make Sure the Categories Span all the Data.* Do not leave out the extreme highs or lows.

3. *For Numerical Categories Make the Category Interval Width the Same for all Categories.* It is all right for the lowest category to indicate all below or equal to a certain value and for the highest category to indicate all above a certain value.

   In Table 3.3 we chose categories of width $5,000, starting with $30,000 and ending with $75,000. We also created "end" categories to show that we have counted everything. Then we counted how many salaries are in each category.

4. *Make the Categories Mutually Exclusive.* That is, there should be no overlap. This ensures that each data point goes into only one category. For example, in the raw data shown in Table 3.2, the value of 70,000 goes in the category >65 K to ≤70 K in Table 3.3.

Looking at Table 3.3, the frequency indicates the number of data points in a given category. As we have mentioned before, for example, there are 10 salaries that are greater than $30,000 and less than or equal to $35,000.

| TABLE 3.4 | FREQUENCY DISTRIBUTION OF SALARIES IN DALLAS WITHOUT SIGNS |
|---|---|

**Frequency Distribution**

| Salary ($000) | Dallas No. of Employees |
|---|---|
| ≤30 | 0 |
| 30–35 | 10 |
| 35–40 | 36 |
| 40–45 | 44 |
| 45–50 | 109 |
| 50–55 | 65 |
| 55–60 | 50 |
| 60–65 | 33 |
| 65–70 | 43 |
| 70–75 | 10 |
| >75 | 0 |
| **Total** | **400** |

For presentation purposes, we sometimes leave off the ">" signs and the "≤" signs to make the table and corresponding histogram look cleaner as in Table 3.4. But you should know the strict definitions when presenting it. Use the style of table that makes sense to you.

We then constructed the corresponding histogram, which is a bar chart type of display of the frequency distribution, shown in Figure 3.2.

On the horizontal axis (x-axis) are the categories, and on the vertical axis (y-axis) is the scale that indicates the number of data points in each category. In this case, it is the number of employees whose salaries are in each category. As we have mentioned before, for example, looking at the tallest bar, there are 109 employees whose salaries are greater than $45,000 and less than or equal to $50,000.

Now you repeat the process for the Greenville salaries, and construct a frequency distribution (Table 3.5) and histogram (Figure 3.3). Anticipating that we will be comparing the two locations, we need to have the same categories.

We now want to compare the two distributions, so we combine them onto one table (Table 3.6) and one chart (Figure 3.4) to facilitate comparison.

You notice the significant difference in the distribution of salaries. These distributions provide supporting evidence for the average salary in Dallas of $52,300 being 21% higher than the average salary in Greenville of $43,200. You seek to discover why the salaries are so different for allegedly identical manufacturing plants. You know that the 50 mile geographic difference between the two locations is too small to explain this.

**TABLE 3.5       FREQUENCY DISTRIBUTION OF
SALARIES IN GREENVILLE**

| Frequency Distribution | |
| --- | --- |
| Salary ($000) | Greenville No. of Employees |
| ≤30 | 0 |
| 30–35 | 78 |
| 35–40 | 93 |
| 40–45 | 83 |
| 45–50 | 63 |
| 50–55 | 33 |
| 55–60 | 35 |
| 60–65 | 10 |
| 65–70 | 0 |
| 70–75 | 0 |
| >75 | 0 |
| **Total** | **395** |

You next examine grades using frequency distributions (Table 3.7) and histograms (Figure 3.5).

You conclude the reason for the difference in the salary distributions is the fact that there is a difference in the grade distribution. For allegedly identical plants you would have expected similar, but not necessarily exactly the same distributions. But this difference is too much.

Perhaps the reason for the high number of transfer requests to go from Greenville to Dallas is the higher grades and concomitant salary for the "same" jobs.

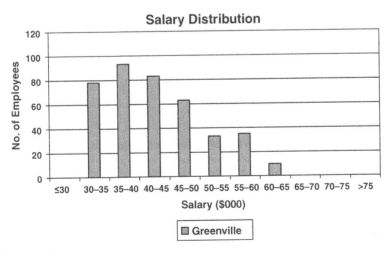

**FIGURE 3.3       HISTOGRAM OF SALARY DISTRIBUTION IN GREENVILLE**

| TABLE 3.6 | FREQUENCY DISTRIBUTIONS OF SALARIES IN DALLAS AND GREENVILLE | |

| | Frequency Distribution | |
|---|---|---|
| Salary ($000) | Dallas No. of Employees | Greenville No. of Employees |
| ≤30 | 0 | 0 |
| 30–35 | 10 | 78 |
| 35–40 | 36 | 93 |
| 40–45 | 44 | 83 |
| 45–50 | 109 | 63 |
| 50–55 | 65 | 33 |
| 55–60 | 50 | 35 |
| 60–65 | 33 | 10 |
| 65–70 | 43 | 0 |
| 70–75 | 10 | 0 |
| >75 | 0 | 0 |
| **Total** | **400** | **395** |

Just accounting for salaries and the underlying grades only, your cause-and-effect model is shown in Figure 3.6.

Other aspects of the initial model need to be investigated. For example, you may have a very strict plant manager in Greenville who is a slave driver and keeps grades and salaries down. Or you may have a very lax plant manager in Dallas who wants to be "liked" by employees by allowing jobs to be graded high with subsequent high salaries.

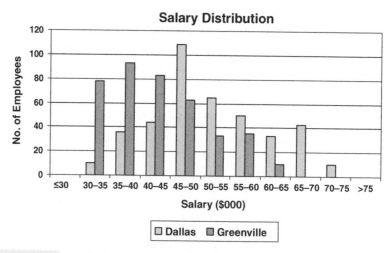

| FIGURE 3.4 | HISTOGRAMS OF SALARY DISTRIBUTIONS IN DALLAS AND GREENVILLE |

**TABLE 3.7    FREQUENCY DISTRIBUTIONS OF GRADES IN DALLAS AND GREENVILLE**

| | Frequency Distribution | |
| --- | --- | --- |
| Grade | Dallas<br>No. of Employees | Greenville<br>No. of Employees |
| 21 | 0 | 17 |
| 22 | 0 | 55 |
| 23 | 23 | 88 |
| 24 | 56 | 77 |
| 25 | 94 | 68 |
| 26 | 57 | 45 |
| 27 | 75 | 35 |
| 28 | 43 | 10 |
| 29 | 52 | 0 |
| Total | 400 | 395 |

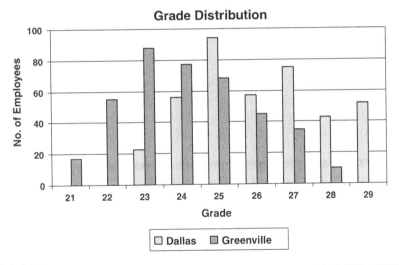

**FIGURE 3.5    HISTOGRAMS OF GRADE DISTRIBUTIONS IN DALLAS AND GREENVILLE**

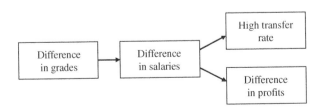

**FIGURE 3.6    CAUSE AND EFFECT MODEL**

But for purposes of illustration, we have used frequency distributions and histograms to uncover a fruitful area, namely grades, for further investigation. Your next action is to conduct a job audit and reevaluation of all the jobs in both plants to ensure fair treatment everywhere, as you really *don't* know if the grades of jobs in Dallas are inflated or the grades of jobs in Greenville are deflated.

## 3.2  COMPARING DISTRIBUTIONS

We have compared the distributions of salaries and grades between the two manufacturing facilities. There are two other aspects of comparing distributions that are important to know: absolute comparison versus relative comparison, and the number of comparisons that is practical on one chart.

### Absolute Comparison and Relative Comparison

Frequency distributions and histograms provide a very powerful way to compare distributions of data sets. In this case, you have to decide whether to use absolute frequencies or relative frequencies. There is no single right answer. When comparing data sets of similar total sizes, the pictures of absolute and relative comparisons will look very much alike. But when comparing data sets of very different total sizes, the pictures will be different.

In Table 3.8, there are two comparisons of the distribution of ages of BPD employees in two locations, Chicago and San Jose. The first comparison is on absolute frequency. The second comparison is on relative frequency.

**TABLE 3.8    ABSOLUTE AND RELATIVE FREQUENCY DISTRIBUTIONS OF AGES IN TWO LOCATIONS**

| Age | Chicago Absolute Frequency No. of Employees | San Jose Absolute Frequency No. of Employees | Chicago Relative Frequency % of Employees | San Jose Relative Frequency % of Employees |
|---|---|---|---|---|
| ≤20 | 0 | 2 | 0 | 4 |
| 21–25 | 5 | 5 | 3 | 10 |
| 26–30 | 10 | 10 | 5 | 20 |
| 31–35 | 20 | 12 | 10 | 24 |
| 36–40 | 30 | 10 | 15 | 20 |
| 41–45 | 35 | 5 | 18 | 10 |
| 46–50 | 45 | 3 | 23 | 6 |
| 51–55 | 30 | 2 | 15 | 4 |
| 56–60 | 10 | 1 | 5 | 2 |
| 61–65 | 10 | 0 | 5 | 0 |
| >65 | 5 | 0 | 3 | 0 |
| Total | 200 | 50 | 100 | 100 |

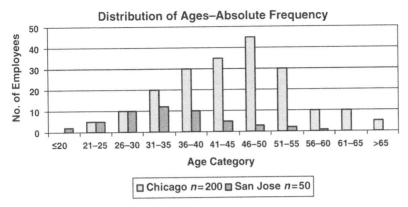

**FIGURE 3.7**    HISTOGRAMS OF ABSOLUTE FREQUENCY DISTRIBUTIONS OF AGES IN TWO LOCATIONS

To calculate the relative frequency for a given category, divide the absolute frequency in that category by the total in the column and multiply by 100 to get a percent. For example, for the category age ≤20 in San Jose, the absolute frequency is 2 employees and the total number of employees in San Jose is 50. The relative frequency for that category is $(2/50)(100) = 4\%$.

We now look at the two histograms. The first is a histogram of the absolute frequencies, shown in Figure 3.7.

We see that there is a younger population in San Jose than in Chicago, and since there are only one-fourth the number of employees in San Jose, their bars in the histogram are a lot shorter than the bars for Chicago.

If you are interested in a comparison of the actual number of employees, then an absolute comparison such as this would be the appropriate one to make.

The second comparison is a histogram of the relative frequencies, shown in Figure 3.8.

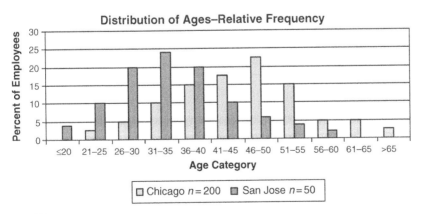

**FIGURE 3.8**    HISTOGRAMS OF RELATIVE FREQUENCY DISTRIBUTIONS OF AGES IN TWO LOCATIONS

In this case you still can see that there is a younger population in San Jose. However, since the percentages for each city have to add up to 100%, the bars are more similar in size.

If you are interested in a comparison where the size of one data set relative to another may inadvertently distort a comparison, then a comparison of relative frequencies such as this would be the appropriate one to make.

If you are not sure which type of comparison is appropriate, then a good rule is to compare both the absolute frequencies and the relative frequencies and then decide.

## Comparing More Than Two Distributions

You can compare the distributions of more than two data sets with histograms. You can easily make sense when comparing two or three sets. Four sets are borderline because it starts to be difficult to distinguish the various bars. More than four is confusing and hard to understand. This is true whether comparing absolute frequencies or relative frequencies.

The following examples show the relative distributions of the most recent performance appraisal ratings of BPD employees in various locations. The first is with three data sets, shown in Figure 3.9, and the second is with five, shown in Figure 3.10.

One can easily distinguish the differences in distributions in the three locations. The distribution for Chicago tends to be oriented toward the high side. The distribution for Los Angeles is centered.

But the display with five distributions is cluttered, and it is almost impossible to distinguish the distributions in different locations.

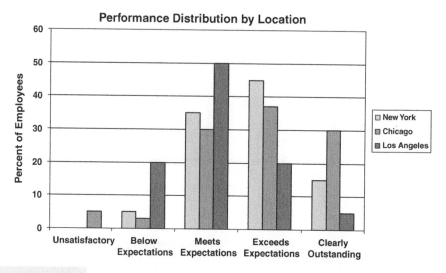

FIGURE 3.9    HISTOGRAMS COMPARING THREE DATA SETS ON ONE CHART

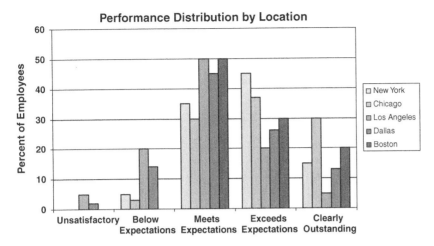

**FIGURE 3.10    HISTOGRAMS COMPARING FIVE DATA SETS ON ONE CHART**

An alternate way to compare multiple distributions is to stack them vertically in one display, shown in Figure 3.11.

With this stacked display, you can easily compare the distributions, as they are all separate from each other.

## 3.3 INFORMATION LOSS AND COMPREHENSION GAIN

A characteristic of frequency distributions and histograms, and indeed of any summary statistic, is that any summarization or categorization of the data necessarily loses the individuality of the data. In other words, at the stage of the frequency distribution, using the example in Table 3.8 of the five employees in Chicago with ages in the category of 21–25, we don't know if the employees are 21, 22, 23, 24, and 25 years old, or 21, 21, 23, 23, and 24 years old, or what. All we know at this stage is that their ages are somewhere from 21 to 25 years old.

So we have lost some detail. But what we have gained by this loss is simplicity and the ability to comprehend and make sense out of the data. This is a necessary step in our quest to organize the data in various fashions to enable us to draw conclusions. As long as the cost of the lost data is less than the benefit of summarization that gives us comprehension, we are on the right track.

At this point, we have moved from the data component to the information component of the Decision Model presented in Chapter 1, Figure 1.2 and repeated here as Figure 3.12.

## 3.4 CATEGORY SELECTION

Changing the number or boundaries of the categories can have a major influence on the conclusions you reach from a histogram. Here is an example where the

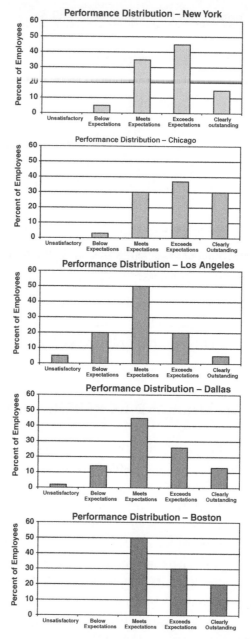

**FIGURE 3.11    HISTOGRAMS COMPARING FIVE DATA SETS ON FIVE STACKED CHARTS**

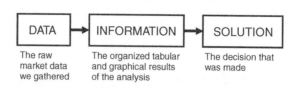

**FIGURE 3.12    DECISION MODEL**

**TABLE 3.9      RAW DATA UNUSED DAYS OF VACATION**

| Unused Days of Vacation | No. of Employees | Unused Days of Vacation | No. of Employees |
|---|---|---|---|
| 1 | 0 | 16 | 7 |
| 2 | 1 | 17 | 9 |
| 3 | 2 | 18 | 10 |
| 4 | 5 | 19 | 27 |
| 5 | 15 | 20 | 21 |
| 6 | 28 | 21 | 41 |
| 7 | 12 | 22 | 35 |
| 8 | 7 | 23 | 15 |
| 9 | 2 | 24 | 10 |
| 10 | 5 | 25 | 13 |
| 11 | 3 | 26 | 4 |
| 12 | 5 | 27 | 0 |
| 13 | 3 | 28 | 3 |
| 14 | 3 | 29 | 3 |
| 15 | 9 | 30 | 0 |

BPD benefits manager was looking at the distribution of unused days of vacation for a group of nonmanagerial employees in the Chicago office. Both tables and charts are based on the same data.

The data are in Table 3.9 and also in Data Set 2.

Table 3.10 and Figure 3.13 have categories of every 10.

This shows a trend that the number of employees with unused vacation days increases as the number of unused vacation days increases.

Now we look at the data with categories of every 1, shown in Table 3.11 and Figure 3.14.

Here, we see that the unused vacation day's distribution has a decidedly different distribution, and the number of employees does not necessarily increase as unused vacation days increase.

These two frequency distributions and histograms violate the "rule" on number of categories (from 5 to 15). However, it does illustrate the point that one should try different categories to uncover any "surprise" distributions in the data.

**TABLE 3.10      FREQUENCY DISTRIBUTION OF UNUSED VACATION DAYS WITH CATEGORIES OF 10**

| Unused Days of Vacation | No. of Employees |
|---|---|
| 1–10 | 77 |
| 11–20 | 97 |
| 21–30 | 124 |

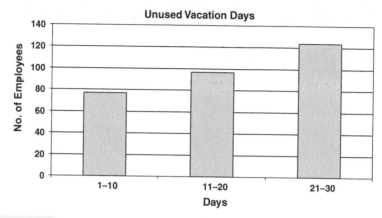

**FIGURE 3.13**    HISTOGRAM OF UNUSED VACATION DAYS WITH CATEGORIES OF 10

**TABLE 3.11**    FREQUENCY DISTRIBUTION OF UNUSED VACATION DAYS WITH CATEGORIES OF 1

| Unused Days of Vacation | No. of Employees | Unused Days of Vacation | No. of Employees | Unused Days of Vacation | No. of Employees |
|---|---|---|---|---|---|
| 1 | 0 | 11 | 3 | 21 | 41 |
| 2 | 1 | 12 | 5 | 22 | 35 |
| 3 | 2 | 13 | 3 | 23 | 15 |
| 4 | 5 | 14 | 3 | 24 | 10 |
| 5 | 15 | 15 | 9 | 25 | 13 |
| 6 | 28 | 16 | 7 | 26 | 4 |
| 7 | 12 | 17 | 9 | 27 | 0 |
| 8 | 7 | 18 | 10 | 28 | 3 |
| 9 | 2 | 19 | 27 | 29 | 3 |
| 10 | 5 | 20 | 21 | 30 | 0 |

## 3.5 DISTRIBUTION SHAPES

Often it is helpful to describe the general shape of a distribution. It helps communicate what the data collectively "look" like. We discuss four typical shapes usually encountered in compensation, along with the Normal distribution.

In a purely theoretical sense, the names used to describe the distributions imply a mathematical precision and exactness. However, reality rarely is this nice, and hence, the names and descriptive adjectives are often used to refer to the *general shapes* of distributions.

A symmetric distribution is one where the left side of the histogram is a mirror image of the right side. Some of the distributions we discuss are symmetric and some are not, as shown in Table 3.12.

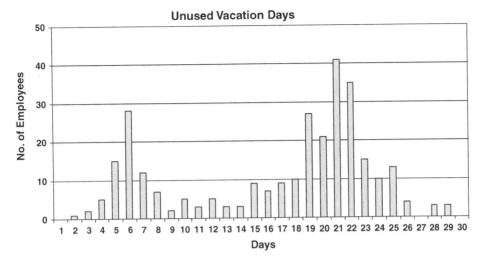

**FIGURE 3.14**    HISTOGRAM OF UNUSED VACATION DAYS WITH CATEGORIES OF 1

## Uniform Distribution

The uniform distribution, also known as the rectangular or flat distribution, is one with the frequencies equal in each category. For example, a theoretical uniform distribution would have exactly the same frequencies in all categories, as in Figure 3.15.

The distributions we encounter will very rarely be this exact. So the inference, for example, for the two uniform distributions shown next, is that the distributions are *approximately* uniform, or *generally* uniform. In Figure 3.16, all the categories are "full." In Figure 3.17 some are empty, but the overall shape is still approximately uniform.

## Bell-Shaped Distribution

A bell-shaped distribution is one with a peak in the middle and tails at both sides. It looks like the cross section of a bell, as shown in Figure 3.18.

**TABLE 3.12    DISTRIBUTIONS WHETHER SYMMETRIC OR NOT**

| Distribution | Symmetric |
| --- | --- |
| Uniform | Yes |
| Bell-shaped | Yes |
| Normal | Yes |
| Skewed | No |
| Bimodal | Sometimes |

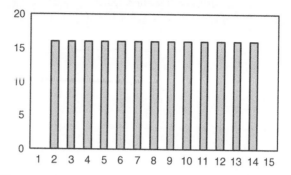

**FIGURE 3.15**     **HISTOGRAM OF THEORETICAL UNIFORM DISTRIBUTION**

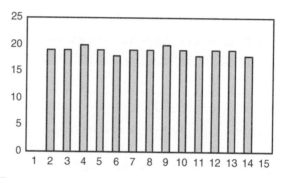

**FIGURE 3.16**     **HISTOGRAM OF UNIFORM DISTRIBUTION—EXAMPLE 1**

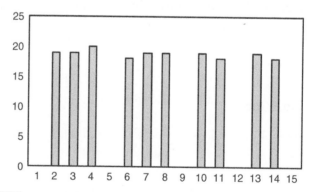

**FIGURE 3.17**     **HISTOGRAM OF UNIFORM DISTRIBUTION—EXAMPLE 2**

Sometimes there are "fat" bells, as in Figure 3.19 and sometimes there are "thin" bells, as in Figure 3.20. Let your imagination be your guide in choosing adjectives.

## Normal Distribution

We mention the Normal distribution here not because we use it or encounter it in compensation, but because it is well known and often misused.

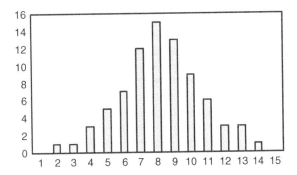

**FIGURE 3.18    HISTOGRAM OF BELL-SHAPED DISTRIBUTION**

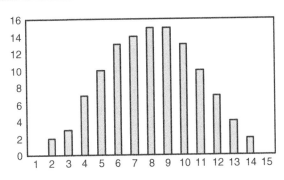

**FIGURE 3.19    HISTOGRAM OF FAT BELL-SHAPED DISTRIBUTION**

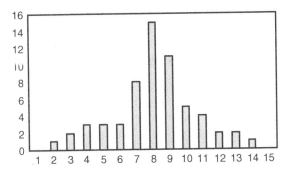

**FIGURE 3.20    HISTOGRAM OF THIN BELL-SHAPED DISTRIBUTION**

There are many bell-shaped distributions—fat ones, skinny ones, tall ones, and short ones. There is one bell-shaped distribution that is particularly useful in statistics: the Normal distribution. The Normal distribution is a symmetric, bell-shaped theoretical distribution shown in Figure 3.21. While the Normal distribution is purely theoretical, there are many situations in nature that are approximately Normal, such as the distribution of heights of women of similar ethnicity, or the distribution of yields of corn per acre. The main uses of the Normal distribution are in statistical inference, such as testing hypotheses or determining confidence intervals.

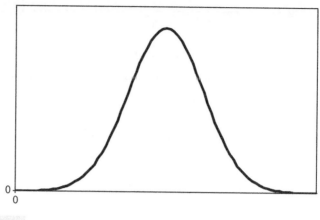

**FIGURE 3.21** NORMAL DISTRIBUTION

In compensation and human resources, however, one popular misuse of the Normal distribution is to force performance rating distributions. For example, in a 5-level rating system, an organization might mandate that 5% of the employees be rated Clearly Outstanding, 15% be rated Exceeds Expectations, 40% be rated Meets Expectations, 15% be rated Below Expectations, and 5% be rated Unsatisfactory. This notion is based on the assumptions that (1) performance ratings "ought" to be Normally distributed and (2) the 5-15-40-15-5 (or any) allocation is Normal.

First of all, there is no *a priori* distribution of performance ratings. The distribution is a function of the wording of the ratings, the manager's ability to assess performance, the work environment, the abilities of the employees, and so on. In other words, there is no "ought" in a performance rating distribution. It "is" what it is in the total context of managing employees.

Furthermore, having discrete categories violates the assumption of Normality because the Normal distribution is a continuous distribution. The allocation illustrated above is symmetric and bell-shaped, but it is not Normal.

The author's opinion as to why this misuse exists is (1) the executives are not comfortable with the managers' abilities to assess performance, and (2) the executives want to control the costs of merit raises. We will discuss this latter point in Chapter 4, *Measures of Location*, when we talk about weighted averages. An organization can meet any salary increase budget with any performance distribution.

This brief discussion on the misuse of the Normal distribution applies to any scheme of forced performance rating distribution, whether Normal or otherwise.

In compensation, you will probably never encounter a pure Normal distribution, even among distributions of salary survey data. But, we mention it here because everyone has heard of it. If you encounter what looks like a Normal distribution, simply say that it is bell-shaped, and unless you conduct certain statistical tests, you really don't know if it is a Normal distribution. More information on the Normal distribution is in the Appendix.

## Skewed Distribution

A skewed distribution is a nonsymmetric distribution with high frequencies on one side, creating a picture of a "hump" or "peak" on that side and low frequencies on the other side, creating a picture of a "tail" on that side. A positively skewed distribution has a peak on the left and a tail off to the right, as shown in Figure 3.22. The tail contains a few data points with high values on the horizontal axis ($x$-axis). A negatively skewed distribution has a peak on the right and a tail off to the left, as shown in Figure 3.23. The tail contains a few data points with low values on the horizontal axis ($x$-axis).

Sometimes your audience will not know where the tail is on a positively or negatively skewed distribution. So it helps to tell them, for example, that "This is a positively skewed distribution, with a tail off to the right." That way there will be no misunderstanding.

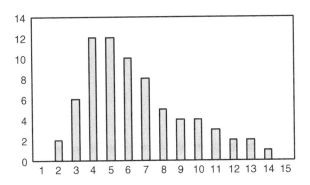

**FIGURE 3.22    HISTOGRAM OF POSITIVELY SKEWED DISTRIBUTION**

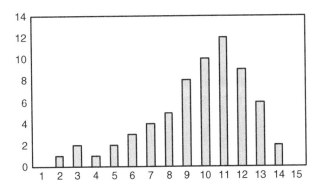

**FIGURE 3.23    HISTOGRAM OF NEGATIVELY SKEWED DISTRIBUTION**

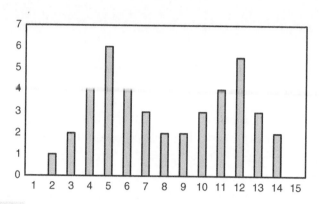

**FIGURE 3.24    HISTOGRAM OF BIMODAL DISTRIBUTION**

## Bimodal Distribution

A bimodal distribution has two modes, or "humps" as shown in Figure 3.24.

In a histogram, the mode is the category with the highest frequency. Again, carrying through the notion of general trends, theoretically there is only one mode here—just one category has the highest frequency—but for purposes of communication, you may want to say this distribution is "approximately" bimodal or "generally" bimodal.

Earlier in this chapter, Figure 3.14 illustrated a distribution of unused vacation days that was approximately bimodal. It had two humps, even though there was only one category with the highest frequency.

**CASE STUDY 4, PART 1 OF 3**

BPD is hiring a new operations manager, and you have been asked to provide information on the competitive pay for that position. You obtain the following survey data on the comparable job in 13 companies from a survey provider.
    The data are in Table 3.13 and also in Data Set 3.

**TABLE 3.13    SALARY SURVEY DATA FOR OPERATIONS MANAGER**

| Company No. | Operations Manager Annual Salary |
|---|---|
| 1 | 97,240 |
| 2 | 103,950 |
| 3 | 107,640 |
| 4 | 110,630 |
| 5 | 114,380 |
| 6 | 119,730 |
| 7 | 124,800 |
| 8 | 129,150 |

**TABLE 3.13**    *(CONTINUED)*

| Company No. | Operations Manager Annual Salary |
|---|---|
| 9 | 136,500 |
| 10 | 140,180 |
| 11 | 146,520 |
| 12 | 157,950 |
| 13 | 176,640 |

You construct a frequency distribution in Table 3.14 and histogram in Figure 3.25 of these data.

**TABLE 3.14**    FREQUENCY DISTRIBUTION OF SALARY SURVEY DATA FOR OPERATIONS MANAGER

| Annual Salary ($000) | No. of Incumbents |
|---|---|
| 90–100 | 1 |
| 100–110 | 2 |
| 110–120 | 3 |
| 120–130 | 2 |
| 130–140 | 1 |
| 140–150 | 2 |
| 150–160 | 1 |
| 160–170 | 0 |
| 170–180 | 1 |

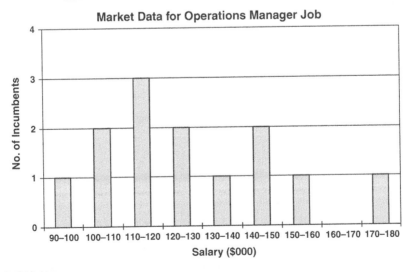

**FIGURE 3.25**    HISTOGRAM OF SALARY SURVEY DATA FOR OPERATIONS MANAGER

You note that the data are skewed positively. In your experience, the majority of jobs have positively skewed distributions of salaries in salary survey data.

This case study will further continue in Chapter 4.

## RELATED TOPICS IN THE APPENDIX

A.3 Types of Numbers

A.9 Normal Distribution

## PRACTICE PROBLEMS

The data for the next four problems are in Table 3.15 and also in Data Set 4.

**3.1** You are conducting a comparative pay study of various job families in a BPD manufacturing facility in Europe. The first population you are studying consists of 30 technicians. Construct an absolute frequency distribution and histogram of their salaries. What is the distribution shape?

**TABLE 3.15    SALARIES OF TECHNICIANS AND STAFF ASSISTANTS**

| Employee No. | Technician Annual Salary (Euros) | Employee No. | Production Assistant Annual Salary (Euros) | Employee No. | Production Assistant Annual Salary (Euros) |
|---|---|---|---|---|---|
| 1 | 43,450 | 1 | 46,260 | 41 | 50,100 |
| 2 | 44,740 | 2 | 46,270 | 42 | 50,110 |
| 3 | 45,050 | 3 | 46,280 | 43 | 50,120 |
| 4 | 45,750 | 4 | 47,660 | 44 | 50,150 |
| 5 | 46,350 | 5 | 47,680 | 45 | 50,160 |
| 6 | 46,840 | 6 | 47,700 | 46 | 50,170 |
| 7 | 47,340 | 7 | 47,760 | 47 | 50,210 |
| 8 | 47,350 | 8 | 48,000 | 48 | 50,240 |
| 9 | 47,650 | 9 | 48,040 | 49 | 50,250 |
| 10 | 48,040 | 10 | 48,260 | 50 | 50,290 |
| 11 | 48,140 | 11 | 48,310 | 51 | 50,310 |
| 12 | 48,240 | 12 | 48,360 | 52 | 51,060 |
| 13 | 48,940 | 13 | 48,660 | 53 | 51,090 |
| 14 | 49,750 | 14 | 48,720 | 54 | 51,100 |
| 15 | 49,840 | 15 | 48,780 | 55 | 51,760 |
| 16 | 49,950 | 16 | 48,790 | 56 | 51,780 |
| 17 | 49,950 | 17 | 48,820 | 57 | 51,810 |
| 18 | 50,040 | 18 | 48,960 | 58 | 51,860 |
| 19 | 50,140 | 19 | 48,960 | 59 | 51,870 |

**TABLE 3.15** *(CONTINUED)*

| Employee No. | Technician Annual Salary (Euros) | Employee No. | Production Assistant Annual Salary (Euros) | Employee No. | Production Assistant Annual Salary (Euros) |
|---|---|---|---|---|---|
| 20 | 50,240 | 20 | 49,030 | 60 | 51,940 |
| 21 | 50,440 | 21 | 49,100 | 61 | 51,960 |
| 22 | 50,540 | 22 | 49,250 | 62 | 51,970 |
| 23 | 50,750 | 23 | 49,330 | 63 | 52,350 |
| 24 | 51,040 | 24 | 49,410 | 64 | 52,370 |
| 25 | 51,340 | 25 | 49,460 | 65 | 52,650 |
| 26 | 51,740 | 26 | 49,550 | 66 | 52,660 |
| 27 | 52,040 | 27 | 49,560 | 67 | 52,680 |
| 28 | 52,240 | 28 | 49,640 | 68 | 52,700 |
| 29 | 52,340 | 29 | 49,660 | 69 | 53,160 |
| 30 | 53,740 | 30 | 49,750 | 70 | 53,210 |
|  |  | 31 | 49,760 | 71 | 53,650 |
|  |  | 32 | 49,860 | 72 | 53,710 |
|  |  | 33 | 49,860 | 73 | 54,250 |
|  |  | 34 | 49,940 | 74 | 54,320 |
|  |  | 35 | 49,950 | 75 | 54,950 |
|  |  | 36 | 49,960 | 76 | 55,030 |
|  |  | 37 | 50,020 | 77 | 55,260 |
|  |  | 38 | 50,040 | 78 | 55,350 |
|  |  | 39 | 50,050 | 79 | 56,550 |
|  |  | 40 | 50,050 | 80 | 56,650 |

**3.2** You also have salaries of 80 production assistants who work in the same manufacturing facility in Europe. Construct an absolute frequency distribution and histogram of their salaries. What is the distribution shape?

**3.3** Compare the absolute distributions of the technician salaries and staff assistant salaries.

**3.4** Compare the relative distributions of the technician salaries and staff assistant salaries.

The data for the next three problems are in Table 3.16 and also in Data Set 5.

**TABLE 3.16   AGES OF EMPLOYEES**

| Employee No. | Age | Employee No. | Age | Employee No. | Age | Employee No. | Age | Employee No. | Age |
|---|---|---|---|---|---|---|---|---|---|
| 1 | 21 | 36 | 34 | 71 | 50 | 106 | 54 | 141 | 70 |
| 2 | 21 | 37 | 34 | 72 | 50 | 107 | 54 | 142 | 70 |
| 3 | 23 | 38 | 34 | 73 | 50 | 108 | 55 | 143 | 70 |
| 4 | 23 | 39 | 34 | 74 | 50 | 109 | 55 | 144 | 70 |
| 5 | 23 | 40 | 34 | 75 | 50 | 110 | 55 | 145 | 70 |
| 6 | 24 | 41 | 34 | 76 | 50 | 111 | 55 | 146 | 71 |
| 7 | 25 | 42 | 35 | 77 | 50 | 112 | 57 | 147 | 71 |
| 8 | 25 | 43 | 35 | 78 | 50 | 113 | 60 | 148 | 71 |
| 9 | 26 | 44 | 35 | 79 | 50 | 114 | 64 | 149 | 72 |
| 10 | 27 | 45 | 37 | 80 | 50 | 115 | 65 | 150 | 72 |

*(continued)*

**TABLE 3.16** (*CONTINUED*)

| Employee No. | Age | Employee No. | Age | Employee No. | Age | Employee No. | Age | Employee No. | Age |
|---|---|---|---|---|---|---|---|---|---|
| 11 | 29 | 46 | 38 | 81 | 50 | 116 | 66 | 151 | 72 |
| 12 | 30 | 47 | 40 | 82 | 50 | 117 | 66 | 152 | 73 |
| 13 | 30 | 48 | 43 | 83 | 50 | 118 | 66 | 153 | 73 |
| 14 | 30 | 49 | 45 | 84 | 50 | 119 | 66 | 154 | 73 |
| 15 | 30 | 50 | 46 | 85 | 51 | 120 | 67 | 155 | 74 |
| 16 | 30 | 51 | 46 | 86 | 51 | 121 | 67 | 156 | 74 |
| 17 | 30 | 52 | 46 | 87 | 51 | 122 | 67 | 157 | 74 |
| 18 | 30 | 53 | 47 | 88 | 51 | 123 | 67 | 158 | 75 |
| 19 | 30 | 54 | 47 | 89 | 51 | 124 | 67 | 159 | 75 |
| 20 | 30 | 55 | 47 | 90 | 52 | 125 | 67 | 160 | 75 |
| 21 | 31 | 56 | 47 | 91 | 52 | 126 | 67 | 161 | 75 |
| 22 | 31 | 57 | 47 | 92 | 52 | 127 | 68 | 162 | 76 |
| 23 | 31 | 58 | 47 | 93 | 52 | 128 | 68 | 163 | 76 |
| 24 | 31 | 59 | 48 | 94 | 52 | 129 | 68 | 164 | 76 |
| 25 | 31 | 60 | 48 | 95 | 53 | 130 | 68 | 165 | 76 |
| 26 | 32 | 61 | 48 | 96 | 53 | 131 | 68 | 166 | 77 |
| 27 | 32 | 62 | 48 | 97 | 53 | 132 | 68 | 167 | 77 |
| 28 | 32 | 63 | 48 | 98 | 53 | 133 | 70 | 168 | 77 |
| 29 | 32 | 64 | 48 | 99 | 53 | 134 | 70 | 169 | 77 |
| 30 | 33 | 65 | 48 | 100 | 54 | 135 | 70 | 170 | 77 |
| 31 | 33 | 66 | 48 | 101 | 54 | 136 | 70 | 171 | 77 |
| 32 | 33 | 67 | 49 | 102 | 54 | 137 | 70 | 172 | 77 |
| 33 | 33 | 68 | 50 | 103 | 54 | 138 | 70 | 173 | 79 |
| 34 | 33 | 69 | 50 | 104 | 54 | 139 | 70 | | |
| 35 | 33 | 70 | 50 | 105 | 54 | 140 | 70 | | |

3.5 You have been asked to ascertain the distribution of ages of the 173 nonmanagerial employees in the same European manufacturing facility. Construct a frequency distribution and histogram using age categories of 20–29, 30–39, 40–49, etc.

3.6 For the same group of employees, construct a frequency distribution and histogram using age categories of 21–30, 31–40, 41–50, etc. Why are these results different from those in Practice Problem 3.5? What point is illustrated?

3.7 For the same group of employees, construct a frequency distribution and histogram using age categories of 21–25, 26–30, 31–35, 36–40, etc. Why are these results different from those in Practice Problem 3.5? What point is illustrated?

3.8 Describe the shapes of the histograms shown in Figure 3.26.

3.9 BPD has three start-up offices in Europe. Initially most of the workforce consisted of contractors. But now the company wants to change the balance to get a higher percentage of employees. Table 3.17 has the data for the branch offices.

Construct a histogram of the percent employees comparing the three locations over a 3 year time period. What might be an issue with this analysis?

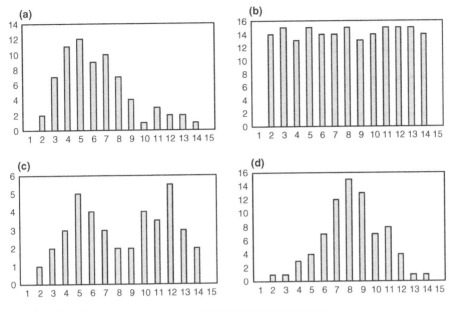

**FIGURE 3.26    HISTOGRAMS OF DIFFERENT DISTRIBUTIONS**

**TABLE 3.17    WORKFORCE POPULATIONS IN THREE LOCATIONS**

| Branch | No. of Contractors | No. of Employees | Total Workforce | % of Employees |
|--------|--------------------|--------------------|-----------------|----------------|
|        | 2011 | 2011 | 2011 | 2011 |
| London | 36 | 12 | 48 | 25.0 |
| Brussels | 36 | 12 | 48 | 25.0 |
| Geneva | 7 | 2 | 9 | 22.2 |
|        | 2012 | 2012 | 2012 | 2012 |
| London | 42 | 15 | 57 | 26.3 |
| Brussels | 36 | 14 | 50 | 28.0 |
| Geneva | 6 | 4 | 10 | 40.0 |
|        | 2013 | 2013 | 2013 | 2013 |
| London | 45 | 18 | 63 | 28.6 |
| Brussels | 36 | 15 | 51 | 29.4 |
| Geneva | 4 | 5 | 9 | 55.6 |

**3.10**   One of your analysts used the same data to produce another histogram, showing visually a dramatic increase for the Geneva office, shown in Figure 3.27. What might be an issue with this presentation?

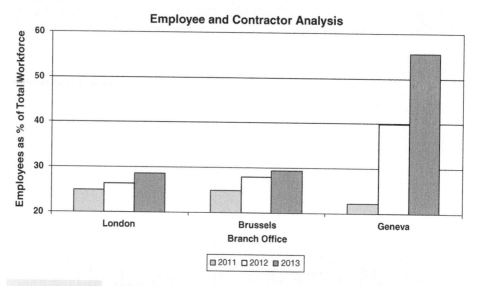

**FIGURE 3.27**    HISTOGRAM OF WORKFORCE POPULATIONS IN THREE LOCATIONS

# CHAPTER 4

# Measures of Location

For a set of data (e.g., salaries, grades, revenues, number of employees), we want to easily summarize and describe the important features that will facilitate an analysis. The most common and useful statistics are called measures of location. A measure of location is one that indicates where a certain part of the data is located. We will discuss two broad categories: measures of central tendency and percentiles.

A measure of central tendency is one that describes the central part of the data, or what is representative, or typical, or expected. We will discuss three measures of central tendency: the mode, the median, and the mean. Along with this, we will discuss four kinds of means: simple, trimmed, weighted, and unweighted.

A percentile is a measure of where other parts of a set of data besides the central tendency are located.

All these measures of location will be defined as we go along in this chapter.

## 4.1 MODE

The mode of a set of data is that value with the highest frequency. In a histogram or bar chart, the mode is the category with the highest frequency.

### Examples

- Your company has more employees in Columbus than in any other city. Columbus is the *mode* city.

---

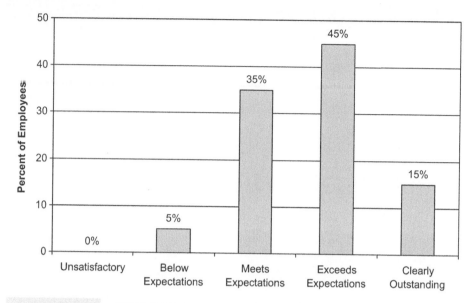

**FIGURE 4.1**    **HISTOGRAM OF PERFORMANCE DISTRIBUTION**

- The accounting department has had the largest number of new hires. The accounting department is the *mode* department.

- There are more employees with ages between 31 and 35 than any other age category. That is the *mode* age category.

- In Figure 4.1, there are more employees rated "Exceeds Expectations" than any other category. That is the *mode* performance category.

Often we don't use the word "mode" in our daily compensation vocabulary. But when we say "This is more" or "This has the most" we are talking about the mode.

Sometimes more than one number will have identical highest frequencies, in which case there is more than one mode. In Figure 4.2, there are two modes with a frequency of six employees: one at 1 year and one at 5 years.

Furthermore, the mode is not necessarily a measure of *central* tendency if the data are skewed. The mode could actually be the highest or the lowest value. In Figure 4.3, the mode is zero training courses completed.

If all data values have the same frequency, then there is no mode. In Table 4.1, each value appears only once, so there is no mode.

## 4.2 MEDIAN

The median of a set of data is a value that half of the data are greater than and half of the data are less than or equal to. If there is an odd number of data points, the

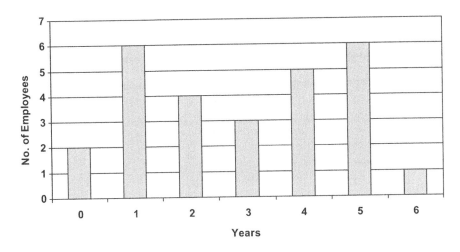

FIGURE 4.2    HISTOGRAM OF TIME IN GRADE

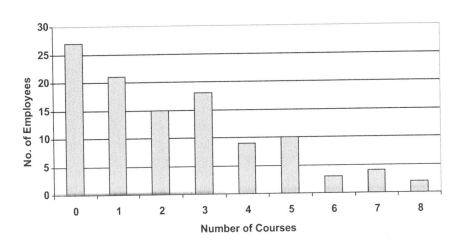

FIGURE 4.3    HISTOGRAM OF NUMBER OF TRAINING COURSES COMPLETED

TABLE 4.1    SALARY DATA WITH NO MODE

| Salary | |
|---|---|
| 21,000 | 25,100 |
| 22,000 | 25,600 |
| 22,500 | 26,000 |
| 23,700 | 26,500 |
| 24,000 | 27,000 |
| 24,300 | 27,100 |

| TABLE 4.2 | AGE OF NEW HIRE—MEDIAN |
|-----------|------------------------|

| Odd No. of Data Points | | Even No. of Data Points | | | Insensitive to Extremes | | |
|---|---|---|---|---|---|---|---|
| | Age of New Hire | | Age of New Hire | | | Age of New Hire | |
| 1 | 22 | 1 | 22 | | 1 | 22 | |
| 2 | 23 | 2 | 23 | | 2 | 23 | |
| 3 | 24 | 3 | 24 | | 3 | 24 | |
| 4 | 24 ←—Median | 4 | 24 ⎱ | 25 Median | 4 | 24 ⎱ | 25 Median |
| 5 | 26 | 5 | 26 ⎰ | (average of | 5 | 26 ⎰ | |
| 6 | 27 | 6 | 27 | middle 2) | 6 | 27 | |
| 7 | 28 | 7 | 28 | | 7 | 28 | |
| | | 8 | 30 | | 8 | 63 | |

median is the value of the middle item when the data are arranged in ascending or descending order, as shown in the first example of Table 4.2. If there is an even number of data points, the median is the average of the two middle values when the data are arranged in ascending or descending order, as shown in the second example.

Since the median is in the middle, it is not affected by data points with extremely high or low values (also known as outliers). In the third example, the median is not affected by the extreme value of 63. That stability and the insensitivity to extreme values are its main positive attributes. However, it does not use all the information represented by the data, but just the information in the middle when the data are ordered ascending or descending. That is its main negative attribute.

## Examples

- Your compensation philosophy is that you want your average pay to be right in the middle of the competitive pay. That is, you want half the competition to pay above you and half to pay below you. The targeted pay level among your competition for your average pay is the *median*.

- Often in comparing the cost of housing in different locations, the *median* home price is used as the relevant statistic to negate the effect of a palatial estate. That means half the houses cost more and half cost less.

## 4.3 MEAN

The mean of a set of *n* numbers is defined as their sum divided by *n*. This is also known as the simple mean, the arithmetic mean, the average, the simple average, and the arithmetic average. When there is no adjective in front of the term mean

**TABLE 4.3    AGE OF NEW HIRE—MEAN AND MEDIAN**

| | Age of New Hire | | | Sensitive to Extremes Age of New Hire | |
|---|---|---|---|---|---|
| 1 | 22 | | 1 | 22 | |
| 2 | 23 | | 2 | 23 | |
| 3 | 24 | | 3 | 24 | |
| 4 | 24 } | 25 Median | 4 | 24 } | 25 Median |
| 5 | 26 | | 5 | 26 | |
| 6 | 27 | | 6 | 27 | |
| 7 | 28 | | 7 | 28 | |
| 8 | 30 | | 8 | (63) | |
| | 25.5 | Mean | | 29.6 | Mean |

or average, it is implied that the mean or average is a simple one. Often the terms "mean" and "average" are used as synonyms.

Using symbols, let

$n =$ the number of data points
$x =$ the data, the individual data points
$\bar{x} =$ the mean, pronounced "x-bar"
$\Sigma =$ "the sum of"

$$\text{Then} \quad \bar{x} = \frac{\Sigma x}{n}$$

The main characteristic of the mean is that it uses all the data points. This is good news and sometimes bad news.

The good news is that it *does* use all the data points and in its own way represents all the numbers under study. The bad news is that as such, it is influenced, or affected by every one of the data points, including any extreme values. If there are extremely high values, the mean is pulled toward the high end, as shown in the second example of Table 4.3. If there are extremely low values, the mean is pulled toward the low end.

Here are two examples in comparing the median and the mean. The first is shown in Figure 4.4.

Suppose there were five people with the indicated heights in the first picture. The median is 70 in. and the mean is 69.4 in. But then the man on the end wanted to be a basketball player and took growth steroids. Now look at the second picture. What happened to the median? Nothing, because it is not influenced by extreme values. What happened to the mean? It went up to 73.2 in., because it is influenced by every data point, including the extreme values.

65   67   70   72   73

Height (in.) Before
Mean = 69.4
Median = 70

65   67   70   72   92

Height (in.) After
Mean = 73.2
Median = 70

**FIGURE 4.4**     **HEIGHT—EFFECT OF HIGH OUTLIER**

170 173 175  183  185
Height (cm.) Before
Mean = 177.2
Median = 175

123 173 175  183  185
Height (cm.) After
Mean = 167.8
Median = 175

**FIGURE 4.5**     **HEIGHT—EFFECT OF LOW OUTLIER**

The second example is shown in Figure 4.5.

Suppose your European headquarters was celebrating its 10th anniversary and was having a multicultural dance troupe perform. The original cast is shown in the first picture, with indicated heights. The median is 175 cm and the mean is 177.2 cm. But the woman on the left got sick and had her daughter stand in for her. Now look at the second picture. What happened to the median? Nothing, because it is not influenced by extreme values. What happened to the mean? It went down to 167.8 cm, because it is influenced by every data point, including the extreme values.

**TABLE 4.4**     **EXAMPLES OF MEANS**

| | |
|---|---|
| Average insurance premium is 13,280 | Average sales is 420,000 |
| Average pay for analysts is 84,000 | Average bonus is 2,600 |
| Average raise is 3.2% | Average number of employees is 127 |
| Average age is 32.6 years | Average number of hits on the HR website is 2,231 per day |
| Average length of service is 7.9 years | Average wait time when calling the Call Center is 2 min 32 sec |
| Average number of accounts is 38 | |
| Average number of bids is 5.4 | |

The mean, or average, is the most widely used measure of central tendency. We use it for almost every variable or measurement we deal with in all areas of compensation and human resources, to describe or to compare sets of data. Here are some examples in Table 4.4.

## 4.4  TRIMMED MEAN

The trimmed mean is the average of the remaining middle data points after trimming, or excluding a certain number of the highest and lowest values.

There are situations where you want to exclude the influence of the extreme values, and at the same time use as much of the remaining information as possible. The median does exclude the influence of the extreme values but it also uses just one or two middle values in its calculation. The trimmed mean is a nice compromise between the mean and the median, between using all the information with the influence of outliers and using too little information without the influence of outliers.

Here are two possible trimming schemes, among others.

- Trim the top and bottom values and take the mean of the remaining data points. You would use this scheme if, in your experience, there were just occasional outliers or spurious data in the type of data you are analyzing, and you wanted to negate their effect.

  If it turns out that there are two identical highest (or lowest) values and you are only trimming the top (or bottom) "one," then just trim one of the values and leave the other one in you calculations.

- Trim the top and bottom 5% of the values and take the mean of the remaining middle 90% of the data points. You would use this scheme if your intent was to just reflect the middle 90% of the data in your analysis and not worry about the top and bottom 5%.

  In any event, you must trim the *same number* of data points from the high side as you do from the low side. Otherwise you will be deliberately biasing the results.

  In addition, when you do trim, you should be transparent with what you did, and report your trimming scheme. So if you trimmed the top and bottom 5% of the raw data, then tell that you did and why you did so.

## 4.5  OVERALL EXAMPLE AND COMPARISON

In Table 4.5, we have 12 salaries, which have been ordered and numbered for convenience.

**TABLE 4.5**    **COMPARISON OF MEASURES OF CENTRAL TENDENCY**

|  | Salary | Measures of Central Tendency | |
|---|---|---|---|
| 1 | 21,000 | Mode | 23,000 |
| 2 | 22,000 | Median | 23,500 |
| 3 | 23,000 | | |
| 4 | 23,000 | $n$ | 12 |
| 5 | 23,000 | Total | 306,000 |
| 6 | 23,000 | Mean | 25,500 |
| 7 | 24,000 | | |
| 8 | 24,000 | | Excluding low and high |
| 9 | 27,000 | $n$ | 10 |
| 10 | 28,000 | Total | 247,000 |
| 11 | 30,000 | Trimmed mean | 24,700 |
| 12 | 38,000 | | |

The salary with the highest frequency is 23,000. There are four employees with that salary. Thus, 23,000 is the *mode*.

There is an even number of data points. Hence, the *median* is the average of the middle two, namely the average of the 6th data point (23,000) and the 7th data point (24,000), or 23,500.

The *mean* is the total, 306,000, divided by the number of data points, 12, giving a value of 25,500.

**TABLE 4.6**    **CHARACTERISTICS OF MEASURES OF CENTRAL TENDENCY**

| Measure | Characteristics |
|---|---|
| Mode | The value or category with the highest frequency<br>Not necessarily a measure of central tendency<br>There can be more than one<br>A set of data might have no mode |
| Median | The value in the middle when the data are ordered<br>A measure of central tendency<br>Uses information only from the middle values<br>Not influenced by extreme values |
| Mean | The sum divided by the number of data points<br>A measure of central tendency<br>Uses all the information<br>Influenced by extreme values |
| Trimmed mean | The average of the middle portion of the data<br>A measure of central tendency<br>Uses most of the information<br>Not influenced by extreme values |

To calculate a *trimmed mean*, we have excluded the lowest- and highest-valued data points (21,000 and 38,000), taken the sum of the remaining middle 10 data points, which is 247,000, and divided by 10, giving a value of 24,700.

Note that the value of the trimmed mean is usually between the mean and the median. Indeed, if you continue trimming more and more data points, you ultimately end up with the median.

## Comparison

Table 4.6 shows a comparison of the four measures of central tendency discussed.

It is important to note that these measures are nothing more and nothing less than their definitions. Which one you use depends on your particular situation—what you are analyzing and the associated context.

---

**CASE STUDY 4, PART 2 OF 3**

You are finding the market pay for the job of operations manager as input to the hiring offer. In Chapter 3 , we had just constructed a frequency distribution and histogram from the salary survey data. The data are in Table 3.13 and also in Data Set 3. The frequency distribution (Table 4.7) and histogram (Figure 4.6) are repeated here.

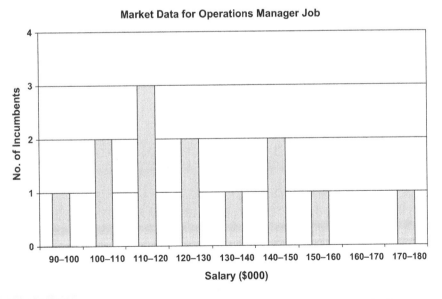

**FIGURE 4.6**   **HISTOGRAM OF SALARY SURVEY DATA FOR OPERATIONS MANAGER**

| TABLE 4.7 | FREQUENCY DISTRIBUTION OF SALARY SURVEY DATA FOR OPERATIONS MANAGER |
|---|---|

| Frequency Distribution | |
|---|---|
| Annual Salary ($000) | No. of Incumbents |
| 90–100 | 1 |
| 100–110 | 2 |
| 110–120 | 3 |
| 120–130 | 2 |
| 130–140 | 1 |
| 140–150 | 2 |
| 150–160 | 1 |
| 160–170 | 0 |
| 170–180 | 1 |

You now calculate measures of central tendency using the formulas presented earlier in this chapter. The ones you are interested in are the mean and the median, as you have found these two measures to be relevant in hiring situations. You calculate them and get the following.

| | |
|---|---|
| Mean | 128,101 |
| Median | 124,800 |

You note that the mean is higher than the median, which is typical for a positively skewed distribution.

This case study will conclude in Chapter 5.

## 4.6 WEIGHTED AND UNWEIGHTED AVERAGE

Technically, these terms apply when you are averaging averages. This takes place quite frequently when determining the average "market pay" for a job or skill, and you have averages from different companies or from different surveys and you have to combine them into a single number. Weighting is a method of assigning importance to a number.

The weighted average is the average of the company averages weighted by number of incumbents. It is mathematically equivalent to the simple average of all the incumbents. It treats each incumbent equally. In salary survey situations, sometimes the weighted average is called the "incumbent-based average." In this case, you are deciding that the averages of companies with many incumbents

**TABLE 4.8     WEIGHTED AND UNWEIGHTED AVERAGE—EXAMPLE 1**

| Company | No. of Incumbents | Average Pay | Total Pay (No.)(Avg.) |
|---------|-------------------|-------------|-----------------------|
| A | 10 | 20,000 | 200,000 |
| B | 20 | 26,000 | 520,000 |
| Total | 30 | | 720,000 |

are more important in identifying the market pay of a job than the averages of companies with just a few incumbents.

The unweighted average is the simple average of the company averages. It treats each company equally. In salary survey situations, sometimes the unweighted average is called the "company-based average." In this case, you are deciding that each company's average is of equal importance to every other company's average in identifying the market pay of a job.

We will illustrate the mathematics with data from a survey of a job with two participating companies. We start with the number of incumbents and the average pay for each company in Table 4.8.

Then we add a column for the total pay of the incumbents in each company. To get the total pay for the incumbents in Company A, multiply the number of incumbents times their average pay.

$$\text{Total pay for Company A} = (10)(20,000) = 200,000$$

A similar calculation is done for Company B. The totals are then added to produce 720,000.

The weighted average is calculated by dividing this total of the total pay by the total number of incumbents.

$$\text{Weighted average} = 720,000/30 = 24,000$$

As a side note, if all 30 individual salaries were available, their total would be 720,000 with an average of 24,000.

The unweighted average is the simple average of the averages, calculated by summing the company averages and dividing by the number of companies.

$$\text{Unweighted average} = (20,000 + 26,000)/2 = 46,000/2 = 23,000$$

---

Technically, the weighted average of a set of numbers (in the example above, the numbers themselves are averages and the weights are the numbers of incumbents) is

$$\text{Weighted average} = \frac{\Sigma(\text{weight})(\text{number})}{\Sigma\text{weight}}$$

**TABLE 4.9** **WEIGHTED AND UNWEIGHTED AVERAGE—EXAMPLE 2**

| Company | No. of Incumbents | Average Pay | Total Pay |
|---|---|---|---|
| A | 15 | 50,000 | 750,000 |
| B | 20 | 45,000 | 900,000 |
| C | 10 | 47,000 | 470,000 |
| D | 55 | 60,000 | 3,300,000 |
| Total | 100 | | 5,420,000 |
| Unweighted average | | 50,500 | |
| Weighted average | | 54,200 | |

If the company with the largest weight pays high, the weighted average will be pulled toward the high side and will be higher than the unweighted average. If the company with the largest weight pays low, the weighted average will be pulled toward the low side and will be lower than the unweighted average. If the company with the largest weight pays in the middle, the weighted average will be pulled toward the middle, and be close to the unweighted average.

This is shown in Table 4.9 with four companies.

Note that company D has the largest weight and also has the highest average pay. Hence, the weighted average is pulled toward the higher side and ends up higher than the unweighted average.

*Just as the simple mean is pulled toward extreme values, so the weighted mean is pulled toward the value with the greatest weight.*

Note in our example that there is an approximate 7% difference between the two means, and this is a significant difference if you are basing a pay decision on the result. What is "significant" is a judgment call. In the author's opinion, anything over 5% in a situation like this is "significant." You should build up your own history file and develop your own definition of "significant." In any event, you have to decide which one to use—the unweighted average or the weighted average—and why.

## Which Measure to Use?

The answer is "It depends." The most common market reference points chosen against which to posture a compensation program are the weighted mean, the unweighted mean, and the median. Following are comments that give a perspective on the different kinds of measures.

*Weighted Mean.* One argument for using the weighted mean is that it reflects equally the number of incumbents in the survey, which, if the companies participating are truly representative of the chosen market, reflects the market value of the job or skill. The number of incumbents represents the number of potential openings for that job or skill. In the extreme, if there were only two companies in the market—one with 10 incumbents and one with 1,000 incumbents—then the one with the 1,000 is determining the market, and the weighted mean would be the measure to use. If all the chosen competitors were participants in the survey, then the weighted mean would make sense. And even if they all were not in the survey, *if the survey is representative of the chosen market, and it is desired to target pay levels at the mean pay of the incumbents in the market, then choose the weighted mean.*

*Unweighted Mean.* One argument for using the unweighted mean is that in the process of getting companies to participate in the survey, there may be a desired company with a large number of incumbents that is not in the survey, and may be unknowingly on the high or low side of the data.

Expanding on the example above where the first company had 10 incumbents and the second company had 1,000 incumbents, suppose there were a third company with 1,000 incumbents that was in the market, but you do not have their data, and you do not know if they are a low, middle, or high payer. You might want to hedge your bet in giving all the weight to the 1,000-incumbent company for which you do have data.

In this case, giving equal weight to each company would make sense to determine the "typical" or "representative" value for that job or skill. In other words, *if you are not sure of the representativeness of the survey of the chosen competition, then choose the unweighted mean.*

*Combination Mean.* Some companies use a combination mean, which is the mean of the weighted mean and the unweighted mean. This gives equal weight to both the number of incumbents and the number of companies. *If you want to balance the importance given to the number of incumbents in each company and the importance given to the companies themselves, then choose the combination mean.*

The weighted mean, unweighted mean, and combination mean are analogous to the U.S. Congress. The weighted mean gives equal importance to each incumbent, so that the company with the most incumbents has the most influence; the House of Representatives gives equal importance to each citizen, so that the state with the most citizens has the most influence. The unweighted mean gives equal weight to each company—each company is of equal importance no matter how many incumbents are in each one; the Senate gives equal weight to each state—each state is of equal importance no matter how many citizens are in each one. The combination mean gives equal weight to both the incumbents and the companies; for a bill to become law, both houses of Congress must pass it, each one of equal status.

*Median.* One argument for using the median is that it is not impacted by extremely high or low values. By its very nature of using all the information, the weighted mean is influenced by both the companies with the highest number of incumbents and the companies that pay very high or very low.

Sometimes all these balance out, but *if you want to avoid the influences of outliers, or just want to reference the middle of the market pay, then choose the median.*

*Trimmed Mean.* One argument for using the trimmed mean is that it is not impacted by extremely high or low values, but it does use all the information in the middle of the data. So *if you want to avoid the influences of outliers but use all the information in the middle, then choose the trimmed mean.*

Whichever measure you use, be sure to have a reason for choosing it and tell your audience the reason. If you decide to use a different measure for a particular job or group of jobs, be sure to disclose the different measure you are using and why. That way, you will be transparent with your statistical methodology and all that can be questioned, then, are your assumptions.

> Your openness will engender credibility not only in your results but also in your honesty and integrity.

## Application of Weighted Averages to Salary Increase Guidelines

If an organization gives raises based on the combination of performance and position in range, compensation will typically develop guidelines for the manager to use. The guidelines are usually in the form of a "merit matrix," so-called because it is a two-dimensional table with performance level in the vertical direction and position in range in the horizontal direction.

The challenge is to develop guidelines that will help the managers stay within the overall organizational salary increase budget. This is done through the use of weighted averages.

This method will allow an organization to control costs without having to have a forced distribution of performance ratings, or levels.

One of BPD's divisions has a $10 million payroll and a 3.2% salary increase budget. That translates to a dollar amount of $320,000. Assume there are five performance levels and three position in range categories. You first create the following template shown in Table 4.10, which you will use more than once.

**TABLE 4.10    TEMPLATE FOR SALARY INCREASE GUIDELINES**

| Performance Level | Position in Range | | |
| :---: | :---: | :---: | :---: |
| | Lower 1/3 | Middle 1/3 | Upper 1/3 |
| 5 | | | |
| 4 | | | |
| 3 | | | |
| 2 | | | |
| 1 | | | |

**TABLE 4.11     TEMPLATE WITH SALARIES ENTERED**

| | Salaries | | |
|---|---|---|---|
| | Position in Range | | |
| Performance Level | Lower 1/3 | Middle 1/3 | Upper 1/3 |
| 5 | 875,000 | 300,000 | 125,000 |
| 4 | 2,850,000 | 1,400,000 | 425,000 |
| 3 | 2,100,000 | 1,100,000 | 350,000 |
| 2 | 290,000 | 125,000 | 60,000 |
| 1 | 0 | 0 | 0 |
| Grand total | | 10,000,000 | |

In each cell, enter the total salaries of the employees whose performance level and position in range are in that cell, shown in Table 4.11.

The next step involves a trial-and-error process. For trial 1 and using the same template, enter the raise percent in each cell you think is appropriate for an employee with that performance level and in that position in range, shown on the left side of Table 4.12.

To obtain the dollars the percents would produce, using the same template multiply the salaries in each cell (Table 4.11) by the raise percent for that cell to get the dollar cell payout, shown in the right side of Table 4.12. For example, for the upper left cell, $(875,000)(6.0\%) = 52,500$. Then add up all the cell payouts to see how the total compares with the budget of 320,000. We then divide that number by the total salaries of 10,000,000 and multiply by 100 to arrive at an overall salary increase percentage.

In the context of weighted averages, the numbers we are averaging are the cell percents and the weights are the cell salaries.

**TABLE 4.12     COMPLETED TEMPLATES FOR SALARY INCREASE GUIDELINES—TRIAL 1**

| | Trial 1 Raise Percent | | | | Trial 1 Raise Dollars | | |
|---|---|---|---|---|---|---|---|
| | Position in Range | | | | Position in Range | | |
| Performance Level | Lower 1/3 | Middle 1/3 | Upper 1/3 | Performance Level | Lower 1/3 | Middle 1/3 | Upper 1/3 |
| 5 | 6.0 | 5.0 | 4.0 | 5 | 52,500 | 15,000 | 5,000 |
| 4 | 5.0 | 4.0 | 3.0 | 4 | 142,500 | 56,000 | 12,750 |
| 3 | 4.0 | 3.0 | 2.0 | 3 | 84,000 | 33,000 | 7,000 |
| 2 | 2.0 | 1.0 | 0.0 | 2 | 5,800 | 1,250 | 0 |
| 1 | 0.0 | 0.0 | 0.0 | 1 | 0 | 0 | 0 |
| | | | | Grand total | | 414,800 | 4.15% |

| TABLE 4.13 | | | COMPLETED TEMPLATES FOR SALARY INCREASE GUIDELINES—TRIAL 2 | | | |
|---|---|---|---|---|---|---|
| **Final Trial Raise Percent** | | | | **Final Trial Raise Dollars** | | |
| | **Position in Range** | | | | **Position in Range** | |
| Performance Level | Lower 1/3 | Middle 1/3 | Upper 1/3 | Performance Level | Lower 1/3 | Middle 1/3 | Upper 1/3 |
| 5 | 5.5 | 4.5 | 3.5 | 5 | 48,125 | 13,500 | 4,375 |
| 4 | 4.0 | 3.0 | 2.0 | 4 | 114,000 | 42,000 | 8,500 |
| 3 | 3.0 | 2.0 | 1.0 | 3 | 63,000 | 22,000 | 3,500 |
| 2 | 0.0 | 0.0 | 0.0 | 2 | 0 | 0 | 0 |
| 1 | 0.0 | 0.0 | 0.0 | 1 | 0 | 0 | 0 |
| | | | | Grand total | 319,000 | 3.19% | |

The results of our first trial are too high, so we revise the raise percents. After several tries, we get the following results in Table 4.13, which meets the budget of 3.2%. As illustrated here, you will seldom get the exact approved budget amount, so it is prudent to come in a little less.

Of course the managers may possibly deviate from these guidelines, or you will give them some flexibility within each cell, but this will give them a start to help them come in at or under budget.

There are no firm rules for developing a merit matrix other than the philosophy on what the organization wants to reward. There are many variations in raise percents that will result in the same answer.

There are other ways to develop salary increase guidelines. This is the most common method, using weighted averages.

## 4.7 SIMPSON'S PARADOX

Simpson's Paradox is a statistical paradox in which the trends observed in several groups are reversed when the groups are combined, or when the trend observed in a combination of groups is reversed when examining the groups separately. When we encounter such a reversal of trends in compensation or human resources, it often comes as a surprise. This paradox is related to the concept of weighted averages.

BPD has a training program for assemblers where they assemble products, and then the products are tested for defects. Each week you give an award to the trainee who has the highest percentage of units assembled with no defects. But you have a concern with the numbers you are getting. For example, in comparing two trainees for the first 2 weeks, Trainee A assembled better each week than Trainee B and she got the award both weeks. But when you look at

**TABLE 4.14    TRAINING RESULTS SIMPSON'S PARADOX**

| | Percent of Units Assembled Without Any Defects | | |
| --- | --- | --- | --- |
| | Week 1 | Week 2 | Overall |
| Trainee A | 30 | 90 | 35 |
| Trainee B | 10 | 60 | 55 |

what they have done in total for both weeks, Trainee B was better, but did not get any award. You don't understand how this can be. Here are the numbers in Table 4.14.

This is an example of what is known as *Simpson's Paradox*, which is a statistical paradox described by E. H. Simpson [6] in 1951 and G. U. Yule [9] in 1903, in which the trends observed in several groups are reversed when the groups are combined.

To understand what is going on, we have to examine the raw data, shown in Table 4.15.

We can see that each employee improved percentage-wise from week 1 to week 2, and that each week, Trainee A did better than Trainee B. But when the numbers are totaled, the 60 units assembled by Trainee B in week 2 overpowered all the other numbers and had the deciding influence on the totals.

Mathematically, we have two things going on. First, "week" is a "lurking variable" that influences in different ways the performances of the two trainees. Trainee B improved much more during the second week than Trainee A for whatever reason. Trainee B improved sixfold while Trainee A improved only threefold.

Second, the sizes of the data sets within each week are different (100 versus 10) for each trainee.

Simpson's Paradox is caused by a combination of a lurking variable and data from unequal-sized groups being combined into a single data set. The unequal

**TABLE 4.15    TRAINING RESULTS SIMPSON'S PARADOX EXPLAINED**

| | | Total No. of Units Assembled | No. of Units Assembled Without Any Defects | % Without Any Defects |
| --- | --- | --- | --- | --- |
| Trainee A | Week 1 | 100 | 30 | 30 |
| | Week 2 | 10 | 9 | 90 |
| | Total | 110 | 39 | 35 |
| Trainee B | Week 1 | 10 | 1 | 10 |
| | Week 2 | 100 | 60 | 60 |
| | Total | 110 | 61 | 55 |

group sizes, in the presence of a lurking variable, can weigh the results such that they may lead to seriously flawed conclusions.

One possible solution in this training program example is to give recognition for each week, and then another recognition for the overall time period, if that is what you want to do.

So the moral of the story here is threefold. First of all decide what you really want to do. In this case, do you want to reward weekly performance, overall performance, or both? Second, be cautious when combining data sets of different sizes from different sources. And third, look at the details in the data so that you thoroughly understand what is going on.

It is important to understand that Simpson's Paradox does not happen every time you combine data sets. But when you have a surprise reversal of results, then investigate to see if the paradox is there.

Simpson's Paradox is such a significant issue and generally not known in the compensation or human resources communities that we will illustrate it with another example.

BPD has a performance rating scale as follows.

| Performance Level | Rating |
|---|---|
| Clearly outstanding | 5 |
| Exceeds expectations | 4 |
| Meets expectations | 3 |
| Below expectations | 2 |
| Unsatisfactory | 1 |

The HR manager was analyzing the ratings given by four departments to compare the 4 or 5 ratings given to men to the 4 or 5 ratings given to women. Here, are the results of the analysis in Table 4.16.

In each department, the percentage of women rated 4 or 5 was greater than the percentage of men rated 4 or 5. For example, in Dept. A, 3 out of 8 women, or 38%, received a 4 or 5, and 4 out of 12 men, or 33%, received a 4 or 5.

**TABLE 4.16    PERFORMANCE LEVELS SIMPSON'S PARADOX EXPLAINED**

| | No. of Women | No. of Men | Total | No. of Women 4 or 5 | No. of Men 4 or 5 | % of Women 4 or 5 | % of Men 4 or 5 |
|---|---|---|---|---|---|---|---|
| Dept. A | 8 | 12 | 20 | 3 | 4 | 38 | 33 |
| Dept. B | 2 | 9 | 11 | 1 | 3 | 50 | 33 |
| Dept. C | 5 | 16 | 21 | 2 | 6 | 40 | 38 |
| Dept. D | 3 | 90 | 93 | 2 | 53 | 67 | 59 |
| **Total** | **18** | **127** | **145** | **8** | **66** | **44** | **52** |

Yet, when looking at the totals, the difference is reversed, where the percentage of women rated 4 or 5 (44%) is less than the percentage of men rated 4 or 5 (52%).

Here again, the reason for this reversal is the particular combination of different numbers of employees in each department (e.g., the large number of men in Dept. D) along with the lurking variable of "department."

In this case, it may be inappropriate to combine the numbers because of the lurking variable.

The reason Simpson's Paradox is so important to be aware of and to understand is that decisions based on the individual group results may be different from and conflict with decisions based on the results of the combined groups. When you discover such a paradox, go back to the beginning of why you are doing the analysis in the first place and confirm or revise the study objectives.

## 4.8 PERCENTILE

A percentile is a value that a given percentage of the data is less than or equal to. For example, the 90th percentile is a value that 90% of the data are less than or equal to. The 50th percentile is a value that 50% of the data are less than or equal to. The 50th percentile is also known as the median.

Sometimes a company will target its average pay or its benefits toward a certain percentile of the market, say the 50th or 60th or 75th. Sometimes a company will target the pay of individuals toward certain percentiles of the market pay for their job based on performance. For example,

| Performance Level | Targeted Percentile |
|---|---|
| Clearly outstanding | 90th |
| Exceeds expectations | 75th |
| Meets expectations | 50th |
| Below expectations | 25th |
| Unsatisfactory | 10th |

Standard percentiles are the 10th, 25th, 50th, 75th, and the 90th. Often these percentiles are denoted by P10, P25, P50, P75, and P90, respectively. The 0th percentile (P0) is the minimum value in a data set, and the 100th percentile (P100) is the maximum value in a data set.

Here are data from a survey of Accounting Assistant salaries. They are sorted in order and numbered from low to high to designate rank along with a histogram in Figure 4.7 to understand the distribution. The data are in Table 4.17 and also in Data Set 6. Replicated values are numbered separately (42,600; 41,500; 40,100; and 39,900). The numbering indicates the rank of the data.

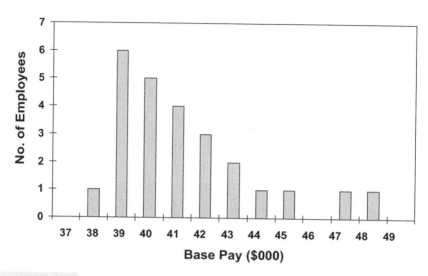

**Distribution of Market Pay
Accounting Assistant**

**FIGURE 4.7**     HISTOGRAM OF ACCOUNTING ASSISTANT SALARIES

To calculate a percentile of a data set you first calculate the rank of the percentile, and then find or calculate the number that corresponds to that rank. We will illustrate this with the Accounting Assistant data.

The rank, or position, of the percentile when the data are ordered and numbered from low to high, including numbering all the repeat values, is given by the following formula. Let,

$p$ = proportion of the percentile
$n$ = number of data points
$r$ = rank of the percentile

Then, $r = pn + (1 - p)$

If the calculated rank is a whole number, then the percentile is the data value with that rank. If the calculated rank is not a whole number and falls between the rank of two data points, then linear interpolation is used to calculate the value of the percentile.

*Example 4.1*

Calculate the 75th percentile. There are 25 data points, and we sort them (already done) and number them from low to high (already done) to get the ranks.

$$\text{Rank of P75} = (0.75)(25) + (1 - 0.75) = 18.75 + 0.25 = 19$$

**TABLE 4.17**  **RAW DATA AND RANKS OF ACCOUNTING ASSISTANT SALARIES**

| Rank | Accounting Assistant Annual Salary |
|------|-----------------------------------:|
| 25   | 48,400 |
| 24   | 47,600 |
| 23   | 45,100 |
| 22   | 44,200 |
| 21   | 43,300 |
| 20   | 43,100 |
| 19   | 42,600 |
| 18   | 42,600 |
| 17   | 42,500 |
| 16   | 41,800 |
| 15   | 41,500 |
| 14   | 41,500 |
| 13   | 41,200 |
| 12   | 40,800 |
| 11   | 40,700 |
| 10   | 40,200 |
| 9    | 40,100 |
| 8    | 40,100 |
| 7    | 39,900 |
| 6    | 39,900 |
| 5    | 39,900 |
| 4    | 39,900 |
| 3    | 39,700 |
| 2    | 39,500 |
| 1    | 38,600 |

The 75th percentile is the value of the 19th data point, namely 42,600. Here the percentile is equal to one of the data values. In this data set, there are two values of 42,600. That is fine. We just want the one with a rank of 19.

*Example 4.2*
Calculate the 90th percentile. There are 25 data points, and we sort them (already done) and number them from low to high (already done) to get the ranks.

$$\text{Rank of P90} = (0.90)(25) + (1 - 0.90) = 22.5 + 0.1 = 22.6$$

The 90th percentile is the value of the 22.6th data point, which doesn't exist. So it is a calculated value six tenths of the way up from the 22nd data value to the 23rd data value.

P90 has a rank of 22.6

22nd data value is 44,200

23rd data value is 45,100

P90 is six tenths of the way from 44,200 to 45,100

$$P90 = 44,200 + (0.6)(45,100 - 44,200)$$
$$= 44,200 + (0.6)(900) = 44,200 + 540 = 44,740$$

The mathematics just performed is called linear interpolation. Here the percentile is a calculated value between two data points. If the two data points in question happen to have the same value, the percentile "between them" is that same value.

## Reverse Percentile

Sometimes you want to know what percentile a number is when compared to a set of other numbers. For example, you might want to know what percentile your company's revenue is compared to your designated competition.

You first find the rank of your revenue, and then use that to calculate the percentile. Let,

$r$ = rank of your revenue
$n$ = number of data points, including you
$p$ = proportion of percentile

Rearranging the previous formula, we get

$$p = (r - 1)/(n - 1)$$

You then multiply $p$ by 100 to convert it to a percentile.

In the example shown in Table 4.18 and Figure 4.8, there are 11 companies that you consider the relevant competition for one of BPD's divisions and you

**TABLE 4.18**  REVENUE OF COMPETITION

| Rank | Revenue ($M) |
| --- | --- |
| 12 | 892 |
| 11 | 710 |
| 10 | 655 |
| 9 | 632 |
| 8 | 601 |
| 7 | 587 |
| 6 | 565 |
| 5 | 532 |
| 4 | 490 |
| 3 | 450 |
| 2 | 450 |
| 1 | 423 |

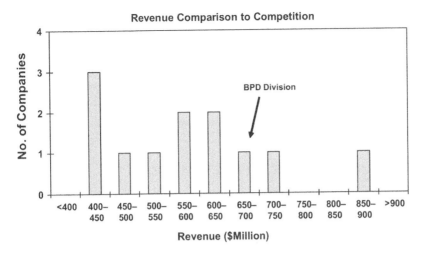

**FIGURE 4.8**   **HISTOGRAM OF REVENUE OF COMPETITION**

want to calculate what percentile BPD is in the group, based on revenue. You place your revenue in the group in the correct ordered position, increasing the total number to 12. Your rank is 10, with revenues of $655 million.

Applying the formula, we get

$$p = (10 - 1)/(12 - 1) = 9/11 = 0.82 \text{ (rounded)}$$

$$(100)(0.82) = 82$$

Your company's revenue is at the 82nd percentile among your competition.

**Estimating Percentiles**   Sometimes you only have the standard percentiles from a salary survey and do not have access to the raw data, but want to determine an "in-between" percentile. Suppose you want to estimate the 60th percentile for a job for which you only have the standard percentiles, shown in Table 4.19.

**TABLE 4.19**   **STANDARD PERCENTILES OF SALARIES**

|  | Salary |
|---|---|
| $P_{10}$ | 81,000 |
| $P_{25}$ | 88,000 |
| $P_{50}$ | 95,000 |
| $P_{75}$ | 105,000 |
| $P_{90}$ | 112,000 |

We will estimate the 60th percentile using linear interpolation. The 60th percentile is between the 50th percentile and the 75th percentile. How far between? It is four tenths of the way up from the 50th to the 75th.

$$(60 - 50)/(75 - 50) = 10/25 = 0.4$$

So we go four tenths of the way from the value of the 50th percentile up to the value of the 75th percentile.

$$P60 \text{ estimate} = 95{,}000 + (0.4)(105{,}000 - 95{,}000) = 95{,}000 + (0.4)(10{,}000)$$
$$= 95{,}000 + 4{,}000 = 99{,}000$$

**Ill-Defined Function**   In mathematics, the percentile is considered an ill-defined function, which is one in which you can arrive at more than one answer that satisfies the definition. For example, in Table 4.18 any value greater than or equal to 565 and less than 587 would satisfy the definition of the 50th percentile, namely that half the data points are less than or equal to it.

Hence, there are a variety of formulas that are used to calculate percentiles. The one presented here is the one that is most commonly agreed-upon, so that we will be consistent in our analyses and presentations.

In the past, the formula for the rank of the $p$th percentile was given as $p(n+1)$, which was the common default in most of the older statistics software. Its advantage is that it is simple. However, there were problems with that formula, because with very small data sets some of the percentiles had ranks outside the data set and thus couldn't be calculated. For example, with $n=5$ and $p=0.90$ then $p(n+1)$ gives a rank of 5.4, which is outside the data set.

The most commonly used spreadsheet software and statistical software programs use the $pn + (1-p)$ formula for rank, which is presented here and which will always give a rank and corresponding percentile within the data set. The two formulas give the same rank for P50. All the other percentiles will be slightly different. You should verify that the software program you are using uses this formula.

## 4.9 PERCENTILE BARS

One useful way to display percentiles, especially when comparing two or more data sets, is with percentile bars. Here, the values of percentiles, usually the "standard" percentiles of P10, P25, P50, P75, and P90 are plotted versus the category for which they were calculated.

To illustrate this, we go to BPD's mining division, and suppose you calculated the standard percentiles of the pay of engineers and geologists along with the respective means, shown in Table 4.20.

TABLE 4.20    STANDARD PERCENTILES AND MEANS OF SALARIES OF ENGINEERS AND GEOLOGISTS

| Percentile | Engineers | Geologists |
|---|---|---|
| 10th | 86,900 | 79,800 |
| 25th | 91,750 | 85,250 |
| 50th | 97,000 | 91,000 |
| 75th | 105,750 | 98,750 |
| 90th | 120,500 | 105,500 |
| Mean | 100,550 | 92,100 |

The percentiles and means are displayed in Figure 4.9.

The legend indicates what percentiles the various parts of the bars correspond to, along with the symbol for the mean.

Comparing percentile bars is similar to comparing histograms. With percentile bars we just show the standard percentiles from a "top down view" of the histograms.

Here, visually we see that the pay of engineers is higher than that of geologists, in all percentiles and in the means. In addition, we can infer something about the distribution shapes.

The distribution of pay for engineers is nonsymmetric and tends to be skewed with the tail on the high side. We see this because the distance between P75 and P90 is more than the distance between P10 and P25, along with the distance between P50 and P75 being more than the distance between P25 and P50. In addition, the mean is higher than the median (P50), being pulled up by high outliers.

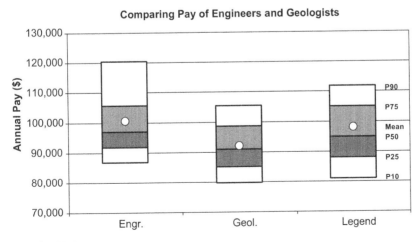

FIGURE 4.9    PERCENTILE BARS OF STANDARD PERCENTILES AND MEANS OF SALARIES OF ENGINEERS AND GEOLOGISTS

The distribution of pay for geologists tends to be symmetric. We see this because the distance between P75 and P90 is about the same as the distance between P10 and P25, along with the distance between P50 and P75 being about the same as the distance between P25 and P50. In addition, the mean is approximately equal to the median (P50).

# PRACTICE PROBLEMS

**4.1**   You have survey data for a call center job family. The data are in Table 4.21 and also in Data Set 7. Calculate the mode, median, simple mean, and trimmed mean. You are interested in the middle 90% of the data, so trim the top and bottom 5% of the data values.

**TABLE 4.21**    CALL CENTER SALARIES

| Incumbent No. | Call Center Annual Salary | Incumbent No. | Call Center Annual Salary | Incumbent No. | Call Center Annual Salary | Incumbent No. | Call Center Annual Salary |
|---|---|---|---|---|---|---|---|
| 1 | 32,000 | 26 | 53,000 | 51 | 65,000 | 76 | 78,000 |
| 2 | 33,000 | 27 | 54,000 | 52 | 65,000 | 77 | 78,000 |
| 3 | 40,000 | 28 | 55,000 | 53 | 66,000 | 78 | 80,000 |
| 4 | 45,000 | 29 | 55,000 | 54 | 67,000 | 79 | 80,000 |
| 5 | 45,000 | 30 | 55,000 | 55 | 68,000 | 80 | 82,000 |
| 6 | 45,000 | 31 | 55,000 | 56 | 68,000 | 81 | 83,000 |
| 7 | 45,000 | 32 | 55,000 | 57 | 68,000 | 82 | 85,000 |
| 8 | 45,000 | 33 | 55,000 | 58 | 68,000 | 83 | 85,000 |
| 9 | 45,000 | 34 | 56,000 | 59 | 68,000 | 84 | 85,000 |
| 10 | 47,000 | 35 | 56,000 | 60 | 68,000 | 85 | 85,000 |
| 11 | 47,000 | 36 | 56,000 | 61 | 68,000 | 86 | 88,000 |
| 12 | 47,000 | 37 | 57,000 | 62 | 69,000 | 87 | 90,000 |
| 13 | 50,000 | 38 | 58,000 | 63 | 69,000 | 88 | 90,000 |
| 14 | 50,000 | 39 | 58,000 | 64 | 70,000 | 89 | 90,000 |
| 15 | 50,000 | 40 | 59,000 | 65 | 70,000 | 90 | 90,000 |
| 16 | 50,000 | 41 | 60,000 | 66 | 70,000 | 91 | 92,000 |
| 17 | 50,000 | 42 | 60,000 | 67 | 70,000 | 92 | 100,000 |
| 18 | 50,000 | 43 | 60,000 | 68 | 70,000 | 93 | 100,000 |
| 19 | 50,000 | 44 | 60,000 | 69 | 70,000 | 94 | 100,000 |
| 20 | 50,000 | 45 | 60,000 | 70 | 72,000 | 95 | 102,000 |
| 21 | 52,000 | 46 | 60,000 | 71 | 75,000 | 96 | 106,000 |
| 22 | 52,000 | 47 | 63,000 | 72 | 75,000 | 97 | 106,000 |
| 23 | 53,000 | 48 | 64,000 | 73 | 75,000 | 98 | 108,000 |
| 24 | 53,000 | 49 | 65,000 | 74 | 75,000 | 99 | 110,000 |
| 25 | 53,000 | 50 | 65,000 | 75 | 75,000 | | |

**4.2**   Just looking at the numerical results of the previous problem, what would you guess is the distribution shape? Why?

**4.3**   For the following survey data in Table 4.22, calculate the weighted mean and the unweighted mean. Which is greater? Why?

**TABLE 4.22    SURVEY DATA**

| Company | No. of Incumbents | Mean Salary (€/y) |
|---|---|---|
| A | 14 | 80,000 |
| B | 12 | 70,000 |
| C | 9 | 90,000 |
| D | 25 | 120,000 |

**4.4**  You have to develop guidelines for salary increases for your BPD division. There are 306 employees with a total annual salary of 15,000,000. You have a salary increase budget of 4.5%. Using the performance rating and position in range data in Table 4.23, develop guidelines which, if managers follow them, will meet the budget. For this problem, do not give any flexibility within cells. Note that there can be more than one correct answer to this problem.

**TABLE 4.23    DATA FOR CALCULATING SALARY INCREASE GUIDELINES**

| | No. of Employees (Position in Range) | | | | Annual Salary (Position in Range) | | |
|---|---|---|---|---|---|---|---|
| Performance Level | Lower 1/3 | Middle 1/3 | Upper 1/3 | Performance Level | Lower 1/3 | Middle 1/3 | Upper 1/3 |
| 5 | 27 | 9 | 5 | 5 | 1,312,500 | 450,000 | 187,500 |
| 4 | 86 | 42 | 14 | 4 | 4,275,000 | 2,100,000 | 637,500 |
| 3 | 63 | 33 | 11 | 3 | 3,150,000 | 1,650,000 | 525,000 |
| 2 | 9 | 5 | 2 | 2 | 435,000 | 187,500 | 90,000 |
| 1 | 0 | 0 | 0 | 1 | 0 | 0 | 0 |

**4.5**  Suppose instead of a 4.5% salary increase budget, you only had a 3.0% salary increase budget. With this small budget, you decide to grant raises based on performance levels only. Using the same data in Table 4.23, develop guidelines which, if managers follow them, will meet the budget. As before, do not give any flexibility within cells. Note that there can be more than one correct answer to this problem.

**4.6**  You are comparing the results of the annual sales successes of the two top BPD sales associates, Jane and Virginia. Complete the comparisons in Table 4.24.

**TABLE 4.24    SALES PERFORMANCE**

| | Jane | | | Virginia | | |
|---|---|---|---|---|---|---|
| | No. of Sales Attempts | No. of Sales Successes | % Success | No. of Sales Attempts | No. of Sales Successes | % Success |
| Old territory | 40 | 20 | | 95 | 44 | |
| New territory | 32 | 8 | | 19 | 3 | |
| Total | | | | | | |

In both old and new territories, Jane has a higher success rate than Virginia. But overall Virginia has a better success rate. Why are the trends reversed?

**4.7**  You have the salaries of 48 assemblers in your Detroit facility. The data are in Table 4.25 and also in Data Set 8. Calculate the mean and the five "standard" percentiles.

**TABLE 4.25   ASSEMBLER SALARIES**

| Employee No. | Assembler Annual Salary | Employee No. | Assembler Annual Salary | Employee No. | Assembler Annual Salary |
|---|---|---|---|---|---|
| 1 | 30,600 | 17 | 29,000 | 33 | 31,700 |
| 2 | 36,600 | 18 | 29,800 | 34 | 33,300 |
| 3 | 28,900 | 19 | 30,200 | 35 | 36,700 |
| 4 | 32,100 | 20 | 31,600 | 36 | 27,700 |
| 5 | 28,900 | 21 | 31,700 | 37 | 29,200 |
| 6 | 29,000 | 22 | 32,400 | 38 | 29,200 |
| 7 | 31,600 | 23 | 28,900 | 39 | 30,500 |
| 8 | 32,300 | 24 | 29,100 | 40 | 34,100 |
| 9 | 28,700 | 25 | 29,700 | 41 | 28,600 |
| 10 | 31,500 | 26 | 30,300 | 42 | 29,200 |
| 11 | 27,600 | 27 | 30,800 | 43 | 30,600 |
| 12 | 29,800 | 28 | 33,200 | 44 | 34,200 |
| 13 | 29,900 | 29 | 28,800 | 45 | 29,000 |
| 14 | 32,200 | 30 | 29,100 | 46 | 30,900 |
| 15 | 28,500 | 31 | 29,300 | 47 | 31,600 |
| 16 | 28,900 | 32 | 30,500 | 48 | 29,000 |

**4.8**   Suppose a new assembler's salary was 32,100. What percentile is that salary with respect to the others in the previous problem?

**4.9**   From a salary survey, you are given the mean and the five standard percentiles for the job of Vice President of Sales, shown in Table 4.26. You want to target her salary toward the 60th percentile. Estimate the 60th percentile. What might be an alternative target?

**TABLE 4.26   SURVEY MEAN AND STANDARD PERCENTILES**

| | Salary |
|---|---|
| No. of Incumbents | 37 |
| Mean | 370,700 |
| P10 | 279,160 |
| P25 | 307,910 |
| P50 | 345,080 |
| P75 | 387,170 |
| P90 | 467,850 |

**4.10**   BPD executives are discussing the company's base pay compensation policy. That is, they are discussing where with respect to the competition they want to target the company's average base pay. The main discussion has been between the mean and the median of the competitive pay.

You have been keeping a history file of surveys. You note that approximately 2/3rd of the jobs in surveys are positively skewed, with the mean being 3.7% above the median on average for those jobs. And approximately 1/3rd of the jobs in surveys are negatively skewed, with the mean being 2.1% below the median on average for those jobs. When all jobs are combined, the mean is 1.8% above the median on average.

What issues are raised by your figures?

# Measures of Variability

## 5.1 IMPORTANCE OF KNOWING VARIABILITY

In this chapter, we will talk about the notion of variability, which tends to be overlooked or ignored in many of our analyses. In the author's opinion, measuring the variability of data is just as important as measuring their location (e.g., means, percentiles) because it gives you a perspective on the spread of the data and may influence the decisions you make.

An important characteristic of most collections of data points is that the values are generally not all alike. Knowledge of the extent to which they vary among themselves is critical in order to evaluate the "goodness" or comfort level of generalizations or decisions you make from the data.

Consider the following question. The average depth in Lake Clifford is only 10 cm. Can you wade across it?

□ Yes. □ No. □ It depends.  Why?

The correct answer is "It depends." It depends on the variability of the depth of the lake. If the depth varied only ±5 cm from the average, then you could wade across the lake. If the depth ranged from 5 cm to 20 m, then you probably could not wade across it.

So the moral here is,

> *Knowledge of the variability of a set of data could influence the decisions you make based on that data.*
>
> *Averages do not tell the whole story.*

*Statistics for Compensation: A Practical Guide to Compensation Analysis,* By John H. Davis
Copyright © 2011 John Wiley & Sons Inc.

The most common uses of measures of variability are as follows:

- To get a sense of how much the data are spread out
- To identify changes in the nature of the data
- To compare data sets
- To identify outliers

These will be illustrated as we discuss different measures of variability.

There are many measures of variability. We will present four of them: standard deviation, coefficient of variation, range, and P90/P10. For the first two, we need to first discuss the notions of a population and a sample.

## 5.2  POPULATION AND SAMPLE

When you are gathering and analyzing data, you need to know whether you have a population or a sample. Think of what it is you are describing or making a decision or inference about. If you have data from ALL the elements (e.g., employees, companies), which you are analyzing or about which you are making inferences, you have a population. If you have data from just SOME of the elements you are analyzing or about which you are generalizing, you have a sample.

> *Population*: That part of reality which you are analyzing and describing or about which you are making inferences or decisions. It is a collection of elements you want to describe or about which you want to make an inference. An element is an object on which a measurement is taken.
>
> *Sample*: Part of and ideally representative of the population from which it was taken.

### Examples of Populations

- If you are analyzing the benefits choices of your employee population and you have all of them in your data set for your analysis, you have a population. The population consists of the data of all your employees.
- If a decision has been made that a certain group of companies is the chosen market reference for pay levels, and you have data from all of them, you have a population. The population consists of the data from the group of companies.

### Examples of Samples and Populations

- If you have conducted an employee opinion survey by sending questionnaires to every fifth employee in a company (e.g., every fifth name, listed alphabetically—this is called a serial sample), you have a sample. The sample is the opinion data you have received from those designated employees. The population consists of the opinions of all of your employees.

- If you are comparing salaries of certain jobs to those of jobs in a particular industry, and you have many but not all of the companies in that industry in your database, you have a sample. The sample is the salary data from the companies in your database. The population consists of the salaries from all companies in that particular industry.

- If you are interested in predicting the outcome of an election, the voters are your population. The sample consists of the voters you interview to ask their preference.

- To determine which television programs have higher viewerships, monitors are attached to the television sets of certain viewers. The population consists of all television viewers. The sample consists of those viewers whose televisions are monitored.

- To determine the effect of certain drugs on rats, certain rats are given the drug and certain rats are given a placebo. The samples consist of those rats given either drugs or placebos. The population consists of all rats with similar characteristics (e.g., weight).

## 5.3  TYPES OF SAMPLES

In general, there are two types of samples random and convenience (nonrandom) To make valid *statistical* inferences from a sample about the population from which the sample was drawn, the sample must be random. A random sample of size $n$ is one in which every possible sample of that size has an equal chance of being selected. One example would be to draw names from a hat. There are other statistically valid sampling procedures, such as stratified sampling, cluster sampling, proportional sampling, serial sampling, and so on. This book does not address these types of samples.

The types of samples we usually encounter in compensation, however, are convenience samples. For example, we try to get every desired company we can in a compensation survey, but typically we do not get all of them. We get those whose compensation practitioners have an interest in the survey and who have time to participate. In these cases, the sample is not random. The best we can say when we generalize from the sample to the population is to make a statement as to its representativeness, based on our judgment.

To illustrate measures of variability, we will use the data in Table 5.1, which are also in Data Set 9. The data are a survey sample of salaries of 15 administrative

TABLE 5.1

**TABLE 5.1**      **RAW DATA OF SALARY SURVEY OF TWO JOBS**

| Annual Pay Administrative Assistant | Annual Pay Accounting Assistant | | | |
|---|---|---|---|---|
| 55,500 | 43,575 | 59,250 | 69,375 | 78,705 |
| 57,765 | 45,495 | 59,400 | 70,035 | 82,605 |
| 59,370 | 47,055 | 59,505 | 70,785 | 83,835 |
| 62,115 | 49,125 | 60,465 | 72,390 | 86,010 |
| 68,940 | 49,920 | 61,335 | 73,530 | 88,200 |
| 69,300 | 50,400 | 64,500 | 74,205 | 89,040 |
| 70,800 | 50,880 | 65,400 | 74,985 | 93,075 |
| 71,700 | 51,765 | 65,700 | 75,555 | 94,380 |
| 71,700 | 54,150 | 66,750 | 76,230 | 101,070 |
| 72,765 | 57,255 | 66,825 | 76,500 | 103,320 |
| 76,440 | 58,425 | 66,885 | 76,905 | 104,460 |
| 81,600 | 58,980 | 68,085 | 77,700 | |
| 83,715 | | | | |
| 87,765 | | | | |
| 89,745 | | | | |

**TABLE 5.2**      **SALARY SURVEY OF TWO JOBS—NUMBERS OF INCUMBENTS AND MEANS**

| | Administrative Assistant | Accounting Assistant |
|---|---|---|
| No. of data points | 15 | 47 |
| Average | 71,948 | 69,660 |

assistants and 47 accounting assistants whose jobs match those at the BPD home office. These are convenience samples. The data are ordered for ease of use.

Our analysis starts with calculating the average (mean) of each group, shown in Table 5.2.

We note that the administrative assistants are paid 2,288, or 3.3% more on average than the accounting assistants.

$$(100)(71,948 - 69,660)/69,660 = (100)(2,288/69,660) = (100)(0.033) = 3.3\%$$

Now we will look at various measures of variability.

## 5.4 STANDARD DEVIATION

The standard deviation is an "average" deviation or distance of all the data points from the mean, ignoring the sign of the deviation (whether above or below the mean.) Its definition is the square root of the "average" squared deviation of the data points from the mean. Which "average" you use depends on whether you have a population or a sample.

Here are the formulas for the standard deviation for a population and for a sample.

Using symbols, let

$n$ = the number of data points
$x$ = the data, the individual data points
$\bar{x}$ = the mean
$s_n$ = the standard deviation if the data is the population
$s_{n-1}$ the standard deviation if the data is a sample from the population

Then,

$$s_n = \sqrt{\frac{\sum (x - \bar{x})^2}{n}}$$

$$s_{n-1}\sqrt{\frac{\sum (x - \bar{x})^2}{n-1}}$$

---

### $n$ AND $n-1$

Why are there two formulas for the standard deviation—one for the population and one for the sample? The answer lies in that part of statistics called statistical inference, where we use sample statistics to estimate population statistics. One characteristic we require of a sample statistic used as an estimator is that it is unbiased—that is, it will not systematically underestimate or overestimate the population statistic. If we use $n$ in the denominator of the sample standard deviation, it would be a biased estimator which is systematically influenced by sample size, tending to underestimate the population standard deviation with small sample sizes. To correct this bias, we substitute $n-1$ for $n$, which can be shown both mathematically and empirically to provide an unbiased estimate of the population standard deviation.

There are other explanations for this, using the concept of degrees of freedom, but they are not presented here.

The sample mean is an unbiased estimator of the population mean, so no adjustments are necessary there.

---

For large data sets, with $n > 30$, the difference between the value of the population standard deviation and the value of the sample standard deviation is less than 2%. As the size of the data set gets larger, the difference becomes negligible.

The larger the standard deviation, the more spread out are the data points. The smaller the standard deviation, the closer together they are. One characteristic of the standard deviation is that since it uses all the data in its

| **TABLE 5.3** | **SALARY SURVEY OF TWO JOBS—MEASURES OF VARIATION SUMMARY WITH STANDARD DEVIATION** | |
| --- | --- | --- |
| | **Administrative Assistant** | **Accounting Assistant** |
| No. of data points | 15 | 47 |
| Average | 71,948 | 69,660 |
| Sample standard deviation | 10,568 | 15,509 |

calculation, it is necessarily impacted by any outliers. Outliers are defined later in this chapter.

Both the sample and the population standard deviations are standard outputs in spreadsheet and statistical software packages.

We calculate the sample standard deviations of the two groups, shown in Table 5.3.

## Interpretations and Applications of Standard Deviation

1. *The Values Themselves.* Looking at the values of the sample standard deviations themselves, we see that for administrative assistants, the data points vary from the mean of 71,948 by about 10,568 on "average." Some of the salaries are above the mean, and some are below the mean. Some are close to the mean and some are far away. But as a ballpark estimate, the distance from the mean averages about 10,568.

    Similarly, for accounting assistants, the data points vary from the mean of 69,660 by about 15,509 on "average."

2. *Comparing Different Data Sets.* We know immediately that the accounting assistant salaries vary from their mean more than the administrative assistant salaries do from their mean, by approximately 5,000 on "average."

**Graphical Illustration of Standard Deviation**   Keeping in mind that a picture is worth a thousand words, let us look at the histograms of the two sets of data, shown in Figures 5.1 and 5.2. Superimposed on each are the locations of the mean and ± one and two standard deviations from the mean.

It can easily be seen that the variability of the administrative assistant salaries is much less than the variability of the accounting assistant salaries.

3. *Comparing the Same Data Set Over Time.* Suppose you examined the data for accounting assistants from last year to this year, shown in Table 5.4.

    You notice a big jump in the standard deviation. This should be a strong indication that you should investigate why. You may have to look at individual data points. For example, were the three accounting assistants with the highest salaries new hires that increased the spread of the data? This

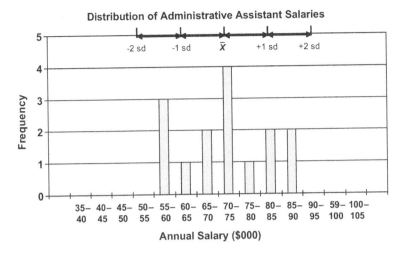

**FIGURE 5.1**    HISTOGRAM OF ADMINISTRATIVE ASSISTANT SALARIES

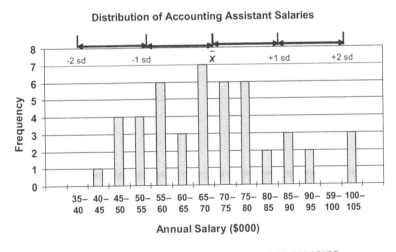

**FIGURE 5.2**    HISTOGRAM OF ACCOUNTING ASSISTANT SALARIES

**TABLE 5.4**    COMPARISON OVER TIME—MEASURES OF VARIATION SUMMARY WITH STANDARD DEVIATION

|  | Accounting Assistant | |
| --- | --- | --- |
|  | Last Year | This Year |
| No. of data points | 44 | 47 |
| Average | 66,180 | 69,660 |
| Sample standard deviation | 11,467 | 15,509 |

illustrates the point that you should understand your data thoroughly and try as best as you can to explain what is going on.

4. *The Two-Sigma Rule.* The Greek letter lower case sigma, $\sigma$, is a symbol used in the field of statistics for standard deviation. So a two-sigma rule means a two standard deviation rule. This interpretation applies a rule that is a consequence of a statistical theorem called Chebyshev's Inequality. From that theorem, we get the two-sigma rule.

*Two-Sigma Rule* —In any distribution of data, at least 75% of the data points will fall within two standard deviations of the mean.

Applying the two-sigma rule first to the administrative assistant data, we calculate the salary values that are two standard deviations from the mean.

$$\text{Two standard deviations below the mean} = 71,948 - (2)(10,568)$$
$$= 50,812$$

$$\text{Two standard deviations above the mean} = 71,948 + (2)(10,568)$$
$$= 93,084$$

The rule states that at least 75% of the administrative assistant salaries will fall between 50,812 and 93,084. Looking at Table 5.1 and Figure 5.1, we see that all of the data points are within these bounds, which of course is at least 75% of them.

For the accounting assistant data, we perform similar calculations.

$$\text{Two standard deviations below the mean} = 69,660 - (2)(15,509)$$
$$= 38,462$$

$$\text{Two standard deviations above the mean} = 69,660 + (2)(15,509)$$
$$= 100,678$$

Looking at Table 5.1 and Figure 5.2, we find that 44 out of the 47 accounting assistant salaries fall between 38,462 and 100,678, or 94% of them, which is at least 75%.

Often in salary surveys, most of the salaries for a benchmark job will fall within two standard deviations of the mean. But in any event, we know at least 75% of them will.

So if all you know about a set of data are the mean and standard deviation, you can calculate limits that will bound at least 75% of the data points.

For example, if you were told that the average salary in a salary survey for the job of a legal assistant was $85,000 with a standard deviation of $10,000, you would know that at least 75% of the salaries were between

85,000 − (2)(10,000) and 85,000 + (2)(10,000), or between $65,000 and $105,000.

5. *z-Scores—Outliers*

An outlier is a data point whose value appears to deviate markedly from other data points in the data set in which it occurs. It is numerically distant from the rest of the data. There is no rigid statistical criterion of what constitutes an outlier. However, z-scores are often used as a basis for identifying possible outliers.

A z-score is the number of standard deviations a particular data value is from the mean. It is calculated as:

$$z = \frac{\text{data value} - \text{mean}}{\text{standard deviation}}$$

Referring to the legal assistant survey above, suppose the salary of a BPD legal assistant was $98,000. Her z-score with respect to the survey is

$$z = \frac{98,000 - 85,000}{10,000} = \frac{13,000}{10,000} = 1.3$$

Her salary is 1.3 standard deviations above the mean.

z-scores are useful in two main situations. The first use is to get a sense of how "far out" a data value is from its mean, with the perspectives of both the two-sigma rule and what you usually expect in similar situations. If it is "too far out" you will consider it an outlier, worthy of further investigation. How far is "too far" is a judgment call.

Suppose you are considering certain companies, designated Competition A, in the manufacturing industry as a possible comparator group for BPD compensation program purposes. You compare the revenue of BPD with those of the group as shown in Table 5.5 and Figure 5.3.

Here BPD's revenue is 3.4 standard deviations above the mean of the comparators, as indicated by a z-score of 3.4. (Raw data and calculations are not shown.) Are we part of this group or not? According to statistical theory, the probability that we are part of this group is very small, since we are that many standard deviations above the mean, all other things being equal. There is no single right answer as to how many standard deviations are too many.

Many practitioners would say that anything more than three standard deviations should be investigated. Often in salary surveys, a z-score with absolute value of 3.0 or more is a flag to investigate to ensure that the job match is valid. Some survey providers use different thresholds, such as z-scores of 2.0 or 2.5.

| TABLE 5.5 | FREQUENCY DISTRIBUTION OF REVENUES OF COMPETITION A | |
| --- | --- | --- |
| **Annual Revenue** | **Competition A Comparators** | **BPD** |
| 5B–10B | | |
| 10B–15B | 1 | |
| 15B–20B | 4 | |
| 20B–25B | 5 | |
| 25B–30B | 3 | |
| 30B–35B | 1 | |
| 35B–40B | | |
| 40B–45B | | 1 |
| 45B–50B | | |
| 50B–55B | | |
| 55B–60B | | |

So if you get a data value with a $z$-score of absolute value of 3.0 or more, be sure you investigate it to be sure that it is valid in the context of your analysis. However, in your data analyses you should build up a history file of your own and develop your own "rules" of what constitutes an outlier.

It appears that BPD is an outlier here, and not really a part of this comparator group. This would raise the issue of the appropriateness of this particular group of companies as a comparator group and might cause you to select another set of companies.

On the other hand, if the comparison looked liked Table 5.6 and Figure 5.4, with another group called Competition B, you would conclude that

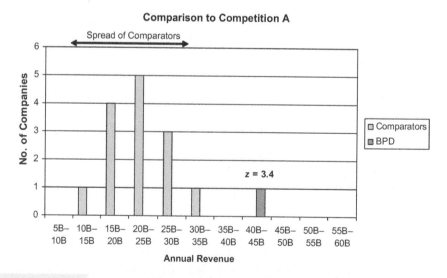

FIGURE 5.3    HISTOGRAM OF REVENUES OF COMPETITION A

| Annual Revenue | Competition B Comparators | BPD |
|---|---|---|
| 5B–10B | 1 | |
| 10B–15B | 2 | |
| 15B–20B | 1 | |
| 20B–25B | 3 | |
| 25B–30B | 2 | |
| 30B–35B | 2 | |
| 35B–40B | 1 | |
| 40B–45B | 1 | 1 |
| 45B–50B | 1 | |
| 50B–55B | 2 | |
| 55B–60B | 1 | |

**TABLE 5.6    FREQUENCY DISTRIBUTION OF REVENUES OF COMPETITION B**

even though you were in the upper half of the group, you definitely were a part of this group, as you are within the spread of the revenues. Your $z$-score is 0.77, well within the group. (Raw data and calculations are not shown.) You are not an outlier.

6. *z-Scores—Merging Data Sets*

The second use of $z$-scores is when comparing data sets that measure the same item but use different scales. An historical example is IQ tests. Many have different scales, but they are converted to $z$-scores and then adjusted to all have a mean of 100 and a standard deviation of 15. That enables them to be easily compared.

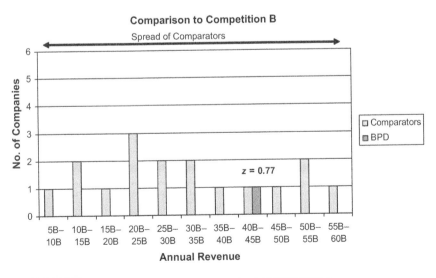

FIGURE 5.4    HISTOGRAM OF REVENUES OF COMPETITION B

**TABLE 5.7     PERFORMANCE DISTRIBUTIONS Z-SCORES**

| Rating | Company A | | | Rating | Company B | | |
|---|---|---|---|---|---|---|---|
| | No. of Employees | % | z-Score | | No. of Employees | % | z-Score |
| 5 | 45 | 15 | 1.7 | 7 | 30 | 6 | 2.1 |
| 4 | 135 | 45 | 0.4 | 6 | 70 | 13 | 1.2 |
| 3 | 105 | 35 | −0.9 | 5 | 230 | 44 | 0.2 |
| 2 | 15 | 5 | −2.2 | 4 | 130 | 25 | −0.7 |
| 1 | 0 | 0 | −3.5 | 3 | 40 | 8 | −1.6 |
| Total | 300 | 100 | | 2 | 20 | 4 | −2.5 |
| | | | | 1 | 0 | 0 | −3.4 |
| Mean rating | | 3.70 | | Total | 520 | 100 | |
| Standard deviation rating | | 0.78 | | | | | |
| | | | | Mean rating | | 4.73 | |
| | | | | Standard deviation rating | | 1.09 | |

One application in the field of human resources is merging two performance appraisal systems in an acquisition situation, each with different scales. Suppose Company A uses a 1–5 scale and Company B uses a 1–7 scale. In each case, the higher the number, the better the performance. There are many ways to merge these two systems. We will just show one using $z$-scores. Here are the starting data in Table 5.7.

To illustrate the calculations, for company A, the mean rating of 3.70 was calculated by taking the mean rating of all 300 employees. Since we have the data on all employees, we have a population. Hence, the standard deviation of 0.78 is a population standard deviation. For the rating of 5, the $z$-score is calculated as $(5 - 3.70)/0.78 = 1.7$. Similar calculations were done for the other ratings in Company A and for all the ratings in Company B. All values are rounded.

Suppose it has been decided to use the 5-level system of Company A. Now we must convert the 1–7 scale of Company B to a 1–5 scale. The scheme we will use is to associate the Company B rating $z$-score for a given rating to the Company A rating with the closest $z$-score, shown in Table 5.8.

Of course this is just a mathematical way to merge two scales, but it does form a nice objective starting point. Other considerations would have to be taken into account, including the wording of the level definitions and management concerns.

**TABLE 5.8    MERGED PERFORMANCE DISTRIBUTIONS BASED ON Z-SCORES**

| Co. B Old Rating | Co. B z-Score | Closest Co. A z-Score | Co. A Rating | Co. B New Rating |
|---|---|---|---|---|
| 7 | 2.1 | 1.7 | 5 | 5 |
| 6 | 1.2 | 1.7 | 5 | 5 |
| 5 | 0.2 | 0.4 | 4 | 4 |
| 4 | −0.7 | −0.9 | 3 | 3 |
| 3 | −1.6 | −2.2 | 2 | 2 |
| 2 | −2.5 | −2.2 | 2 | 2 |
| 1 | −3.4 | −3.5 | 1 | 1 |

## 5.5 COEFFICIENT OF VARIATION

The standard deviation is a measure of the *absolute* variation of data values about their mean. The coefficient of variation (CV) is a measure of the *relative* variation of data values about their mean in terms of a percent. It allows a nice comparison of the variability of data sets that vary in magnitude.

The CV is an "average" percent *of* the mean of the data points *from* the mean. It is calculated by dividing the standard deviation by the mean, and expressing the ratio as a percent.

$$CV = \frac{\text{Standard deviation}}{\text{Mean}} \times 100$$

Using data from a sales survey in which BPD participated we get the following in Table 5.9.

For the VP Sales position, CV = (26,200/162,300)(100) = 16.1%. A similar calculation is done for the Sales Associate data.

In this example, the standard deviation of the salaries of VP Sales, in absolute dollars, is almost four times that of the salaries of Sales Associates. However, in relative terms, they are both about the same—the standard deviation is approximately 16% of the mean for both jobs. Hence, on a relative basis the variability is about the same.

Like the standard deviation, the CV is necessarily impacted by outliers since it is based on calculations involving all the data points.

**TABLE 5.9    COEFFICIENT OF VARIATION FOR VP SALES DATA**

| | VP Sales | Sales Associate |
|---|---|---|
| Mean | 162,300 | 42,560 |
| Standard deviation | 26,200 | 6,660 |
| CV | 16.1% | 15.6% |

**TABLE 5.10** SALARY SURVEY OF TWO JOBS—MEASURES OF VARIATION SUMMARY WITH CV

|  | Administrative Assistant | Accounting Assistant |
|---|---|---|
| No. of data points | 15 | 47 |
| Average | 71,948 | 69,660 |
| Sample standard deviation | 10,568 | 15,509 |
| Coefficient of variation | 14.7% | 22.3% |

Continuing with the example of administrative assistant and accounting assistant salaries, we have Table 5.10.

## Interpretations and Applications of Coefficient of Variation

1. *The Values Themselves.* As with the standard deviation, the smaller the CV, the less relative variation there is of the data points about the mean. And the larger the CV, the larger the relative variation there is of the data points about the mean.

   In this example, the salaries of the administrative assistants vary on average by about 14.7% of their mean, and the salaries of the accounting assistants vary on average by about 22.3% of their mean.

2. *Comparing Different Data Sets.* In this example, we see that on a relative basis, the salaries of the accounting assistants vary more than the salaries of the administrative assistants.

3. *Comparing the Same Data Set Over Time.* We continue to compare the data for the accounting assistants from last year to this year, shown in Table 5.11.

   Here too, we notice a jump in the coefficient of variation. Repeating what we said before, this should be a strong indication that you should investigate why. In looking at individual data points, for example, you may find the three accounting assistants with the highest salaries are new hires and skewed the distribution positively. Again, you should understand your data thoroughly and try as best as you can to explain what is happening.

**TABLE 5.11** COMPARISON OVER TIME—MEASURES OF VARIATION SUMMARY WITH CV

|  | Accounting Assistant | |
|---|---|---|
|  | Last Year | This Year |
| No. of data points | 44 | 47 |
| Average | 66,180 | 69,660 |
| Sample standard deviation | 11,467 | 15,509 |
| Coefficient of variation | 17.3% | 22.3% |

4. *The Two-Sigma Rule.* For the administrative assistants, we know that at least 75% of the salaries fall within 29.4% of their mean (2)(14.7%). For the accounting assistants, we know that at least 75% or the salaries fall within 44.6% of their mean (2)(22.3%).

## 5.6  RANGE

The range is a "quick and dirty" measure of the variability, and is easy to calculate. It is the difference between the maximum and the minimum.

$$\text{Range} = \text{maximum} - \text{minimum}$$

Here are the ranges for our two groups, shown in Table 5.12.

We see that the range of the accounting assistant salaries is almost twice the range of the administrative assistant salaries. This is evident from the two histograms shown earlier.

Counterbalancing the ease of calculation of the range is that it is defined only by extreme values, and thus greatly impacted by them. You may want a measure that is not influenced by just one value that happens to be very high or very low.

### Interpretations and Applications of Range

1. *The Values Themselves.* As with the other measures of variation, the smaller the range, the less variation there is of the data points and the larger the range, the larger the variation there is of the data points. The range of the administrative assistant salaries is 34,245 and of the accounting assistant salaries is 60,885.

2. *Comparing Different Data Sets.* In this example with the range of the accounting assistant salaries almost twice that of the administrative assistant salaries, we can conclude the accounting assistant data are much more varied than the administrative assistant data.

**TABLE 5.12    SALARY SURVEY OF TWO JOBS—MEASURES OF VARIATION SUMMARY WITH RANGE**

| | Administrative Assistant | Accounting Assistant |
|---|---|---|
| No. of data points | 15 | 47 |
| Average | 71,948 | 69,660 |
| Sample standard deviation | 10,568 | 15,509 |
| Coefficient of variation | 14.7% | 22.3% |
| Maximum | 89,745 | 104,460 |
| Minimum | 55,500 | 43,575 |
| Range | 34,245 | 60,885 |

| TABLE 5.13 | COMPARISON OVER TIME—MEASURES OF VARIATION SUMMARY WITH RANGE | |
|---|---|---|
| | **Accounting Assistant** | |
| | **Last Year** | **This Year** |
| No. of data points | 44 | 47 |
| Average | 66,180 | 69,660 |
| Sample standard deviation | 11,467 | 15,509 |
| Coefficient of variation | 17.3% | 22.3% |
| Maximum | 93,000 | 104,460 |
| Minimum | 42,750 | 43,575 |
| Range | 50,250 | 60,885 |

3. *Comparing the Same Data Set Over Time.* We continue to compare the data for accounting assistants from last year to this year, shown in Table 5.13.

   We note that there is a large jump in the range from last year to this year, and went on to investigate if the net increase of three accounting assistants is responsible for it.

   At this point, the three measures of variation have one property in common: they are all impacted by extreme values. What we need is a measure of variability than is insulated from that impact, yet uses a lot of the information.

   The analogy here is like the trimmed mean when compared to the mean. It is not impacted by extreme values but it uses a lot of information. However, we don't want a measure analogous to the median, because it only uses information from the very middle of the data.

   The compromise presented here is to calculate a measure of variability of the middle 80% of the data, excluding the top and bottom 10%. We call it P90/P10.

## 5.7  P90/P10

The term P90/P10 is the ratio of the 90th percentile to the 10th percentile. It is pronounced "*p* ninety to *p* ten."

$$P90/P10 = \frac{P90}{P10}$$

This is a measure of the spread of the middle 80% of the data. As such, P90/P10 is not impacted by outliers. It doesn't matter what the top 10% and the bottom 10% of the data are doing.

One could just take the range from P10 to P90, which would be an absolute measure. In the author's experience a relative measure is more useful, and that is what the ratio does. It provides nice comparisons of data set variability where the data values in the sets differ in magnitude. This is similar to calculating the

**TABLE 5.14   SALARY SURVEY OF TWO JOBS—MEASURES OF VARIATION SUMMARY WITH P90/P10**

|                           | Administrative Assistant | Accounting Assistant |
|---------------------------|--------------------------|----------------------|
| No. of data points        | 15                       | 47                   |
| Average                   | 71,948                   | 69,660               |
| Sample standard deviation | 10,568                   | 15,509               |
| Coefficient of variation  | 14.7%                    | 22.3%                |
| Maximum                   | 89,745                   | 104,460              |
| Minimum                   | 55,500                   | 43,575               |
| Range                     | 34,245                   | 60,885               |
| P90                       | 86,145                   | 90,654               |
| P10                       | 58,407                   | 50,208               |
| P90/P10                   | 1.47                     | 1.81                 |

range spread of a graded salary structure, where the range spread is the percent that the maximum is more than the minimum. A common range spread is 50%, and is easy to interpret for both low salary grades and for high salary grades.

You may encounter another measure based on percentiles, called the interquartile range. It is the range from P25 to P75. This is an absolute measure of variability of the middle 50% of the data. In the author's experience, this ignores too much information. The data have been trimmed too severely.

Table 5.14 is our completed table with all four measures of variation we have discussed.

## Interpretations and Applications of P90/P10

1. *The Values Themselves.* The smaller the P90/P10, the smaller is the spread of the middle 80% of the data. The larger the P90/P10, the larger is the spread of the middle 80% of the data.

   If we MOTH the P90/P10 ratio, we get a percentage spread of the middle 80% of the data. (MOTH is not a formal statistics term, but it is easy to remember and very useful.)

   $$\text{MOTH} = \text{Minus One Times Hundred}$$

   In this example, the spread of the middle 80% of the administrative assistant salaries is 47%. That is, the 90th percentile is 47% more than the 10th percentile.

   $$\text{Percent spread} = (1.47 - 1)(100) = (0.47)(100) = 47\%$$

   Similarly, the spread of the middle 80% of the accounting assistant salaries is 81%. That is, the 90th percentile is 81% more than the 10th percentile.

2. *Comparing Different Data Sets.* Based on this measure we see that the variability of the accounting assistant salaries is more than the variability of the

**TABLE 5.15** COMPARISON OVER TIME—MEASURES OF VARIATION SUMMARY WITH P90/P10

| | Accounting Assistant | |
| --- | --- | --- |
| | Last Year | This Year |
| No. of data points | 44 | 47 |
| Average | 66,180 | 69,660 |
| Sample standard deviation | 11,467 | 15,509 |
| Coefficient of variation | 17.3% | 22.3% |
| Maximum | 93,000 | 104,460 |
| Minimum | 42,750 | 43,575 |
| Range | 50,250 | 60,885 |
| P90 | 85,320 | 90,654 |
| P10 | 49,470 | 50,208 |
| P90/P10 | 1.72 | 1.81 |

administrative assistant salaries. The middle 80% of the accounting assistant salaries are spread out more than the middle 80% of the administrative assistant salaries.

3. *Comparing the Same Data Set Over Time.* We make the final comparison for the accounting assistant salaries from last year to this year in Table 5.15.

We note a small jump in the P90/P10 from last year to this year. Based on your experience and perspective, this may be a reasonable change for this statistic.

4. *Input to Decision on Salary Structure Range Spread.* In deciding the range spread for a job or for a structure, a survey P90/P10 offers a useful perspective. It indicates the range spread of the middle 80% of the market data; that is, of the bulk of the market data. If you have a P90/P10 say, of 1.50 for a surveyed job, that means that the range spread of the middle 80% of the data is 50%, and would be a justification for a 50% range spread for that job inside your company. You would be targeting your structure to the middle mass of survey data while ignoring the extremes.

## 5.8 COMPARISON AND SUMMARY

Table 5.16 is a comparison of the four measures of variability presented here.

The use in compensation of measures of variability is in its infancy compared to the use of measures of location. However, if you want to gain an understanding of data, there is no shortcut. You must measure variability and track it to build up your own trends and rules of thumb so that you not only gain an understanding of the data and underlying situations but also are able to identify when data sets are more varied or less varied than you expect.

Since all four of these measures are easy to compute, start tracking them. Build up your own experience portfolio with all four measures of variation.

**TABLE 5.16   COMPARISON AND CHARACTERISTICS OF MEASURES OF VARIATION**

| Standard Deviation | CV | Range | P90/P10 |
|---|---|---|---|
| Absolute measure of variability—the "average" absolute distance from the mean | Relative measure of variability—the "average" percent distance from the mean | Absolute measure of the difference between the lowest and highest values | Relative measure of the spread of middle 80% of data—the 90th percentile as a multiple of the 10th percentile |
| Uses all the data points | Uses all the data points | Uses only the maximum and minimum | Uses two data points as influenced by nearby points |
| Impacted by outliers | Impacted by outliers | Greatly impacted by outliers | Not impacted by outliers |

You may find that the moral described in the beginning of this chapter is true: knowledge of the variability of a set of data could influence the decisions you make based on that data. Averages do not tell the whole story.

**CASE STUDY 4, PART 3 OF 3**

You are finding the market pay for the job of manager operations as input to the hiring offer. In the two previous chapters, we had just constructed a frequency distribution and histogram, along with calculating two measures of central tendency, the mean and the median. They are repeated here along with the raw data in Table 5.17 (also in Data Set 3), and in Table 5.18 and Figure 5.5.

**TABLE 5.17   SALARY SURVEY DATA FOR OPERATIONS MANAGER**

| Company | Annual Salary |
|---|---|
| 1 | 97,240 |
| 2 | 103,950 |
| 3 | 107,640 |
| 4 | 110,630 |
| 5 | 114,380 |
| 6 | 119,730 |
| 7 | 124,800 |
| 8 | 129,150 |
| 9 | 136,500 |
| 10 | 140,180 |
| 11 | 146,520 |
| 12 | 157,950 |
| 13 | 176,640 |

| TABLE 5.18 | FREQUENCY DISTRIBUTION OF SALARY SURVEY DATA FOR OPERATIONS MANAGER |
|---|---|

| Annual Salary ($000) | No. of Incumbents |
|---|---|
| 90–100 | 1 |
| 100–110 | 2 |
| 110–120 | 3 |
| 120–130 | 2 |
| 130–140 | 1 |
| 140–150 | 2 |
| 150–160 | 1 |
| 160–170 | 0 |
| 170–180 | 1 |

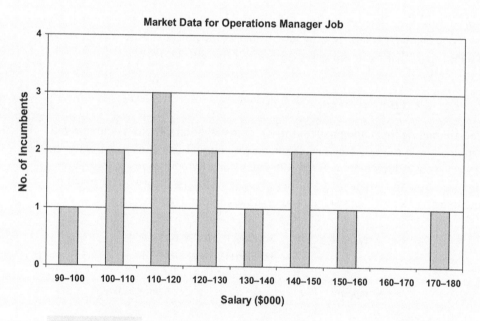

| FIGURE 5.5 | HISTOGRAM OF SALARY SURVEY DATA FOR OPERATIONS MANAGER |
|---|---|

You now calculate various measures of variability and add them to your two measures of location, shown in Table 5.19.

Based on your "history file" of variability of survey data, the CV and the P90/P10 are in line with what you would expect. Indeed, the range from the 10th percentile to the 90th percentile is almost 50%, which is the range spread for the salary structure for this job.

**TABLE 5.19   MEASURES OF LOCATION AND VARIABILITY FOR SALARY SURVEY DATA FOR OPERATIONS MANAGER**

| No. of Incumbents | 13 |
|---|---|
| Mean | 128,101 |
| Median | 124,800 |
| Sample standard deviation | 22,972 |
| CV | 17.9% |
| Low | 97,240 |
| High | 176,640 |
| Range | 79,400 |
| P90 | 155,664 |
| P10 | 104,688 |
| P90/P10 | 1.49 |

Since the standard deviation is the "average" deviation of the data points from the mean, you calculate the values that are $\pm$ one standard deviation from the mean, shown in Table 5.20.

**TABLE 5.20   BASIS FOR SALARY OFFER**

| Mean $-$ 1 sd | 105,129 |
|---|---|
| Mean $+$ 1 sd | 151,073 |

These values are very close to the values of P10 and P90. This gives another perspective on deciding the range of pay to offer.

Since BPD's compensation policy is to target its average pay toward the average in the market, you feel comfortable in recommending, after rounding liberally, that the salary offer be between 105,000 and 155,000. The hiring manager will use this information along with bonus opportunities, other perquisites, and perceived candidate qualifications in crafting a final offer.

# PRACTICE PROBLEMS

**5.1**   You are analyzing the last 12 months of starting salaries of the IT professionals in the Middle East division, and you have the data on all the employees. Do you have a population or a sample?

**5.2**   Your executives have identified the European companies it considers its competition for executive talent in Europe. You have data from many of the companies. Do you have a population or a sample?

| TABLE 5.21 | NETWORK ASSISTANT SALARIES |
|---|---|

| Incumbent No. | Network Assistant Annual Salary |
|---|---|
| 1 | 47,100 |
| 2 | 50,060 |
| 3 | 52,500 |
| 4 | 55,600 |
| 5 | 57,000 |
| 6 | 58,000 |
| 7 | 59,000 |
| 8 | 60,500 |
| 9 | 62,000 |
| 10 | 66,200 |
| 11 | 69,000 |
| 12 | 69,100 |
| 13 | 69,300 |
| 14 | 69,600 |
| 15 | 69,600 |
| 16 | 70,100 |
| 17 | 70,500 |
| 18 | 71,800 |
| 19 | 72,200 |
| 20 | 72,400 |
| 21 | 74,500 |
| 22 | 76,900 |
| 23 | 79,100 |
| 24 | 81,800 |
| 25 | 84,600 |
| 26 | 86,700 |
| 27 | 88,900 |
| 28 | 92,000 |
| 29 | 94,600 |
| 30 | 96,700 |

**5.3** You are conducting an attitude survey on salaries among all employees in the United States, and decide to take a sample by putting their names in an electronic hat, electronically mixing them up, and drawing 10% of the names. Those whose names are selected will receive the attitude questionnaire. What kind of a sample do you have?

**5.4** The BPD executives in India have identified 45 companies in Bangalore, Mumbai, and Delhi as the competition for software engineers. You contact all of them to participate in a compensation study, and 30 of them agree. What kind of a sample do you have?

**5.5** For the survey sample of Network Assistant salaries shown in Table 5.21 and also in Data Set 10, construct a frequency distribution and histogram, and complete Table 5.22.

**5.6** Interpret the results of Practice Problem 5.5.

**5.7** Apply the two-sigma rule to the data in Practice Problem 5.5. Is it true here?

**5.8** You receive a bonus of $9,000. You know that the average bonus received in your company among 49 employees is $6,000 with a standard deviation of $1,200. What is the $z$-score of your bonus?

**TABLE 5.22**   **TEMPLATE FOR MEASURES OF VARIABILITY**

**No. of data points**

Mean
Median
Sample standard deviation
Coefficient of variation
Maximum
Minimum
Range
P90
P10
P90/P10

**5.9**  You are given that the mean of a survey for a job is 110,000 and that the standard deviation is 12,000. Using the two-sigma rule, what can you say about where most of the data are?

**5.10**  You had a sales contest for the 24 sales reps in the Middle East division and the 30 sales reps in the Europe division. The purpose of the contest was to obtain new customers for your products. Summary data are shown in Table 5.23. Ethan, a sales rep in Kuwait, and Jack, a sales rep in Italy, were comparing their individual results, and they each obtained 130 new customers.

Would you say that their results were equivalent? Calculate the respective $z$-scores and interpret the results.

**TABLE 5.23**   **SALES CONTEST SUMMARY**

|  | Middle East | Europe |
|---|---|---|
| No. of reps | 24 | 30 |
| New customers |  |  |
| Mean | 100 | 150 |
| Standard deviation | 10 | 20 |
|  | Ethan | Jack |
| New customers | 130 | 130 |

# Model Building

This chapter will go into more detail on model building, expanding on the basic framework presented in Chapter 1, *Introduction*.

## 6.1 PRELUDE TO MODELS

The remaining statistical techniques discussed in this book will be presented in the framework of models and model building. Models are how we conceptually relate different variables to one another. Indeed, the most important, practical, and sophisticated statistics used in compensation are models.

Here is where we are going.

- *Model Building Concepts*
  Chapter 6    *Model Building*

- *"Pure" Statistical Models*
  Chapter 7    *Linear Model*
  Chapter 8    *Exponential Model*
  Chapter 9    *Maturity Curve Model*
  Chapter 10    *Power Model*
  
  Each of these four chapters will use a small case study to illustrate the particular model.

- *Application of Models*
  Chapter 11    *Market Models and Salary Survey Analysis*
  
  This will include a large case study for our example company BPD to analyze both external salary surveys and internal employee data to (1)

identify the company's market positions, (2) develop salary structures, and (3) develop salary increase budgets. This process is a major annual effort and perhaps the most important one conducted by compensation professionals. There are five types of market model analyses discussed.

Chapter 12   *Integrated Market Model: Linear*
Chapter 13   *Integrated Market Model: Exponential*
Chapter 14   *Integrated Market Model: Maturity Curve*
Chapter 15   *Job Pricing Market Model: Group of Jobs*
Chapter 16   *Job Pricing Market Model: Power Model*

- *Multiple Linear Regression*

Chapter 17   *Multiple Linear Regression* (advanced topic)

This describes models and techniques to use when there is more than one $x$-variable simultaneously affecting the $y$-variable. The terms $x$-variable and $y$-variable are defined below.

## 6.2  INTRODUCTION

A variable is a measured characteristic of an item of interest that varies among items. The measure might be pay of an employee, age of an employee, revenue of a company, and so on. Up to now we have been dealing with just single variables. For example, in our daily work we examine and describe the distribution, measures of location, and measures of variability of salaries, ages, grades, position in range, time in grade, revenue, and so on. Each of these is a single variable. We examine each one all by itself.

Sometimes we use the terms "variable" and "factor" interchangeably.

In each of these examples, the data vary. Salaries vary, ages vary, grades vary, and so on.

A key question to ask is "Why do the data vary?" Is there something that relates to the variable of interest that might help us answer why it varies? That "something" is another variable that will help us answer why. So to answer the main question, we now turn to the *relationship* between variables and work with two or more variables at a time.

As a perspective, it is important to answer the question "Why do we want to know why data vary?"

For example, we are often concerned with paid time off. We observe that paid time off varies for different employees. Not everyone takes the same paid time off. We want to know *why* paid time off varies. We want to find out *what factors* influence the amount of paid time off taken and *how* and *to what degree*.

Why do we want to know this? There are typically three general reasons why we want to know why paid time off varies. They all are related to making better decisions.

- We want to *understand* paid time off. For example, if we observe that paid time off taken is mainly vacation and not personal days, we can make better policy decisions on the balance between different components of paid time off that better suit employee needs.

- We want to *predict* paid time off. For example, if we observe that paid time off varies by salary level, then we can predict or estimate the average paid time off taken by someone with a salary of, say, $50,000 and make better decisions on budgeting for benefits costs.

- We want to *control* paid time off. For example, we want to establish paid time off policies that relate to the market so our programs are competitive, enabling us to make better decisions on the paid time off we offer.

Another example involves pay. We observe that pay varies for a given job or skill. Not everyone is paid the same. We want to know *why* pay varies, beyond just speculating why. We want to find out *what factors* influence the pay for a given job or skill and *how* and *to what degree*.

Why do we want to know this? As stated before, there are typically three general reasons why we want to know why pay varies. They all are related to making better decisions.

- We want to *understand* pay. For example, if we observe that pay for programmers varies by the particular programming language they know, we can use that information to make better decisions on pay differentials among programmers.

- We want to *predict* pay. For example, if we observe that pay varies by experience, we can predict, or estimate the average pay for someone with, say, 5 years of experience and make better decisions on hiring rates.

- We want to *control* pay. For example, we want to establish pay structures that relate to the market, so our pay levels are competitive but not too low or too high, enabling us to make better decisions on the range of pay we offer.

A major objective in many statistical investigations is to establish relationships which make it possible to describe or predict one variable in terms of others. For example, studies are made to predict pay levels of programmers in terms of experience and programming languages, a CEO's pay in terms of the annual sales of the company, and insurance premiums in terms of age.

Outside of human resources, studies are made to predict the potential sales of a new cell phone in terms of its price, a runner's race time in terms of the air temperature and humidity, and a patient's weight in terms of the number of weeks he or she has been on a diet.

Although, it is desirable to be able to predict one quantity *exactly* in terms of others, this is seldom possible, and in most instances we have to be satisfied in predicting averages. Thus, we may not be able to predict *exactly* how much money Paul will make 5 years after graduating from college, but given suitable data we can predict the *average* income of a college graduate in terms of the number of years he has been out of college. Similarly, we can at best predict the *average* yield of a given variety of corn in terms of the rainfall and temperature in June, and we can at best predict the *average* performance of entering college students in terms of their IQ's.

In seeking to answer the questions of what factors influence a given variable, how they influence it, and to what degree they influence it, we will use the framework of model building, which is a general commonsense process used to solve problems.

Model building is simply a way of applying the scientific method.

## 6.3 SCIENTIFIC METHOD

The scientific method is the process by which we endeavor to construct an accurate, reliable, consistent, and nonarbitrary representation of the world. Recognizing that personal and cultural beliefs influence both our perceptions and our interpretations of phenomena, we aim through the use of standard procedures and criteria to minimize those influences when developing a theory.

The scientific method has four steps. (Some authors list a different number of steps, but they are consistent with each other and all follow the same general path.)

1. Observe and describe a phenomenon.
2. Formulate a hypothesis to explain the phenomenon that is consistent with what you have observed. This often takes the place of a mathematical equation.
3. Use the hypothesis to predict the results of new observations.
4. Perform experimental tests of the predictions by independent experimenters and properly performed experiments to verify the hypothesis.

If all goes well, then the hypothesis becomes a theory. If there are conflicting data that are not supported by the hypothesis, then a new hypothesis is proposed and the process is repeated.

To illustrate the scientific method, following is an example of turnover in a BPD accounting department, also discussed later.

1. The manager observes high voluntary turnover in her department.

2. She attributes the turnover to salaries that are low relative to the market and identifies a relationship between turnover and relative salaries.

3. She predicts turnover using that relationship.

4. The manager of compensation uses her relationship to predict turnover and finds that there are unexplained resignations of high-salaried employees. He proposes additional factors that influence turnover.

## 6.4  MODELS

Let us now turn to models. Models are not a whole new way of solving problems. We have all been doing it for years.

A model is an abstraction of reality used to solve problems in an objective manner.

A model has three components.

- The problem—the variable of interest
- Critical factors that impact the problem—other variables
- The relationships between the factors and the problem, and among the factors themselves

For example, a map of getting from the airport to downtown is a model. The map simplifies reality. Omitted are all the details of traffic lights, the volume of traffic, other people, all the shops and stores, the quality of the road, the weather, the trees, the buildings, and so on. The map abstracts just those parts of reality and their relationships that are essential to the problem at hand, namely getting from the airport to downtown.

The power of a model in solving a problem comes precisely from its *not corresponding* to reality *except* in those details pertinent to the problem at hand. In the strictest sense, we find the solution to a real problem by first finding out what will solve the problem in our abstract model, based on experience, research, and analysis, and then impose the same solution on reality. We never know if our models are correct until we apply them to reality.

In the abstract you followed the roads on the road map to drive from the airport to downtown. Turn left here, go right there, and drive 3 miles north, all on the map. Then, having solved the problem in the abstract, you applied the same solution to reality as you got in the car and actually turned left and went right and drove 3 miles north. The fact that you got to your destination indicates

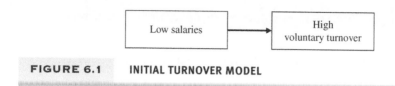

**FIGURE 6.1**    **INITIAL TURNOVER MODEL**

that the map, that is, the model, abstracted the essential parts of reality and with the appropriate relationships between them, helped you go from the airport to downtown.

In this example, the problem was getting from the airport to downtown. The critical factors were the major roads, route indicators, and so on. The relationships were how the roads linked the airport to downtown.

However, our models don't always work. Consider again, for example, the manager of the accounting department mentioned earlier that is experiencing high turnover. She comes to Compensation and says that the people say they are leaving for higher pay. Her model is shown in Figure 6.1.

If you accepted her model, then the solution would be to raise salaries. And if the model was correct and you raised salaries, then voluntary turnover would abate. But if the model was not correct, then you would have spent money inappropriately. Indeed, there may be other factors that influence voluntary turnover. A new and more complete model might be as shown in Figure 6.2, using the model-building framework described in Chapter 1, *Introduction*, and illustrated in Figure 1.4.

It is always instructive to identify those factors over which we have control and those we don't. And within each of those two groups, some factors we have the data on or can get the data on, and some we can't get the data on or are simply not measurable quantitatively.

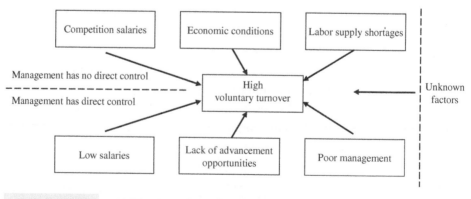

**FIGURE 6.2**    **COMPLETE TURNOVER MODEL**

Controllable variables are those over which we have direct control as managers. It is important to include these in our models because they can become powerful tools if found to influence our problem. For example, if advancement opportunity is such a variable, we can directly reduce turnover through a change in policy.

Known uncontrollable variables are those about which we have information but are not under our direct control as managers. For example, knowledge of economic conditions may well explain a great deal of our turnover, but we have little or no direct control over the variable. From a strategic perspective, we must treat this variable as a constraint—something we have to work around in designing our human resources policies.

Unknown factors are those factors that influence the problem, but about which we have no information. These are the variables that make our models imperfect and incomplete. In many cases, we will find that these influences are more important than known factors in influencing problems such as turnover. There is little managers can do to deal with these influences.

Sometimes we even create a new variable from two other ones, such as relative salaries, where we divide competition salaries by BPD salaries.

$$\text{Relative salaries} = \text{Competition salaries}/\text{BPD salaries}$$

Other factors could be involved in this example, but we want to reinforce three points about models.

- Our models don't always work. The manager's model may not have abstracted the key factors that influenced the problem.
- Our models can be used to suggest or direct strategies to solve the problem. In this example, we may investigate the advancement opportunity situation.
- Our models provide a framework within which the problem can be systematically considered, analyzed, and discussed. For example, what other factors would you consider in the above problem?

**CASE STUDY 5, PART 1 OF 3**

The engineering manager of BPD needs to hire some mid-level individual contributor (nonmanagerial) engineers with 5–10 years of experience, and wants to know what the market pay is so he knows the range of pay to offer. We will approach this problem using a model building process.

**The five steps of model building**

*1. Specify the problem, or issue.* In this step you do two additional things.
- Describe briefly why you are working on this problem. What is driving the problem? Why it is important enough to be spending your time on it? When you make your conclusions and recommendations, the drivers of the problem are what you lead off with.
- Describe your terms mathematically. For example, if you are looking at turnover, you need to define the numerator (e.g., number of regrettable quits during the last 12 months) and the denominator (e.g., the number of relevant employees at the beginning of the year)

*2. Generate critical factors which may explain or impact the problem.* This is the brainstorming step. It is important to realize that you can only develop a model with quantitative data that you have available.
- Some of the factors you will have data on, such as time in your company, or performance level, or number of training courses taken, or pay.
- Some of the factors you may have to gather data on, such as salary survey data, or years of education.
- Some of the factors you know influence the problem but you are unable to get the data, such as managerial effectiveness, or number of openings in other companies.
- Some of the factors may be unknown, yet to be discovered.

*3. Identify the relationship, if any, between each factor and the problem.* This step has two parts.
- Gather the data. You must ensure that the data are valid and "cleaned up." Otherwise your conclusions are meaningless.
- Plot the points.

*4. Quantify the relationship and analyze.* This step includes calculating the coefficients of a mathematical equation that describes the relationship.

*5. Evaluate the model.* Here, we will look at statistical criteria as well as common sense.

**FIGURE 6.3    THE FIVE STEPS OF MODEL BUILDING**

## 6.5 MODEL BUILDING PROCESS

There are five general steps in the model building process. These should be viewed as common sense and things you normally do anyway. What these five steps represent, is an explicit, disciplined approach to the business of problem solving using the context of model building. It is simply an application of the scientific method.

As a perspective, it should be kept in mind that the attempt to formulate a problem in terms of a mathematical model forces one into understanding what question is really being asked. A careful examination of available resources and data sometimes produces the conclusion that it is not possible to answer the question with those particular resources and data.

It should be emphasized that model building is an art. To make it as objective as possible we follow a rational process, as shown in Figure 6.3.[1]

---

[1] I am indebted to George T. Milkovich for the five steps of this model building approach.

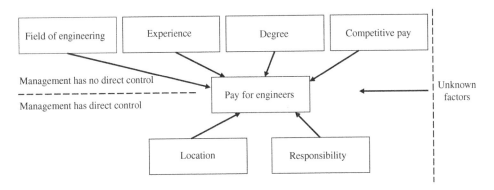

**FIGURE 6.4    MODEL OF FACTORS AFFECTING PAY OF ENGINEERS**

Our mathematical models that we create describe relationships between the variables. They don't describe the variables themselves. Rather they describe relationships.

Models describe relationships.

To illustrate this process we will continue with Case Study 5.

*Step 1. Specify the Problem, or Issue.* The problem is what influences the pay of individual contributor engineers.

- The reason we want to know is that we want to be able to predict the average pay for experienced hires. This will help our recruiters have a good idea what to offer.
- We are interested in annual base salary.

*Step 2. Generate Critical Factors Which May Explain or Impact the Problem.* Here you might want to get with some engineering managers or some engineers themselves to help identify potential factors you want to put in your model. Following are some example factors, shown in Figure 6.4.

- Field of engineering (chemical, mechanical, electrical, etc.)
- Years of experience as an engineer
- Degree (BS, MS, PhD)
- What other companies pay for similar jobs
- Location
- Responsibility level and complexity of work

For purposes of illustration, we will focus on the years of experience of BS chemical engineers.

*Step 3. Identify the Relationship, if any, Between the Factor and the Problem.* This step involves gathering the data and plotting points. It is very important that you ensure the data are valid. The computer adage "garbage in, garbage out" applies here just as strongly.

## Plotting Points

**A Quick Review**    Although you may be using software to plot the data, it is important that you understand the fundamentals of plotting so you know what the software is doing.

- A *plot*, also known as a graph, a chart, a scatter graph, a scatter chart, a scatter plot, and a scattergram, is a visual display showing the relationship between two variables.

- On a plot, there are two *axes*: an $x$-axis (also known as the abscissa), which is the horizontal axis and a $y$-axis (also known as the ordinate), which is the vertical axis.

- A *variable* is simply a measured characteristic of an item of interest, such as the pay of a chemical engineer, or the years of experience of a chemical engineer. In order for the characteristic to be a variable, it must vary. If all chemical engineers had 4 years of experience, then experience would not be a variable.

  ○ The $y$-variable, also known as the *dependent variable*, is the problem variable we are working on.

  ○ The $x$-variable, also known as the *independent variable*, is the factor that we think helps explain the variability of the problem variable.

- A *data point* consists of the measured characteristics of one item of interest. In our example with two variables, a data point consists of the years of experience and the pay of a particular chemical engineer. It is expressed as an ordered pair of values enclosed by parentheses and separated by a comma, such as (10, 110,000). This particular chemical engineer has 10 years of experience and is paid $110,000. In an ordered pair, the $x$-variable is always first and the $y$-variable is always second (alphabetical order). This is an accepted convention in plotting.

- When *plotting a data point*, you locate the $x$-value of the data point on the $x$-axis and locate the $y$-value of the data point on the $y$-axis, and extend those values into the plot area, and where they meet is where the data point is plotted. This is shown in Figure 6.5.

**Which Variable is $x$ and Which Variable is $y$?**    Typically the problem variable is $y$ and the critical factor variable is $x$. In this case, pay is the $y$-variable and experience is the $x$-variable. If you are ever in doubt, then substitute the

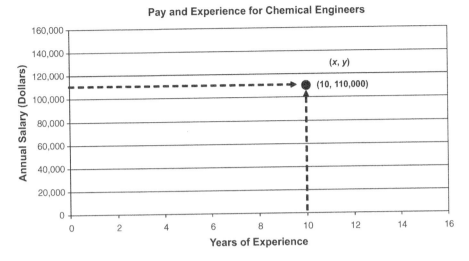

**Pay and Experience for Chemical Engineers**

**FIGURE 6.5      PLOTTING A DATA POINT**

names of the variables in any of the following statements in an order that makes sense to you and that describes the problem you are working on.

I want to find out if ———— is influenced by ————.

I want to determine if ———— is dependent upon ————.

I want to find out if ———— can be predicted by ————.

I want to describe ———— as a function of ————.

I wonder if ———— depends on ————.

The variable that you put in the first blank is the *y*-variable and is the one that is on the vertical axis. The variable that you put in the second blank is the *x*-variable and is the one that is on the horizontal axis.

We will illustrate this using the first statement with our example.

"I want to find out if *pay* is influenced by *experience*."

"I want to find out if *experience* is influenced by *pay*."

The sentence that makes sense and that describes our problem is the first one. We want to find out if *pay* is influenced by *experience*. This means that pay is the *y*-variable and experience is the *x*-variable. Based on our experience in compensation, the second statement does not make sense.

In describing what is being plotted, the usual way of stating it is "*y* versus *x*" or "*y* plotted against *x*." In this example, you would say "pay versus experience" or "pay plotted against experience." This will infer that the first term in your statement is *y* and the second one is *x*. This way of stating what is being plotted is

| TABLE 6.1 | PAY AND EXPERIENCE FOR CHEMICAL ENGINEERS |
|---|---|
| **Experience (Years)** | **Salary** |
| 3 | 80,000 |
| 4 | 80,128 |
| 4 | 94,000 |
| 5 | 98,304 |
| 6 | 118,272 |
| 6 | 100,000 |
| 7 | 87,296 |
| 7 | 132,000 |
| 8 | 114,560 |
| 9 | 138,368 |
| 10 | 128,512 |
| 10 | 108,000 |
| 11 | 135,860 |
| 12 | 164,352 |
| 13 | 122,624 |
| 14 | 153,088 |

not, however, universal, so to ensure there is no confusion, state which variable is the $y$-variable and which is the $x$-variable. In this example, you would say that pay is the $y$-variable, or dependent variable, and experience is the $x$-variable, or independent variable.

Now we return to the model building or our case study. You gather the data and plot the points. The data are in Table 6.1 and also in Data Set 11.

The plotting can be done easily with spreadsheet and most statistical software programs, and is shown in Figure 6.6.

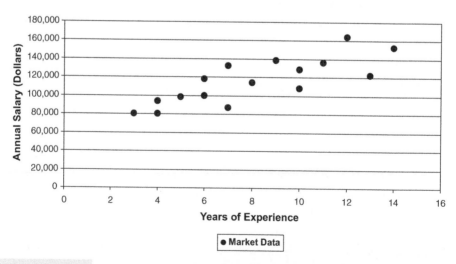

FIGURE 6.6   PAY AND EXPERIENCE FOR CHEMICAL ENGINEERS

After you plot the points, you look at the picture and ask the following questions.

### Is There a Relationship?

- No. Then try other factors.
- Yes. Then ask *What kind?*
  - *Positive* (high values of $y$ go with high values of $x$, low values of $y$ go with low values of $x$) or *negative* (high values of $y$ go with low values of $x$, low values of $y$ go with high values of $x$)
  - *Linear* (the general trend of the points can be described by a straight line) or *nonlinear* (the general trend of the points can be described by a curved line)
  - *Perfect* (the points fall on a perfectly straight line or a nice smooth curved line) or *not perfect* (they don't)

Before we answer these questions, consider the following examples in Figure 6.7.

The answers to the questions are given in Table 6.2.

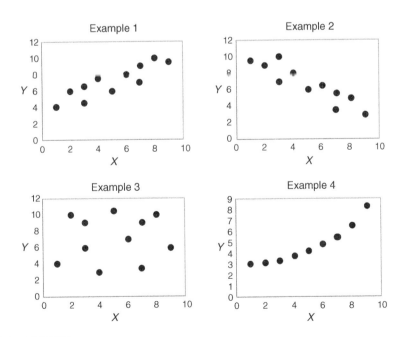

**FIGURE 6.7    FOUR TYPES OF RELATIONSHIPS**

| TABLE 6.2 | FOUR TYPES OF RELATIONSHIPS | |
|---|---|---|
| Example | Is There a Relationship? | If So, What Kind? |
| 1 | Yes | Positive, linear, and not perfect |
| 2 | Yes | Negative, linear, and not perfect |
| 3 | No | |
| 4 | Yes | Positive, nonlinear, and perfect |

So, our plot in Figure 6.6 looks like Example 1. The relationship is positive, linear, and not perfect.

Our goal is to go from a display of the points on a chart to a line that describes the trend we see by doing the following steps.

1. Plot the data.

2. Look at the plot to decide what kind of relationship is there between the two variables.

3. Describe that relationship with an equation of a line (straight or curved) so we can describe or predict.

## Functional Forms

Let us look at some possible relationships. We show two figures of functional forms that describe various relationships. If the relationship suggested by the plot of your data looks like one of these forms, then the associated mathematical formula will be the one that may be appropriate to use to describe the relationship.

There are two groups of functional forms. The first group in Figure 6.8 consists of polynomial models, which are models using the equation of a polynomial. A polynomial equation has powers of $x$ in it, such as $y = a + bx + cx^2$. The second group in Figure 6.9 consists of logarithmic transformation models, which are models where one or more of the variables has been transformed into logarithms, such as Log $y = a + bx$.

We have presented 11 models in the charts for completeness. The most common models encountered in compensation are two polynomial models—linear and maturity curve and two logarithmic transformation models—exponential and power. These four common models will be discussed in subsequent chapters.

The power model doesn't lend itself to drawing a line on a regular plot. The plot is like a comet whose head is crashing at the origin with many data points crowded together and the tail is scattered across the "sky" of the chart. The discussion in Chapter 10, *Power Model*, will elaborate on this.

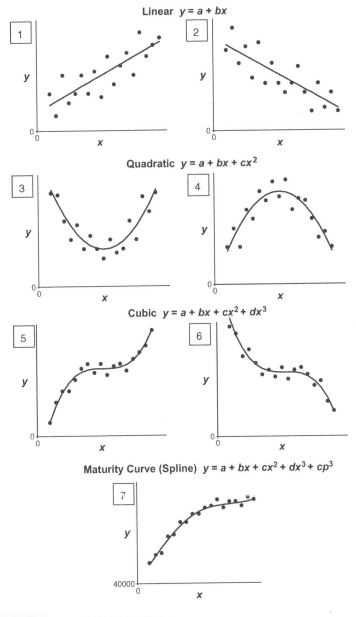

**FIGURE 6.8    POLYNOMIAL MODELS**

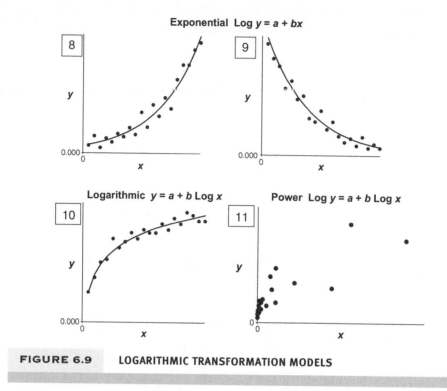

**FIGURE 6.9**     **LOGARITHMIC TRANSFORMATION MODELS**

Sometimes more than one functional form will describe the relationship adequately. In the example in Figure 6.10, there is a relationship, it is positive, nonlinear, and not perfect.

The trend could be described by exponential model 8, the right-hand side of quadratic model 3, or the right-hand side of cubic model 5, as shown in Figure 6.11. The mathematics automatically uses that part of a particular model that is appropriate.

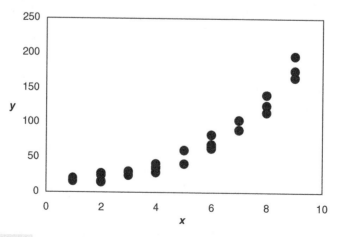

**FIGURE 6.10**     **EXAMPLE RELATIONSHIP 1**

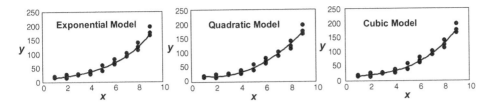

**FIGURE 6.11    POSSIBLE MODELS FOR EXAMPLE RELATIONSHIP 1**

Which one you select depends on the particular problem you are working on. Perhaps there is a context that would favor the use of an exponential model, such as developing a pay structure.

In addition, there is nothing sacred about a single model to describe your data. In the example in Figure 6.12, you could use quadratic model 4 to describe the relationship or you could use two linear models after segregating the data. Use linear model 1 with a positive slope to describe the left side of the data and linear model 2 with a negative slope to describe the right side of the data. This is shown in Figure 6.13.

As previously mentioned, subsequent chapters will discuss the four most common models encountered in compensation:

- *Linear Model* (Chapter 7)
- *Exponential Model* (Chapter 8)
- *Maturity Curve Model* (Chapter 9)
- *Power Model* (Chapter 10)

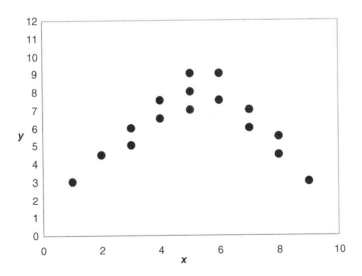

**FIGURE 6.12    EXAMPLE RELATIONSHIP 2**

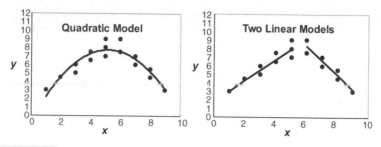

**FIGURE 6.13    POSSIBLE MODELS FOR EXAMPLE RELATIONSHIP 2**

*Step 4. Quantify the Relationship and Analyze.* Once we have decided which model may be appropriate to describe the relationship shown by the plot of the data, the next step is to develop the equation that corresponds to the model chosen. That is, we want to describe the general trend we see with a mathematical equation. We will derive the equation based on the data using the method of least squares.

## Method of Least Squares

The method of least squares is, for a given set of data and a given model, fitting the model to the data by choosing the coefficients of the model equation that minimize the sum of squared deviations of the data points from the model. The resulting equation produces what is called the least squares line. This is illustrated with a linear model in Figure 6.14

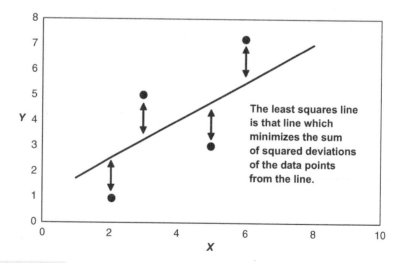

**FIGURE 6.14    LEAST SQUARES LINE**

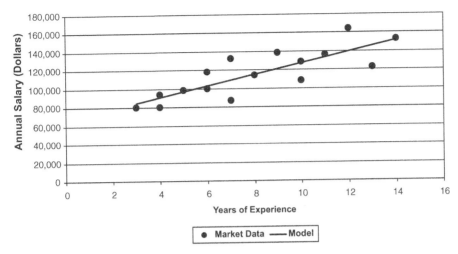

**FIGURE 6.15**    **LEAST SQUARES LINE FOR PAY AND EXPERIENCE FOR CHEMICAL ENGINEERS**

We say that we "fit the line to the data," or "fit the model to the data." That is, the data points come first and our model comes second. The model is hierarchically dependent on the data.

Since the least squares line is entirely data-determined, it is objective and credible, which makes it widely accepted. The method of least squares is what is used in spreadsheet and statistical software programs to fit models to data. Using "eyeball" lines would open the door to challenges of one's objectivity.

We use the term *regression* to describe the process of fitting a model to data using the method of least squares.

Continuing with our example in Figure 6.6 of chemical engineers to illustrate the model building process, we enter the raw data into a spreadsheet or statistical software program and have it produce the equation of the least squares line. We then draw the least squares line drawn on our chart, shown in Figure 6.15.

The equation of the resulting least squares line is

$$y' = 66,681 + 6,112x$$

or

$$\text{Predicted average pay} = 66,681 + (6,112)(\text{Years of Experience})$$

How to interpret and use such an equation is discussed in Chapter 7, *Linear Model*, along with using the symbol $y'$.

*Step 5. Evaluate the Model*

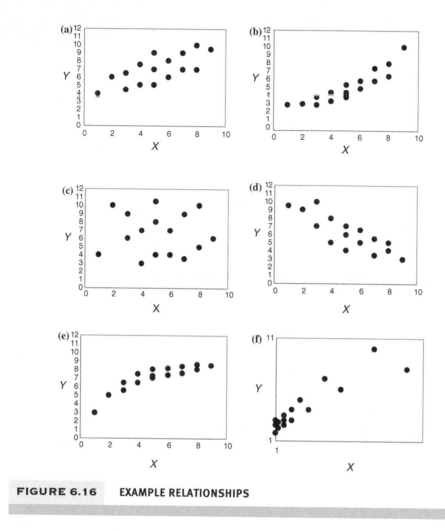

**FIGURE 6.16** **EXAMPLE RELATIONSHIPS**

There are various statistical and commonsense criteria with which to evaluate the goodness of the model. These are also discussed in Chapter 7.

In summary, the five steps of the model building process provide us an objective method of solving a problem. Subsequent chapters will continue to illustrate this.

Case Study 5 is continued in Chapter 7, *Linear Model.*

# PRACTICE PROBLEMS

**6.1** Why are we interested in why data vary?

**6.2** What is a model? What are its three components?

**6.3** You are studying production bonuses for outdoor environmental projects. Your team has developed a model that includes years of service, training, and team assignment. Identify the components of the model.

**6.4**   What are the five steps in model building?

**6.5**   You have expanded upon the model of production bonuses for outdoor environmental projects, and now have the following factors: years of service, training, hourly pay, managerial effectiveness, temperature, precipitation, team assignment, and external distractions (e.g., wild animals). Make a schematic diagram of a model, indicating which factors are under management control and which are not. Which factors are you most likely to have measureable data on or can get measureable data on?

**6.6**   You have a manager in Dubai with 8 years of experience and who is in grade 37. You are modeling grade versus experience. Express the data point for this manager as an ordered pair.

**6.7**   You are modeling the PTO (paid time off) taken by employees and the age of the employees. Which variable is the $x$-variable (independent variable) and which variable is the $y$-variable (dependent variable)?

**6.8**   You are modeling sales performance, and have two variables: commissions and sales. Which variable is the $x$-variable (independent variable) and which variable is the $y$-variable (dependent variable)?

**6.9**   What kinds of relationships are shown in the graphs in Figure 6.16?

**6.10**   Describe the method of least squares. What are its main advantages?

CHAPTER 7

# Linear Model

## 7.1 EXAMPLES

If the plot of two variables indicates that a straight line can describe the observed trend, then we fit a linear model or straight line to the data. This process is called simple linear regression. The term "simple" means that there is just one $x$-variable. Here are two examples of linear models.

The first model, in Figure 7.1, was used to make a decision on the salary range for the BPD VP of Human Resources.

The second model, in Figure 7.2, was used to make decisions on targeting BPD'communication efforts to lower-paid employees on the benefits of contributing to a 401(k).

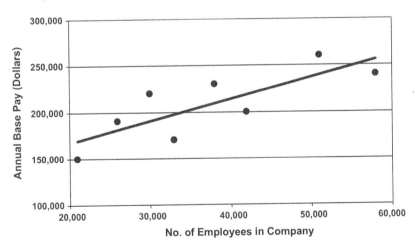

**FIGURE 7.1** LINEAR MODEL OF MARKET DATA FOR VP HUMAN RESOURCES

*Statistics for Compensation: A Practical Guide to Compensation Analysis,* By John H. Davis
Copyright © 2011 John Wiley & Sons Inc.

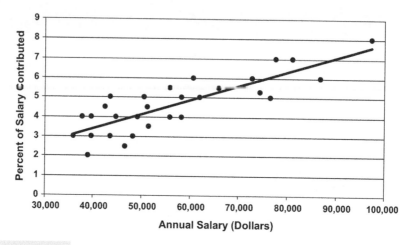

**FIGURE 7.2    LINEAR MODEL OF EMPLOYEE CONTRIBUTION TO 401(K)**

### CASE STUDY 5, PART 2 OF 3

Recall from Chapter 6 that the engineering manager of BPD needs to hire some mid-level chemical engineers with 5–10 years of experience and wants to know what the market pay is, so he knows the range of pay to offer. This statement summarizes the first two steps in the model building process—to specify the problem or issue (what is the pay for mid-level chemical engineers?) and to generate the critical factors that may impact the problem (for this example, we focus on experience of BS chemical engineers).

You know that pay varies with experience, so you collect survey data that include both salary and experience. We will use the same data from Chapter 6

**TABLE 7.1    PAY AND EXPERIENCE FOR CHEMICAL ENGINEERS**

| x (Years of Experience) | y (Salary) |
| --- | --- |
| 3 | 80,000 |
| 4 | 80,128 |
| 4 | 94,000 |
| 5 | 98,304 |
| 6 | 118,272 |
| 6 | 100,000 |
| 7 | 87,296 |
| 7 | 132,000 |
| 8 | 114,560 |
| 9 | 138,368 |
| 10 | 128,512 |
| 10 | 108,000 |
| 11 | 135,860 |
| 12 | 164,352 |
| 13 | 122,624 |
| 14 | 153,088 |

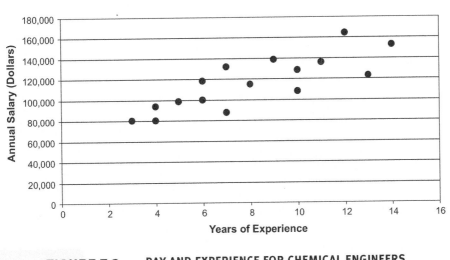

**FIGURE 7.3    PAY AND EXPERIENCE FOR CHEMICAL ENGINEERS**

that were used to illustrate the model building process. The data and corresponding plot are repeated here in Table 7.1 and Figure 7.3. The data are also in Data Set 11.

After you plot the points, you look at the picture and conclude that there is a relationship that is positive, linear, and not perfect. This completes step 3 of the model building process, which is to identify the relationship, if any, between the factors and the problem. Hence, a straight line or linear model would be appropriate to fit to the data.

## 7.2 STRAIGHT LINE BASICS

Since we will be fitting a straight line to the data, we first need to review straight lines. The equation for a straight line is

$$y = a + bx$$

where $a$ is the intercept and $b$ is the slope.

An example equation is

$$y = 40,000 + 5,000x$$

■ *Intercept is the value of y when x is zero.* In this example, the intercept is 40,000, the value of $y$, when $x$ is zero.

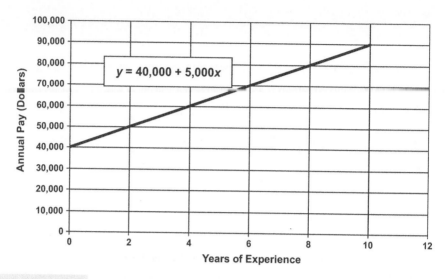

**FIGURE 7.4     STRAIGHT LINE**

■ *Slope is the change in y for a unit change in x.* The slope is the change in $y$ when the value of $x$ increases by 1. In this example, slope is 5,000, which is the change in $y$ when the value of $x$ increases by 1. The slope is also the change in $y$ divided by the corresponding change in $x$.

$$b = (y_2 - y_1)/(x_2 - x_1)$$

where $(x_1, y_1)$ and $(x_2, y_2)$ are any two points on the line.

The terms $a$ and $b$ are called the coefficients of the equation $y = a + bx$. In the example equation, the values 40,000 and 5,000 are the coefficients.

Figure 7.4 shows the straight line plotted on a chart.

In this example, $x$ is the years of experience and $y$ is the annual pay. The equation is

$$y = 40,000 + 5,000x.$$

The intercept is 40,000, which is the value of $y$ when $x$ is zero, and the slope is 5,000, which is the change in $y$ when $x$ increases by 1.

## Interpretations of Intercept and Slope

There are two interpretations for each term.

### Mathematical Interpretations

- The *mathematical interpretation* of the intercept is that $y$ has a value of 40,000 when $x$ is zero.
- The *mathematical interpretation* of the slope is that $y$ increases by 5,000 when $x$ increases by a value of 1.

### Problem Interpretations

- The *problem interpretation* of the intercept is that the pay is 40,000 for zero years of experience.
- The *problem interpretation* of the slope is that pay increases by 5,000 for each additional year of experience.

Note that the mathematical interpretations are simply the definitions of the terms. To get the problem interpretations, just substitute what $x$ is for $x$ and what $y$ is for $y$ in the mathematical interpretations.

## Using the Equation

To use this equation, suppose you want to know the pay for 10 years of experience. Simply rewrite the equation, substitute in 10 for $x$, and calculate $y$.

$$y = 40,000 + (5,000)(10) = 40,000 + 50,000 = 90,000$$

If you are given an equation and want to draw the line on a chart, simply solve the equation for two points, plot the points, and draw the line through them.

For example, using this same equation, suppose you used the values of $x = 0$ and $x = 8$. Simply substitute the values of $x$ in the equation and solve for the corresponding $y$. The first one is easy, because the value of $y$ is the intercept when $x$ is zero.

$$x_1 = 0$$
$$y_1 = \text{intercept} = 40,000$$
$$(x_1, y_1) = (0, \ 40,000)$$
$$x_2 = 8$$
$$y_2 = 40,000 + (5,000)(8) = 40,000 + 40,000 = 80,000$$
$$(x_2, y_2) = (8, \ 80,000)$$

Now plot these two points and draw the line through them, as shown in Figure 7.5.

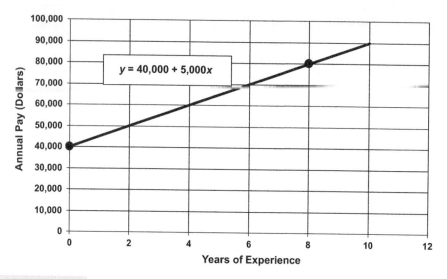

**FIGURE 7.5**     **STRAIGHT LINE THROUGH TWO POINTS**

Sometimes different symbols are used.

$$y = a + bx$$
$$y = mx + b$$
$$y = ax + b$$
$$y = b_0 + b_1 x$$

In all of these, the symbol or letter standing alone is the intercept and the symbol or letter next to the $x$ is the slope. We will use $y = a + bx$.

It should be noted that although there is always a problem interpretation of the slope, sometimes there is no problem interpretation of the intercept. Rather, the intercept is just a locater for the line. Consider the following equation, where $y$ is bonus as a percent of salary and $x$ is grade. The eligibility for bonuses starts at grade 22.

$$y = -15.0 + 1.2x$$

This is plotted in Figure 7.6.

What this means is that each higher grade gets an additional bonus of 1.2% of salary. That is the problem interpretation of the slope. But there is no problem interpretation of the intercept because the organization does not have a grade zero. The value of $-15.0$ is simply a locater for the line.

Note that zero is not shown on the $x$-axis. What we are showing here is just the relevant range of $x$-values—relevant to the problem at hand.

In this last example we have a negative intercept. That is perfectly acceptable. Each combination of different values of $a$ and $b$ represents a unique straight line,

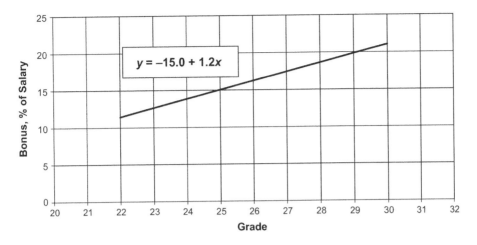

**FIGURE 7.6**    STRAIGHT LINE OF BONUS PERCENT AND GRADE

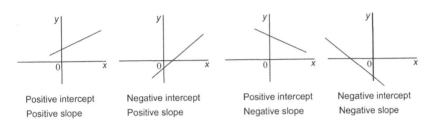

**FIGURE 7.7**    FOUR POSSIBLE COMBINATIONS OF INTERCEPT AND SLOPE

whether the values are positive or negative. In general, straight lines can have four combinations of positive and negative intercepts and slopes, as shown in Figure 7.7.

## 7.3 FITTING THE LINE TO THE DATA

This is step 4 of the model building process, where we quantify the relationship we observe and analyze it. We continue with the chemical engineering data.

We will use the least squares method to derive the equation of the straight line from the data. This is called linear regression or, for just one $x$-variable, simple linear regression.

The formulas for the slope and intercept of the least squares line are as follows:

$$\text{slope } b = \frac{\sum (x - \bar{x})(y - \bar{y})}{\sum (x - \bar{x})^2}$$

$$\text{intercept } a = \bar{y} - b\bar{x}$$

The slope and intercept are also known as regression coefficients.

It is not necessary to be able to use these formulas or any other regression formulas presented in the rest of the book by hand or on a hand calculator, as in almost all cases the practitioner will load the raw data into a spreadsheet or statistical software that will have the formulas imbedded "behind the scenes" and which will produce the regression results. What is important is that you know there are rigorous and theoretically sound formulas that produce the results.

Before we go on, there is one final note on terminology. Sometimes it is helpful in symbols to distinguish between the raw, original data and the predicted values that the least squares equation produces. We do that as follows, using the symbol $y'$ ("$y$ prime").[1]

| Symbol | Meaning |
| --- | --- |
| $y$ | $y$-value of the data point |
| $y'$ | $y$-value predicted or produced from the model |

Now, we find the equation of the least squares line and plot that line on the chart with the data points, both of which can be done with spreadsheet or statistical software.

In this example, the resulting equation is $y' = 66{,}681 + 6{,}112x$, and the plot is shown in Figure 7.8.

This line makes sense of the data points. It summarizes and describes the observed trend and allows us to identify or predict a single value for the given years of experience. For example, if we had a candidate with 10 years of experience, we would substitute 10 for $x$ in the equation and get the following prediction result.

$$y' = 66{,}681 + (6{,}112)(10) = 66{,}681 + 61{,}120 = 127{,}801$$

## What We Are Predicting

Regression predicts *average values*, not the actual ones. We are predicting the *average* pay for a given experience level. Some of the actual pay is higher than the predicted value, and some of the actual pay is lower than the predicted value. So when using the equation, it is important to use the word "average" in our statements.

---

[1] Often in calculus, the symbol $y'$ indicates the first derivative of $y$ with respect to $x$, $dy/dx$. That is not what is meant here. In some statistics texts, the symbol $\hat{y}$ ("$y$ hat") is used for the predicted value.

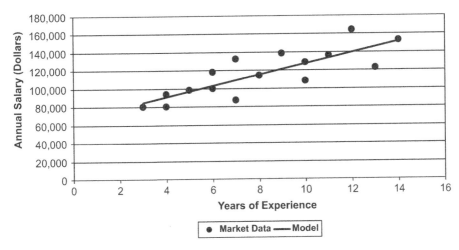

**FIGURE 7.8    PAY AND EXPERIENCE OF CHEMICAL ENGINEERS WITH LINEAR MODEL**

Hence, in our example, the predicted *average* pay for a chemical engineer with 10 years of experience is 127,801.

### Interpretations of Intercept and Slope

- The *mathematical interpretation* of the intercept is that $y'$ has a value of 66,681 when $x$ is zero.
- The *mathematical interpretation* of the slope is that $y'$ increases by 6,112 when $x$ increases by a value of 1.
- The *problem interpretation* of the intercept is that the predicted average annual pay is $66,681 for zero years of experience.
- The *problem interpretation* of the slope is that predicted average annual pay increases by $6,112 for each additional year of experience.

As a side note, the least squares line always goes through the point of the grand mean, $(\bar{x}, \bar{y})$. In this case, the line goes through the point (8.1, 115,960), where 8.1 is the mean of the years of experience and 115,960 is the mean of the annual pay.

## 7.4 MODEL EVALUATION

We have completed step 4 of the model building process, which is to quantify the relationship and analyze. Finally, step 5 is to evaluate the model, where we evaluate the "goodness" of the model. By "goodness" is meant how well the model corresponds to reality, reality being the observed data. We will discuss five types of evaluations.

- The appearance or look of the line through the data points
- Coefficient of determination $r^2$
- Correlation $r$
- The standard error of estimate (SEE)
- Common sense

## Appearance

The first evaluation is visual and qualitative, and is based on your perception of the chart. Does the line go through the "middle" of the data and does it "look like" it describes the general trend? In this case it appears to do a good job of both, just by looking at it. The line does indeed accurately describe the trend.

## Coefficient of Determination

The second evaluation is quantitative, and is called the coefficient of determination or $r^2$. It is a quantitative measure of the strength of association between the two variables. Mathematically, the value of $r^2$ is the proportion of variability of one variable that can be attributed to the variability of the other variable.

In this example, the coefficient of determination is a direct measure of how much of the total variation in pay can be attributed to, or "explained by," the relationship with experience. The value of the coefficient of determination is the proportion of the variation in pay that is explained by experience.

The formula for the coefficient of determination is

$$r^2 = 1 - \frac{\text{SS residual}}{\text{SS total}} = 1 - \frac{\sum (y - y')^2}{\sum (y - \bar{y})^2}$$

where SS residual is "sum of squares residual" and SS total is "sum of squares total."

The total variability of the dependent variable is measured by SS total. The reference for this measure of variability is the mean $\bar{y}$. The model has "explained" some of that variability, but not all of it. The variability not explained is measured by SS residual. The reference for this measure of variability is the predicted values $y'$.

As a technical note, there is a process called partitioning the sum of squares, wherein the total variation of the $y$-variable, SS total, is partitioned into two parts: the variation explained by the model, called SS regression, and the variation that is not explained by the model, called SS residual (sometimes called SS error or the sum of squares error).

SS regression
SS residual
———————————
SS total

As a further note, the SS residual *is* the sum of squares that has been minimized by the least squares criterion.

For linear regression, a mathematically equivalent formula for $r^2$ is

$$r^2 = \frac{\left(\sum (x - \bar{x})(y - \bar{y})\right)^2}{\left(\sum (x - \bar{x})^2\right)\left(\sum (y - \bar{y})^2\right)}$$

The values of $r^2$ range from 0.0 to 1.0. The closer it is to a value of 1, the stronger is the relationship. Keep in mind that we are trying to explain the *variability* of the y-variable as it is associated with the x-variable, and not the values themselves of the y-variable, and that $r^2$ indicates the strength of that association. In this example, we are trying to explain the variability of pay rather than the pay itself. This is a subtle difference, but an important difference nonetheless.

For our example, from the spreadsheet or statistical software we get

$$r^2 = 0.67, \text{ or } 67\%.$$

In the pie chart in Figure 7.9 representing the total variability of pay, the variability explained by the model, that is, by experience, is represented by the portion called regression. The variability not explained by the model is represented by the portion called residual. The residual is the variability that is "left over" after the regression and that the model did not explain.

We already concluded from viewing the plot that the relationship was not perfect. A value of $r^2$ less than 1.0 confirms that observation. A legitimate question to ask at this point is "Is our model invalid since it did not explain everything

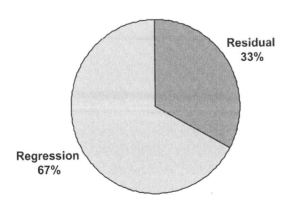

**FIGURE 7.9**    **VARIABILITY COMPONENTS**

perfectly?" The answer is that our model is indeed valid, but in addition to experience, there may be other factors that influence the variability of pay that we identified in step 2 of the model building process, such as degree level, responsibility level, and so on. Perhaps one of those factors or even a combination of factors would produce a higher $r^2$. Combinations of factors are discussed in Chapter 17.

Sometimes it is asked, "What is a good coefficient of determination?" "How big should it be before I can believe the results of a model?"

The answer is, "It depends." It depends on the context of the problem. There is no *a priori* standard for a good $r^2$.

If I am designing a bridge as an engineer or if I am testing out a new medicine as a pharmaceutical researcher, I would like my $r^2$ to be way over 90%. But if I am a social scientist examining human behavior, a value for $r^2$ of 50% would make me very happy, because I finally found a relation with *something*.

Usually when we analyze a data set, we just take whatever $r^2$ we get. It is what it is. In compensation modeling, for example, values over 95% and values as low as 10% abound. Obviously, the higher the value of $r^2$, the more comfort we have with the resulting model.

The advice for the compensation practitioner is to build up a history file of values of $r^2$ to determine what values one might expect in the situations being modeled. When you get a similar value, you conclude that things are as they have been, but when a value appears that varies greatly, an investigation should be made to determine why it is different.

### Interpretations of Coefficient of Determination

- The *mathematical interpretation* of the coefficient of determination is that 67% of the variability in $y$ can be attributed to the relationship with $x$.
- The *problem interpretation* of the coefficient of determination is that 67% of the variability in pay can be attributed to the relationship with experience.

## Correlation

The third evaluation is a quantitative number, but it has only a qualitative interpretation, and is called the correlation or correlation coefficient or $r$.

The value of the correlation is the square root of the coefficient of determination, having the same sign as the sign of the slope of the least squares line. Its values range from $-1.0$ to $+1.0$. The closer the values are to a value of $-1.0$ or $+1.0$, the stronger is the relationship.

$$r = \sqrt{r^2}$$

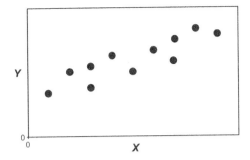

**FIGURE 7.10    POSITIVE CORRELATION**

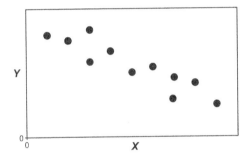

**FIGURE 7.11    NEGATIVE CORRELATION**

Figure 7.10 shows a positive correlation. The high values of $y$ go with the high values of $x$. The low values of $y$ go with the low values of $x$.

Figure 7.11 shows a negative or inverse correlation. The high values of $y$ go with the low values of $x$. The low values of $y$ go with the high values of $x$.

Figure 7.12 shows no correlation. There is no relation between the values of $y$ and the values of $x$.

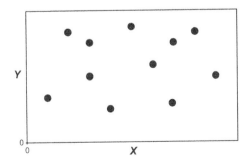

**FIGURE 7.12    NO CORRELATION**

So all you can say about correlation is strictly qualitative. That is why the coefficient of determination is a much more powerful measure of the association. But we mention correlation here because the term is so widely used.

In our example, the correlation is positive and has a value of 0.82. That is all that you can say about correlation. It does not give a quantitative interpretation of the strength of the relationship as the coefficient of determination does.

## Standard Error of Estimate

The fourth evaluation is quantitative and is the standard error of estimate, or SEE. Sometimes it is just called the standard error.

The standard error of estimate has a similar meaning with respect to the least squares line that the standard deviation has with respect to the mean of one variable. Recall that the standard deviation represents an "average" deviation of the values of the data points from the mean. Here, the SEE represents an "average" deviation of the actual $y$-values from the predicted values, or the line.

The formula for the standard error of estimate is

$$\text{SEE} = \sqrt{\frac{\sum (y - y')^2}{n - p}}$$

where $p$ is the number of coefficients in the least squares equation.

In this example, SEE has a value of 15,141.

### Interpretations of Standard Error of Estimate

- The *mathematical interpretation* of SEE is that on average, the actual $y$ values vary from the predicted average $y$ values by about 15,141.

- The *problem interpretation* of SEE is that, on average, the actual pay levels vary by about \$15,141 from the predicted average values of pay. Some are close, some are far away; some are positive, some are negative. But, on average, that difference is about 15,141. So when we predict average pay, we know that the actual pay varies by about ±15,141 on average from the predicted average values.

The lower the value of SEE, the more precise are the predictions made with the model.

## Common Sense

The fifth evaluation is qualitative and is common sense. This is where you have to use your knowledge of the situation and its context as a perspective. In this case, does it make sense for experience to influence pay? And if it does, does

**TABLE 7.2    SUMMARY OF INTERPRETATIONS AND EVALUATIONS**

| Term or Evaluation Criteria | Symbol | Mathematical Interpretation | Problem Interpretation or Evaluation |
|---|---|---|---|
| Appearance | | | Goes through middle of data and describes trend nicely |
| Intercept | $a$ | $y'$ has a value of 66,681 when $x$ is zero | The predicted average pay is $66,681 for zero years of experience |
| Slope | $b$ | $y'$ increases by 6,112 when $x$ increases by a value of 1 | The predicted average pay increases by $6,112 for each additional year of experience |
| Coefficient of determination | $r^2$ | 67% of the variability in $y$ can be attributed to the relationship with $x$ | 67% of the variability in pay can be attributed to the relationship with experience |
| Correlation | $r$ | +0.82—strong association between $y$ and $x$ | +0.82—strong association between pay and experience |
| Standard error of estimate | SEE | On average, $y$ varies from $y'$ by 15,141 | On average, the actual pay varies from the predicted average pay by about $15,141 |
| Common sense | | | Makes sense with respect to pay levels and increases with experience |

it make sense for no experience to be worth about $66,681 per year and for each additional year of experience to be worth about $6,112 annually? For this example, we will assume a positive answer to these questions, based on our experience.

**Perspective**    You will note in all these evaluation criteria that the basis for deciding "good" or "bad" is the context of the analysis coupled with your subject matter expertise, your experience, and your expectations. There is no *a priori* standard for what is good.

## 7.5  SUMMARY OF INTERPRETATIONS AND EVALUATION

Table 7.2 summarizes the interpretations and evaluations of our example.

## 7.6  CAUTIONS

There are four major statistical cautions about the use of these models.

1. *A High Association Between Variables Does Not Imply Cause and Effect Relationships.* We often use measures of association such as correlation and coefficient of determination to help us find cause and effect relationships.

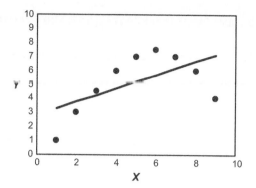

**FIGURE 7.13   FITTING STRAIGHT LINE TO CURVILINEAR DATA**

But in itself, high association does not prove a thing. It is simply a measure of how well two sets of numbers just happen to go together. We often look for cause and effect, but the numbers do not prove a thing.

For example, every time your wristwatch shows 12 o'clock, the bell in the tower rings 12 times. Does this mean that your wristwatch caused the tower bell to ring? Likewise, every time the bell in the tower rings 12 times, your wristwatch shows 12 o'clock. Does this mean that the tower bell caused your wristwatch to indicate 12 o'clock? The answer is "no" to both questions. We must investigate why the numbers go together. In this case, your watch and the bell are set independently to the same time standard. That is why they have a perfect correlation with each other.

2. *Coefficient of Determination and Correlation Measure the Strength of Linear Relationships Only.* Mathematically, these two terms were derived based on the assumption of a linear relationship. You can fit a straight line to curvilinear data, as shown in Figure 7.13, and calculate $r^2$ and $r$, but the interpretation will not be as valid as you would like.

3. *Be Cautious When Using the Model to Make Predictions Too Far Away from the Data Points from Which the Model Was Derived.* The validity of the model is strongest within the range of the data points from which it was derived. If you use the model to predict values too far away from the range of the data points, it may not reflect the true situation, as the relationship may not hold. Figures 7.14 and 7.15 illustrate this. If we use the model from the original data points to predict far away from those data points, we would be overpredicting.

The original data points and model are shown in Figure 7.14.

More data points and the original model are shown in Figure 7.15.

This does not mean do not do it. Sometimes we have to extrapolate from our model, such as predicting health care costs for the next 3 years.

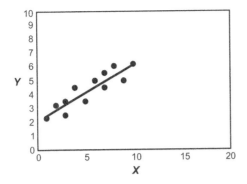

**FIGURE 7.14    ORIGINAL DATA AND LINEAR MODEL**

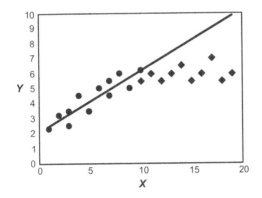

**FIGURE 7.15    MORE DATA AND ORIGINAL LINEAR MODEL**

It just means be cautious and ensure that the original relationship between the variables still holds.

4. *Have Sufficient Data Points to Be Comfortable with Your Model and the Conclusions.* This raises the issue of how many data points are sufficient. For simple linear regression, at least 10–20 data points should be sufficient in most cases to identify any trend or relationship (or lack thereof) present between the two variables.

Sometimes you only have a few data points, as is often the case when we are modeling executive pay as a function of sales. You have to use what you have, but in such cases, your conclusions start to become tenuous. They may be perfectly valid, but the comfort level is not as high.

To summarize these cautions,

Check your assumptions!

**CASE STUDY 5, PART 3 OF 3**

Since you have concluded that your model is a good one, you prepare Table 7.3 for the engineering manager to use as a guide for making offers to experienced chemical engineers.

**TABLE 7.3** **GUIDE FOR SALARY OFFERS**

| Years of Experience | Market Pay for Chemical Engineers | | |
| --- | --- | --- | --- |
| | Average − SEE | Predicted Average | Average + SEE |
| 4 | 75,988 | 91,129 | 106,270 |
| 5 | 82,100 | 97,241 | 112,382 |
| 6 | 88,212 | 103,353 | 118,494 |
| 7 | 94,324 | 109,465 | 124,606 |
| 8 | 100,436 | 115,577 | 130,718 |
| 9 | 106,548 | 121,689 | 136,830 |
| 10 | 112,660 | 127,801 | 142,942 |
| 11 | 118,772 | 133,913 | 149,054 |

For each year of experience, you show the predicted average pay plus and minus the standard error of estimate, that is, plus and minus the "average" deviation from the predicted average. For example, for 4 years of experience, the predicted average pay is $91,129 \pm 15,541$, resulting in a suggested hiring range of 75,988–106,270.

## 7.7 DIGGING DEEPER

Often in analyzing a problem you should explore the data in more depth. This is because the trends that appear on the surface may mask other trends that are in different parts of the data. In such a case you need to segregate the data set and explore the different subsets separately.

One method to do this segregated analysis is to plot two (or more) sets of data on the same chart. That is, you want to plot two sets of $y$-variables versus the same $x$-variable.

We will illustrate this in an analysis of raises given to 30 employees in two departments as a function of their performance ratings. The data are shown in Table 7.4 and also in Data Set 12.

First, we will plot percent raise ($y$) versus performance rating ($x$), since we want to find out if the percent raise is influenced by performance rating. This is shown in Figure 7.16.

**TABLE 7.4    RAISES AND PERFORMANCE OF EMPLOYEES IN TWO DEPARTMENTS**

| Employee | Emp. No. | Department | Performance Rating | Percent Raise |
|----------|----------|------------|--------------------|----------------|
| Allen | 1 | A | 4 | 4.7 |
| Bert | 2 | B | 4 | 4.0 |
| Clifford | 3 | A | 4 | 4.5 |
| Daphne | 4 | A | 3 | 4.5 |
| Ed | 5 | B | 3 | 4.0 |
| Ellen | 6 | A | 3 | 3.8 |
| Flannery | 7 | B | 4 | 4.0 |
| Flora | 8 | A | 3 | 4.0 |
| George | 9 | B | 4 | 4.0 |
| Howard | 10 | A | 5 | 5.0 |
| Iris | 11 | A | 4 | 5.0 |
| Jane | 12 | A | 5 | 6.0 |
| Juan | 13 | B | 5 | 4.0 |
| Kelly | 14 | A | 2 | 2.0 |
| Lana | 15 | A | 3 | 3.6 |
| Marlena | 16 | B | 2 | 4.0 |
| Nan | 17 | A | 4 | 4.0 |
| Oscar | 18 | A | 3 | 3.2 |
| Paul | 19 | B | 3 | 4.0 |
| Quincy | 20 | A | 5 | 4.5 |
| Ramero | 21 | A | 4 | 4.9 |
| Stacy | 22 | B | 5 | 4.0 |
| Tanya | 23 | A | 4 | 3.6 |
| Ura | 24 | B | 3 | 4.0 |
| Victor | 25 | A | 2 | 2.5 |
| Virginia | 26 | A | 4 | 4.1 |
| Will | 27 | A | 3 | 3.0 |
| Xavier | 28 | B | 4 | 4.0 |
| Yasha | 29 | A | 5 | 5.5 |
| Zoe | 30 | A | 4 | 4.8 |

The data look pretty consistent, and are what one might expect. There is a relationship that is generally linear and positive. Everything looks all right, as it shows our managers are rewarding performance appropriately.

You decide to dig deeper into the data and examine whether there are any differences between the two departments. Here in Figure 7.17 we plot the two departments on one chart, but with different symbols.

From this chart it is obvious that the manager of Department A did differentiate raises based on performance, while the manager of Department B gave everyone a 4.0% raise, and did not differentiate raises based on performance at all. This display highlights an important situation that needs investigating.

This is a nice example where graphical displays are very powerful tools in data analysis and where numbers raise issues.

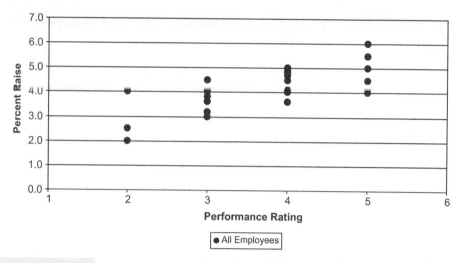

FIGURE 7.16    RAISES AND PERFORMANCE OF EMPLOYEES IN TWO DEPARTMENTS

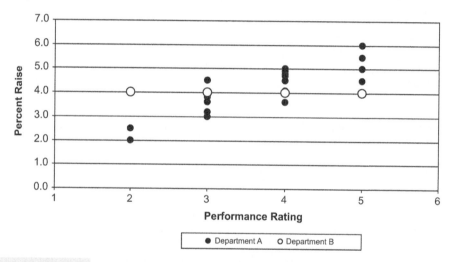

FIGURE 7.17    RAISES AND PERFORMANCE OF EMPLOYEES SEPARATED BY
DEPARTMENT

The moral of this story is that when you have data, keep digging into it to see what you may find. Be a data explorer. It could be very interesting!

## 7.8 KEEP THE HORSE BEFORE THE CART

𝔄 𝔣𝔞𝔟𝔩𝔢. You are the new human resources manager for the information technology division of your organization, and are doing an analysis of unused paid time off (PTO). You are interested in finding out if the more experienced employees accumulate more unused PTO, as your organization is considering

**TABLE 7.5    UNUSED PTO AND EXPERIENCE OF EMPLOYEES IN FOUR DEPARTMENTS**

| Application Development | | Data Center Operations | | Strategic Solutions | | Logistics | |
|---|---|---|---|---|---|---|---|
| Years of Experience | Unused PTO | Years of Experience | Unused PTO | Years of Experience | Unused PTO | Years of Experience | Unused PTO |
| 10 | 80.4 | 4 | 31.0 | 12 | 81.5 | 8 | 65.8 |
| 8 | 69.5 | 9 | 87.7 | 5 | 57.3 | 8 | 57.6 |
| 13 | 75.8 | 6 | 61.3 | 14 | 88.4 | 8 | 77.1 |
| 9 | 88.1 | 13 | 87.4 | 4 | 53.9 | 8 | 88.4 |
| 11 | 83.3 | 14 | 81.0 | 8 | 67.7 | 8 | 84.7 |
| 14 | 99.6 | 8 | 81.4 | 9 | 71.1 | 8 | 70.4 |
| 6 | 72.4 | 11 | 92.6 | 10 | 74.6 | 8 | 52.5 |
| 4 | 42.6 | 7 | 72.6 | 13 | 127.4 | 19 | 125.0 |
| 12 | 108.4 | 10 | 91.4 | 6 | 60.8 | 8 | 55.6 |
| 7 | 48.2 | 5 | 47.4 | 7 | 64.2 | 8 | 79.1 |
| 5 | 56.8 | 12 | 91.3 | 11 | 78.1 | 8 | 68.9 |

**TABLE 7.6    REGRESSION RESULTS**

| Department | Linear Model | $r^2$ | SEE |
|---|---|---|---|
| Application Development | $y' = 30.0 + 5.00x$ | 0.67 | 12.4 |
| Data Center Operations | $y' = 30.0 + 5.00x$ | 0.67 | 12.4 |
| Strategic Solutions | $y' = 30.0 + 5.00x$ | 0.67 | 12.4 |
| Logistics | $y' = 30.0 + 5.00x$ | 0.67 | 12.4 |

some version of a "use it or lose it" policy. There are four departments in the division and you gather the following data in Table 7.5. The data are also in Data Set 13.[2] You also want to know if there are differences between departments on unused PTO.

Since it is so easy to have your software do a regression analysis, you enter the data and obtain the following results in Table 7.6.

"What!" you exclaim. "The employees in all four departments have the *same* relationship between experience and unused PTO. Everything is the same: the intercept, the slope, the coefficient of determination, and the standard error of estimate. That is really something!" You evaluate the models with $r^2$, $r$, SEE, and common sense, and they look like good models. You can hardly wait to report this surprising revelation that everything is same in all the four departments. But you hesitate, because having identical results just seems so strange. And then something nagging in your subconscious reminds you of the five steps of the model building process.

[2] These data are a modification of what is known as Anscombe's Quartet. See Anscombe, F. J. (1973) Graphs in Statistical Analysis, *American Statistician*, **27**: 17–21.

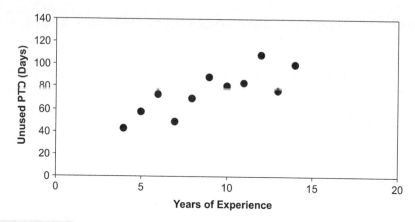

**FIGURE 7.18**   UNUSED PTO AND EXPERIENCE OF EMPLOYEES IN APPLICATIONS DEVELOPMENT

"Let's see," you say to yourself. "The first step is to state the problem. The second step is to brainstorm factors that might influence the problem. The third step is to gather the data and plot them to identify the relationship, if any. The fourth step is to quantify the relationship with a regression analysis. And the fifth step is to evaluate the model. Hmmm. I have done steps 1, 2, 4, and 5. Perhaps I should do step 3 before I get too excited and report my results."

So you plot the data for each of the four departments. The first one, Applications Development, is shown in Figure 7.18.

You say, "Well, this one looks ok. It is the type of plot I would expect for a linear relationship. There is a relationship; it is positive, linear, and not perfect. So far, so good."

Next is Data Center Operations, shown in Figure 7.19.

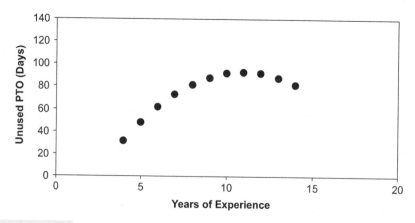

**FIGURE 7.19**   UNUSED PTO AND EXPERIENCE OF EMPLOYEES IN DATA CENTER OPERATIONS

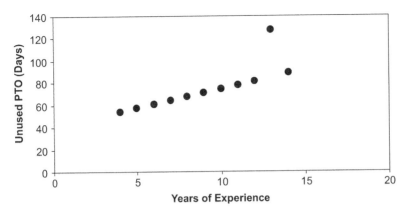

**FIGURE 7.20**   UNUSED PTO AND EXPERIENCE OF EMPLOYEES IN STRATEGIC SOLUTIONS

"Hmmm. There is a relationship, it is positive, but it is nonlinear and it looks perfect. Maybe I should not attempt to fit a straight line to this obvious curvilinear data. And besides, even in my limited experience, *nothing* is perfect, whether it is linear or not. I need to look into this."

This is followed by Strategic Solutions, shown in Figure 7.20.

"Wow. This one is really weird. Except for one outlier, it is a perfect straight line. I should investigate what is going on in this department."

Finally, you look at Logistics, shown in Figure 7.21.

"And this one is even weirder. Everyone has 8 years of experience except the one outlier. I need to check my data. I suspect an error somewhere, or maybe something else. I will find out."

𝕿𝖍𝖊 𝕸𝖔𝖗𝖆𝖑 𝖔𝖋 𝖙𝖍𝖊 𝕾𝖙𝖔𝖗𝖞. So, how can everything be the "same" when they are all different? Well, that can happen, and this example drives home the point that you should *always* plot the data first before you conduct your regression analysis.

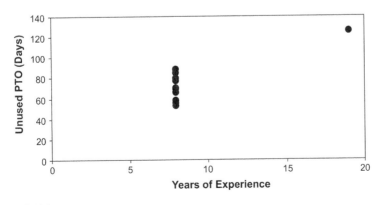

**FIGURE 7.21**   UNUSED PTO AND EXPERIENCE OF EMPLOYEES IN LOGISTICS

A plot validates (or invalidates) the assumption that the relationship between the two variables is linear. In other words, a plot helps you check your assumptions.

If the plot is linear, then fit a linear model to it and all the interpretations you make for a linear model will be valid.

If the plot shows the relationship is not linear, then do not fit a linear model to it. If you do fit a linear model, the interpretations we have made for a linear model will be tenuous at best.

## RELATED TOPICS IN THE APPENDIX

A.10   Linear Regression Technical Note
A.11   Formulas for Regression Terms

## PRACTICE PROBLEMS

**7.1**   The training manager for the BPD New York manufacturing plant has come to you with a proposal to have a reward or recognition tied to productivity that would encourage workers to take more training courses. She wants to know the effectiveness of the amount of training that the production workers have taken. The plant has a productivity measure of units per day for each employee. You collect and verify the productivity and training data for each of the 30 employees, shown in Table 7.7. The data are also in Data Set 14.

Determine the type of relationship there is by plotting the data.

### TABLE 7.7   PRODUCTIVITY AND TRAINING

| Emp. No. | No. of Courses | Units per Day | Emp. No. | No. of Courses | Units per Day |
|---|---|---|---|---|---|
| 1 | 0 | 18 | 16 | 3 | 58 |
| 2 | 0 | 20 | 17 | 3 | 60 |
| 3 | 0 | 25 | 18 | 3 | 65 |
| 4 | 0 | 27 | 19 | 3 | 70 |
| 5 | 0 | 35 | 20 | 3 | 72 |
| 6 | 1 | 25 | 21 | 3 | 85 |
| 7 | 1 | 32 | 22 | 4 | 70 |
| 8 | 1 | 40 | 23 | 4 | 80 |
| 9 | 1 | 48 | 24 | 4 | 85 |
| 10 | 2 | 37 | 25 | 4 | 90 |
| 11 | 2 | 40 | 26 | 5 | 80 |
| 12 | 2 | 45 | 27 | 5 | 90 |
| 13 | 2 | 50 | 28 | 5 | 92 |
| 14 | 2 | 55 | 29 | 5 | 95 |
| 15 | 2 | 70 | 30 | 5 | 100 |

**TABLE 7.8    EXERCISE FACILITY USAGE AND AGE**

| Age | % Employees Using Exercise Facilities |
|---|---|
| 21 | 63 |
| 22 | 45 |
| 23 | 55 |
| 24 | 41 |
| 25 | 58 |
| 26 | 38 |
| 27 | 45 |
| 28 | 43 |
| 29 | 32 |
| 30 | 47 |
| 31 | 20 |
| 33 | 37 |
| 35 | 25 |
| 37 | 42 |
| 39 | 21 |
| 40 | 30 |
| 42 | 12 |
| 43 | 17 |
| 45 | 26 |
| 47 | 2 |
| 49 | 11 |
| 51 | 0 |
| 52 | 8 |
| 54 | 12 |
| 55 | 1 |
| 57 | 3 |
| 59 | 0 |

**7.2**  If the relationship is linear, calculate the equation of the least squares line, the coefficient of determination, the correlation, and the standard error of estimate. Also, report the average number of courses taken and the average units per day produced.

**7.3**  Provide mathematical and problem interpretations of the intercept, slope, coefficient of determination, correlation, and standard error of estimate.

**7.4**  Draw the least squares line on the plot.

**7.5**  Use this equation to estimate the average productivity when three training courses are taken.

**7.6**  The profit margin increases dramatically when productivity is at least 80 units per day. Use the prediction equation to calculate how many training courses it would take to achieve this level of productivity.

**7.7**  What are the four major cautions in simple linear regression?

**7.8**  What is the main purpose of plotting the data before conducting a regression?

**7.9**  BPD has arrangements with various health facilities such as gyms, swimming pools, and fitness centers for the use of its employees. The BPD benefits manager comes to you and asks for some help in analyzing the use of these facilities by BPD employees, as there is a concern that only the younger employees use them. He gathers the data, verifies them, and gives them to you for your analysis. The data are in Table 7.8, and also in Data Set 15.

Conduct a complete analysis. Include the rationale for deciding which is the dependent variable and which is the independent variable, the plot, the identification of the type of relationship (if any), and if there is a relationship develop an equation that describes the trend. Interpret the various regression results.

What is your conclusion regarding whether only the younger employees use the health facilities?

**7.10** Why is there no problem interpretation for the intercept in the previous exercise?

# CHAPTER 8

# Exponential Model

## 8.1 EXAMPLES

If the plot of two variables indicates that an exponential model can describe the observed trend, then we fit an exponential model to the data. An exponential model is one where $y$ increases at an ever-increasing rate as $x$ increases, or decreases at an ever-decreasing rate as $x$ increases. Here are four examples.

Figure 8.1 is a model that was used to make decisions on grade midpoints that reflected the market.

Figure 8.2 is a model that was used to make decisions on recruiting budgets.

Figure 8.3 is a model that was used to justify the requirement of at least four training courses for assemblers.

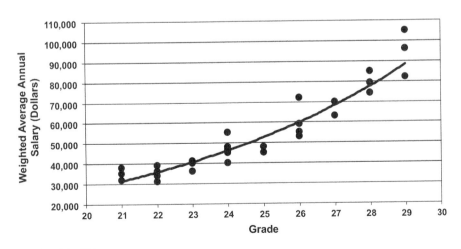

**FIGURE 8.1**   **EXPONENTIAL MODEL OF SURVEY AVERAGES AND BPD GRADES**

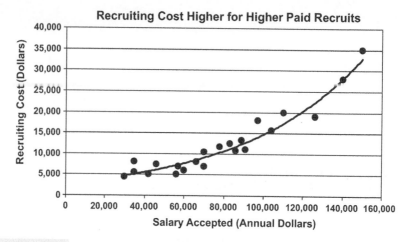

**FIGURE 8.2** **EXPONENTIAL MODEL OF RECRUITING COST AND SALARY ACCEPTED**

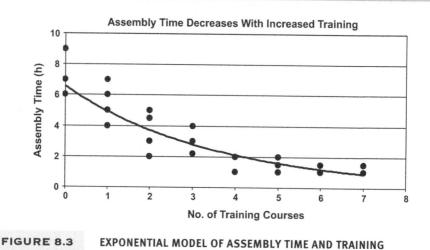

**FIGURE 8.3** **EXPONENTIAL MODEL OF ASSEMBLY TIME AND TRAINING**

Figure 8.4 is a model that was used to help decide the strategy on balancing pay with benefits.

## 8.2 LOGARITHMS

Since the exponential model is one of the logarithmic transformation models, we need to first review the logarithms.

Logarithms are like any other mathematical language. We start with a problem in "regular algebraic" language. We translate it into "logarithmic" language, solve the problem in that language, and then translate the solution back into "regular algebraic" language where we can then understand the solution. This is like having a problem in English, translating it into Spanish, solving the problem in

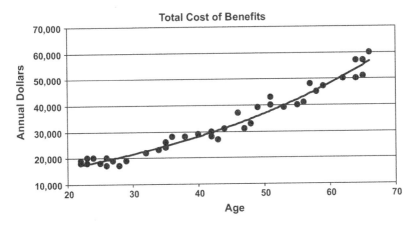

**FIGURE 8.4    EXPONENTIAL MODEL OF BENEFITS COST AND AGE**

Spanish, and then translating the solution back into English where we can then understand the solution.

A logarithm, or log for short, is the exponent or power to which a base is raised to produce a certain value. A common base is 10. For example, if we were to ask as to what power 10 must be raised to produce a value of 1,000, the answer is 3, since

$$10^3 = 1,000$$

This is a mathematical statement in common algebraic language. The language of logarithms, on the other hand, is just another way of saying the same thing. In terms of logarithms, we would say that the logarithm to the base 10 of 1,000 is 3. Mathematically expressed, it is

$$\log_{10} 1,000 = 3$$

Analogously, we have gone from English to Spanish.

Sometimes a logarithm to the base 10 is called a common logarithm. Often the shorthand notation for a common logarithm is simply "log," and we would say the log of 1,000 is 3. Thus we get

$$\log 1,000 = 3$$

where the subscript 10 is not displayed.

You cannot take the logarithm of zero or of a negative number.

| TABLE 8.1 | LOGARITHMIC AND ALGEBRAIC IDENTITIES |
|---|---|
| **Identity** | **Example** |
| $10^{\log x} = x$ | $10^{\log 1,000} = 10^3 = 1,000$ |
| $\log x^k = k \log x$ | $\log 100^2 = 2 \log 100$ |
| $\log ab = \log a + \log b$ | $\log (100)(1,000) = \log 100 + \log 1,000$ |
| $\log (a/b) = \log a - \log b$ | $\log (1,000/10) = \log 1,000 - \log 10$ |
| $10^{(a+b)} = (10^a)(10^b)$ | $10^{(2+3)} = (10^2)(10^3)$ |
| $10^{ab} = (10^a)^b = (10^b)^a$ | $10^{2*3} = 100^3 = 1,000^2 = 1,000,000$ |
| Bases | There are other bases used for logarithms, such as base 2 or base 8. In mathematics, science, and engineering, there is a widely used base denoted by the letter $e$, and logarithms to the base $e$ are called natural logarithms, often denoted by "ln." The value of $e$ is approximately 2.718281828459 |

## Antilogs

Antilogs are the inverse mathematical transformation of a logarithm—from logarithmic language back to algebraic language (Spanish back to English). It consists of raising 10 to the power of the log. Two examples are shown here.

| Logarithms | Antilogs |
|---|---|
| $\log 1,000 = 3$ | $10^3 = 1,000$ |
| $\log 500 = 2.69897$ | $10^{2.69897} = 500$ |

Table 8.1 contains some technical notes about logarithms.

## Scales

Logarithms are used to change the scales or metrics of numbers. Most of the scales we use for plotting and modeling are linear scales. A *linear scale* is one in which a *constant difference* in values on the scale is represented by a constant distance (e.g., centimeters), no matter where you are on the scale. The exploded scale in Figure 8.5 shows the details of a linear scale, where the numbers are evenly spaced. The distance in centimeters between 1 and 2 is the same as the distance between 2 and 3, between 3 and 4, and so on.

A *logarithmic scale* is one in which a *constant ratio* is represented by the same distance no matter where you are on the scale. The exploded scale in Figure 8.6 shows the details of a logarithmic scale, where the numbers are geometrically spaced. The distance in centimeters between 1 and 2 is the same as the distance

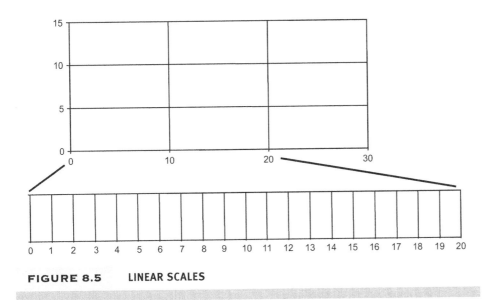

**FIGURE 8.5    LINEAR SCALES**

between 2 and 4, 3 and 6, 4 and 8, 10 and 20, and so on, all representing a ratio of 1:2.

In this illustration, both the *x*- and *y*-axes are in log scales. We call this a log–log plot. When only one axis is transformed to logs, it is called a semilog plot.

## Why Logarithms?

There are two main situations in compensation where logarithmic transformations are useful in graphical representations and in modeling.

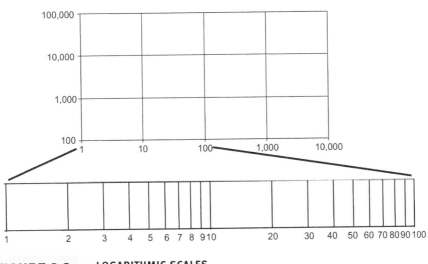

**FIGURE 8.6    LOGARITHMIC SCALES**

The first situation is when the data show a relationship represented by one of the logarithmic transformation models shown in the functional forms display in Figure 6.9. In these situations, one or both the variables are transformed to logarithms in the model. The survey averages and grade model at the beginning of this chapter is an exponential model, where the $y$-variable was changed to logarithm to develop the model.

The second situation is when you have data that vary by orders of magnitude. An order of magnitude between two numbers is when one is ten times the other. This happens often in executive pay models, relating pay to company sales, where both variables vary by orders of magnitude. In this case, you have a power model and would transform both variables into logarithms. This is discussed in Chapter 10.

## 8.3 EXPONENTIAL MODEL

We will illustrate the developing of an exponential model with an example that is common in compensation, namely, developing a model that describes the market pay in terms of internal grades. This is the starting point to identify an organization's market position for pay, develop a salary increase budget, and establish market-based pay ranges. These aspects are discussed in later chapters. For now, we will just develop the model.

---

### CASE STUDY 6

You have survey data from several surveys on matched jobs in the grades shown in Table 8.2 for the IT Department of BPD Company. The data are also given in Data Set 16. Duplicate titles are due to data from different surveys. The market pay values are aged weighted average survey data. Your task is to develop a market model that relates the external pay to the internal grades. The reader may wish to enter these data into a computer with spreadsheet or statistical software and follow along.

**TABLE 8.2    SALARY SURVEY DATA OF GRADE AND MARKET PAY**

| Job No. | Company Job Title | Grade | Market Pay |
|---------|-------------------|-------|------------|
| 1 | Department Assistant | 31 | 46,719 |
| 2 | Scheduling Coordinator | 31 | 52,365 |
| 3 | Field Rep. | 31 | 58,399 |
| 4 | Database Analyst | 32 | 56,453 |
| 5 | Sr. Field Rep. | 32 | 61,319 |

**TABLE 8.2** *(CONTINUED)*

| Job No. | Company Job Title | Grade | Market Pay |
|---------|-------------------|-------|------------|
| 6 | Marketing Analyst | 32 | 60,062 |
| 7 | Stock Coordinator | 32 | 48,000 |
| 8 | HR Coordinator | 33 | 53,380 |
| 9 | Accountant | 33 | 56,761 |
| 10 | Programmer | 33 | 59,689 |
| 11 | Programmer | 33 | 65,474 |
| 12 | Sr. Technical Rep. | 34 | 60,800 |
| 13 | Sr. Market Analyst | 34 | 66,514 |
| 14 | HR Rep. | 34 | 68,493 |
| 15 | Financial Analyst | 34 | 69,970 |
| 16 | Sr. Accountant | 34 | 74,133 |
| 17 | Sr. Programmer | 34 | 76,123 |
| 18 | Sr. Programmer | 34 | 78,704 |
| 19 | Operations Coordinator | 35 | 77,661 |
| 20 | Operations Coordinator | 35 | 81,600 |
| 21 | Product Specialist | 35 | 84,800 |
| 22 | Programmer Specialist | 35 | 89,523 |
| 23 | Sr. Operations Coordinator | 36 | 84,273 |
| 24 | Sr. Product Specialist | 36 | 92,476 |
| 25 | Sr. Programmer Specialist | 36 | 102,155 |
| 26 | Lead Programmer | 36 | 102,630 |
| 27 | Sr. Operations Analyst | 37 | 107,501 |
| 28 | Sr. Operations Analyst | 37 | 112,086 |
| 29 | Programmer Consultant | 37 | 101,357 |
| 30 | Sr. Marketing Specialist | 38 | 108,319 |
| 31 | Operations Specialist | 38 | 112,935 |
| 32 | Sr. Financial Consultant | 38 | 113,200 |
| 33 | Sr. HR Consultant | 38 | 117,414 |
| 34 | Sr. Programmer Consultant | 38 | 126,933 |

You plot the market pay as the $y$-variable versus grade as the $x$-variable and get the plot shown in Figure 8.7.

You compare this plot with the models in the functional form displays in Figures 6.8 and 6.9. There is a slight upward curve to the trend. It looks like either a linear model or an exponential model could describe the trend. Since you are building a model that will ultimately be used to help establish salary ranges, you decide to use an exponential model, as the resulting model will have a constant percent progression between grades. This feature is discussed later.

Mathematically, the exponential model has the equation

$$\log y = a + bx$$

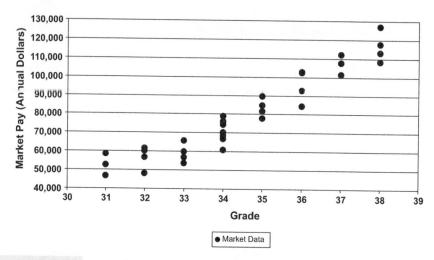

**FIGURE 8.7** SURVEY AVERAGES AND BPD GRADES

To calculate the coefficients of the equation, first transform market pay to log market pay as shown in Table 8.3, and then regress log market pay as the $y$-variable against grade as the $x$-variable.

Using software to conduct the regression, the resulting equation is

$$\log y' = a + bx$$

where

$$a = 3.0217$$
$$b = 0.053929$$

and

$$r^2 = 0.915$$
$$r = 0.956$$
$$\text{SEE} = 0.0371$$

**TABLE 8.3** GRADE AND LOGARITHM OF MARKET PAY

| Job No. | Company Job Title | Grade | Market Pay | Log Market Pay |
|---|---|---|---|---|
| 1 | Department Assistant | 31 | 46,719 | 4.6695 |
| 2 | Scheduling Coordinator | 31 | 52,365 | 4.7190 |
| 3 | Field Rep. | 31 | 58,399 | 4.7664 |
| 4 | Database Analyst | 32 | 56,453 | 4.7517 |
| ⋮ | ⋮ | ⋮ | ⋮ | ⋮ |
| ⋮ | ⋮ | ⋮ | ⋮ | ⋮ |
| 32 | Sr. Financial Consultant | 38 | 113,200 | 5.0538 |
| 33 | Sr. HR Consultant | 38 | 117,414 | 5.0697 |
| 34 | Sr. Programmer Consultant | 38 | 126,933 | 5.1036 |

Substituting values for $a$ and $b$ in the equation, we get, recalling that we are predicting averages in regression,

$$\log y' = 3.0217 + 0.053929x, \text{ or}$$

$$\text{predicted average log market pay} = 3.0217 + 0.053929 \text{ grade}$$

We next calculate the predicted values and plot them on the chart. For this, take the antilogs and use the following formula to calculate the predicted average market pay for various values of grades.

$$y' = 10^{(a+bx)}$$
$$y' = 10^{(3.0217+0.053929x)}$$

For example, to calculate the predicted average market pay for grade 35, substitute 35 for $x$ and get

$$y' = 10^{(3.017+0.053929)(35)} = 10^{(3.0217+1.887515)} = 10^{4.909215} = 81,136$$

Another mathematically equivalent approach is to first calculate the log of the predicted average market pay and then raise 10 to that power.

$$\log y' = a + bx$$

$$\log y' = 3.0217 + (0.53929)(35) = 3.0217 + 1.887515 = 4.909215$$

Then,

$$y' = 10^{4.909215} = 81,136$$

This may look complicated, but spreadsheet and statistical software perform these calculations with ease.

We get the following results given in Table 8.4, which we plot on the chart as a line in Figure 8.8. On a linear plot, (i.e., a plot with linear scales) the line is curved.

If this model is used to establish the midpoints of the grades in a salary structure, a very nice aspect of the exponential model is that it results in a constant percentage midpoint progression. To determine that, first calculate the ratio of one grade midpoint (i.e., the predicted average market pay) to the one below using the following formula:

$$\text{ratio} = 10^{\text{slope}} = 10^b$$

Substituting, we get

$$\text{ratio} = 10^{0.053929} = 1.132$$

| TABLE 8.4 | GRADE AND PREDICTED AVERAGE MARKET PAY |
|-----------|----------------------------------------|

| Grade | Predicted Average Market Pay |
|-------|------------------------------|
| 31 | 49,374 |
| 32 | 55,902 |
| 33 | 63,293 |
| 34 | 71,662 |
| 35 | 81,136 |
| 36 | 91,864 |
| 37 | 104,009 |
| 38 | 117,761 |

This means that the ratio of one midpoint to the one below it is 1.132. Finally, MOTH it to convert from a ratio to a percent difference.

So we have $(1.132 - 1)(100) = (0.132)(100) = 13.2\%$

There is a constant 13.2% increase in midpoint from one grade to the next. That is, from grade 31 to 32, there is a 13.2% increase and from grade 32 to 33, there is a 13.2% increase, and so on. Another way of stating this is that the midpoint of grade 32 is 13.2% more than the midpoint of grade 31, the midpoint of grade 33 is 13.2% more than the midpoint of grade 32, and so on.

## 8.4 MODEL EVALUATION

### Appearance

Visually, this is a very satisfactory model. The (curved) line goes through the data and satisfactorily describes the observed trend.

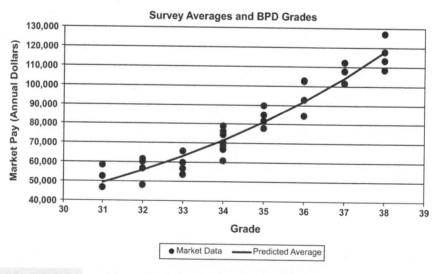

| FIGURE 8.8 | PREDICTED AVERAGE MARKET PAY AND GRADE |
|------------|----------------------------------------|

## Coefficient of Determination

Mathematically, $r^2$ is 0.915, which means that we have explained 91.5% of the variability of log market pay with grade. In terms of the original metric, we need to do some additional mathematics.

What we really want is an equivalent $r^2$ for market pay and grade. You can calculate it using the definition of coefficient of determination and substituting the values of pay (actual pay, average pay, and predicted average pay) in the original metric (dollars) in the following equation:

$$r^2 = 1 - \frac{\text{SS residual}}{\text{SS total}} = 1 - \frac{\sum (y - y')^2}{\sum (y - \bar{y})^2}$$

Recall that SS residual is a measure of the variation of the $y$-variable that is *not* explained by or attributed to the model, and SS total is a measure of the total variation of the $y$-variable about its mean.

On calculating, you get a value of 0.927. We can say that 92.7% of the variation in market pay for matched jobs can be attributed to the grade of those jobs. This compares with the $r^2$ between log market pay and grade of 91.5%. Usually the two values are close, but you cannot predict which one will be higher. Also, if you do not have access to such data to calculate the "real" coefficient of determination, you just report the 91.5% for the log of pay and say that the $r^2$ for pay in the original metric will in all likelihood be in the same neighborhood.

## Correlation

The correlation is the square root of $r^2$ with a positive sign (since the trend is positive) and has a value of +0.956 for log market pay and grade (or +0.963, using the $r^2$ value for market pay and grade). As discussed previously, this is a qualitative measure of the association and its value indicates a very strong association.

## Standard Error of Estimate

Recall that the standard error of estimate (SEE) is a measure of variability of the data points from the predicted values and is the "average" distance of the data points from the line. In the regression output, we see that SEE is 0.0371. Keep in mind that this is in terms of logs. Mathematically, we can say that "on average" the actual log of pay varies from the predicted average log of pay by about 0.0371. Using some logarithmic identities shown earlier, we get on "average"

$$\log y - \log y' = \text{SEE}$$
$$\log (y/y') = \text{SEE}$$
$$y/y' = 10^{\text{SEE}}$$
$$y/y' = 10^{0.0371} = 1.089$$

**TABLE 8.5    SUMMARY OF MODEL EVALUATION**

| Evaluation Criteria | Symbol | In Terms of Log Market Pay | In Terms of Market Pay |
|---|---|---|---|
| Appearance | | | Goes through middle of data and describes trend nicely |
| Coefficient of determination | $r^2$ | 91.5% of variability of log market pay associated with grade | 92.7% of variability of market pay associated with grade |
| Correlation | $r$ | +0.956—strong association | +0.963—strong association |
| Standard error of estimate | SEE | 0.0371 is "average" deviation of log market pay from predicted average log market pay | 8.9% is "average" deviation of market pay from predicted average market pay |
| Common sense | | | Makes sense with respect to pay levels and 13.2% progression from one grade to the next |

After MOTHing, this means that on "average" the actual pay values vary by about 8.9% from the predicted average pay values.

## Common Sense

In comparing the model with our experience, it makes sense that the progression from one grade to the next is 13.2%.

## Summary of Evaluation

A summary of the model evaluation is shown in Table 8.5.

Overall this is a good model, and it very nicely represents the predicted average market pay for a given grade. You decide to use this as the starting point to identify BPD's market position for pay, develop a salary increase budget, and establish market-based pay ranges.

## RELATED TOPICS IN THE APPENDIX

A.12   Logarithmic Conversion

# PRACTICE PROBLEMS

**8.1** The employment manager asks for your help in developing a recruiting budget for the coming year. He wants to predict the average recruiting costs as a function of the salary accepted. He furnishes you the data that you verify in Table 8.6 from the current year's recruiting program, which are also given in Data Set 17.

**TABLE 8.6    RECRUITING COST AND SALARY ACCEPTED**

| Salary Accepted | Recruiting Cost | Salary Accepted | Recruiting Cost |
|---|---|---|---|
| 30,000 | 4,500 | 78,000 | 11,700 |
| 35,000 | 5,500 | 83,000 | 12,450 |
| 35,000 | 8,000 | 86,000 | 10,800 |
| 42,000 | 5,000 | 89,000 | 13,350 |
| 46,000 | 7,500 | 91,000 | 11,000 |
| 56,000 | 5,000 | 97,000 | 18,000 |
| 57,000 | 7,000 | 104,000 | 15,600 |
| 60,000 | 6,000 | 110,000 | 20,000 |
| 66,000 | 8,000 | 126,000 | 18,900 |
| 70,000 | 7,000 | 140,000 | 28,000 |
| 70,000 | 10,500 | 150,000 | 35,000 |

  Determine the type of relationship by plotting the data. State your rationale for deciding which variable is a dependent variable and which is an independent one.

**8.2** If the relationship is exponential, calculate the equation of the least squares line, the coefficient of determination, the correlation, and the standard error of estimate. Also report the average starting salary accepted and the average recruiting cost.

**8.3** Draw the least squares line on the plot.

**8.4** Provide a complete evaluation of the model.

**8.5** Calculate the predicted average recruiting cost for various levels of salary accepted. The employment manager will use this to develop the recruiting budget.

**8.6** The training manager for the BPD New York manufacturing plant comes to you again, thinking that perhaps a policy requiring a certain number of training courses would be better than offering an incentive for employees to voluntarily take the training. Perhaps there would be a reward or recognition at the completion of a certain number of courses. She brings data showing the assembly time for 23 employees with different numbers of training courses completed. You verify the assembly and training data, shown in Table 8.7. The data are also in Data Set 18.

**TABLE 8.7 TRAINING COURSES AND ASSEMBLY TIME**

| Number of Training Courses | Assembly Time (hours) |
|---|---|
| 0 | 9 |
| 0 | 7 |
| 0 | 6 |
| 1 | 4 |
| 1 | 7 |
| 1 | 6 |
| 1 | 5 |
| 2 | 3 |

(*continued*)

**TABLE 8.7**     *(CONTINUED)*

| Number of Training Courses | Assembly Time (hours) |
|:---:|:---:|
| ? | 4.5 |
| 2 | 5 |
| 2 | 2 |
| 3 | 4 |
| 3 | 3 |
| 3 | 2.2 |
| 4 | 2 |
| 4 | 1 |
| 5 | 1.5 |
| 5 | 1 |
| 5 | 2 |
| 6 | 1.5 |
| 6 | 1 |
| 7 | 1.5 |
| 7 | 1 |

Determine the type of relationship by plotting the data. State your rationale for deciding which variable is a dependent variable and which is an independent one.

**8.7**   If the relationship is exponential, calculate the equation of the least squares line, the coefficient of determination, the correlation, and the standard error of estimate. Also report the average number of training courses taken and the average assembly time.

**8.8**   Draw the least squares line on the plot.

**8.9**   Provide a complete evaluation of the model.

**8.10**   Calculate the predicted average assembly time for various levels of the number of training courses.

# CHAPTER 9

# Maturity Curve Model

## 9.1 MATURITY CURVES

A maturity curve is a description of the relationship between pay and experience for people with similar educational backgrounds and doing similar types of work, such as nonsupervisory BS engineers conducting research or first-level supervisory PhD scientists. Sometimes when there are sufficient data, it is possible to break it down by discipline, such as electrical engineering or chemistry.

In job worth hierarchy terms, experience is a surrogate for levels of expertise, responsibility, and impact on the organization.

Maturity curves have their largest use in modeling market pay for employees engaged in technical work such as engineering, and in particular in research and development, although they are also used in many other areas. One reason for using maturity curves to describe the market for technical disciplines is that job matching is difficult because sometimes the incumbents are working on proprietary or classified projects and it is not possible to share descriptions of the actual work they are doing. Another reason is the culture of the organizations that have used maturity curves for many years. Those organizations are comfortable with them [2].

If the plot of two variables indicates that a maturity curve model could describe the observed trend, then we fit a cubic or a spline model to the data. A spline model is a double cubic model joined smoothly at a point called the knot or join. Here are two examples of maturity curves.

The model in Figure 9.1 was used to help make decisions on salary increase budgets.

In this plot, each data point represents an individual engineer, where the *x*-value of the point is the experience and the *y*-value of the point is the pay.

*Statistics for Compensation: A Practical Guide to Compensation Analysis,* By John H. Davis
Copyright © 2011 John Wiley & Sons Inc.

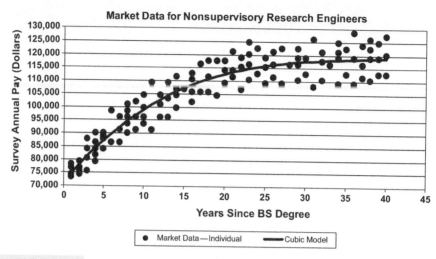

For example, the data point (11, 109,200) represents an engineer with 11 years of experience since BS degree (YSBS) who is making $109,200 annually.

Typically after about 20 years of experience, the curves go up very slowly. A possible reason their upward movement slows down is that some of the higher paid employees, who got that way due to higher performance, were promoted to the next level and are not in the particular database any more.

The model in Figure 9.2 was used to establish pay level targets for different performance levels.

In this plot, each data point represents a calculated percentile of the salaries of all the scientists having the same experience level. For example, the data point

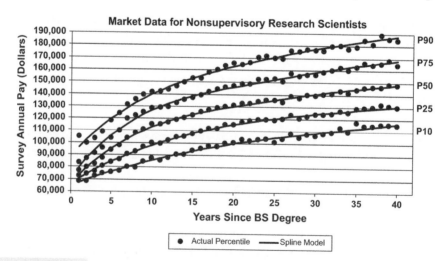

(8, 122,400) represents the 75th percentile of the annual pay of all scientists in the database with 8 years of experience since BS degree. The value of the 75th percentile is $122,400.

As a technical note, if you regress on the raw data, you will get a lower $r^2$ than if you regress on the average or a percentile because calculated measures of location have less variability.

Sometimes the range from P10 to P90 is used as a basis for a salary range. In addition, the various percentiles are often used as targets for different levels of performance.

Most of the time a cubic model will adequately describe the relationship between pay and experience. However, if the horizontal part of the plot starts to curve upward for high values of YSBS, then a spline model may be a better fit. You should try both to see which one works better for your situation.

The cubic model has the following equation:

$$y = a + bx + cx^2 + dx^3$$

The spline model has the following equation:

$$y = a + bx + cx^2 + dx^3 + ep^3$$

where

$$p = 0, \text{ if } x < k, \quad \text{and} \quad x - k, \text{ if } x \geq k$$

$$k = \text{knot, usually having a value of } 20$$

In both equations, the dependent variable $y$ is the pay and the independent variable $x$ is the years of experience, measured as years since the BS degree.

We will illustrate developing a maturity curve model with an example for the research and development (R&D) division of BPD, which has many engineers. This is the starting point to identify the organization's market position for pay, develop a salary increase budget, and establish market-based pay ranges. These aspects will be discussed in a later chapter. For now, we will just develop the model.

---

**CASE STUDY 7**

You have survey data for R&D engineers in your industry that show the average pay as it relates to years since bachelors degree. Your task is to develop a market model that relates the pay to YSBS.

Our starting point is the data in Table 9.1, where we have survey averages for each year of experience as measured by YSBS. The reader may want to enter these data into a computer with statistical software and follow along. The data are also in Data Set 19.

**TABLE 9.1** **SURVEY AVERAGE ANNUAL PAY AND YSBS**

| YSBS | Survey Average Annual Pay | YSBS | Survey Average Annual Pay | YSBS | Survey Average Annual Pay |
|---|---|---|---|---|---|
| 1 | 81,627 | 14 | 113,626 | 27 | 127,050 |
| 2 | 88,774 | 15 | 118,641 | 28 | 123,768 |
| 3 | 84,200 | 16 | 110,191 | 29 | 128,157 |
| 4 | 94,207 | 17 | 115,102 | 30 | 125,587 |
| 5 | 100,447 | 18 | 125,752 | 31 | 125,272 |
| 6 | 97,513 | 19 | 126,930 | 32 | 118,790 |
| 7 | 101,678 | 20 | 127,470 | 33 | 132,857 |
| 8 | 97,026 | 21 | 123,457 | 34 | 128,953 |
| 9 | 109,973 | 22 | 117,894 | 35 | 123,981 |
| 10 | 99,664 | 23 | 126,807 | 36 | 135,920 |
| 11 | 107,484 | 24 | 118,578 | 37 | 125,478 |
| 12 | 112,429 | 25 | 128,093 | 38 | 120,195 |
| 13 | 106,743 | 26 | 123,121 | 39 | 125,993 |
| | | | | 40 | 134,823 |

Each data point represents the calculated average pay of all engineers with the same experience level. For example, the data point (14, 113,626) represents the average annual pay of $113,626 for all engineers with 14 years of experience as measured by YSBS.

## 9.2 BUILDING THE MODEL

We plot the data and get the following chart in Figure 9.3.

For purposes of illustration, we will develop both a cubic model and a spline model. For both we will use multiple linear regression (MLR) mathematics, with the following "translations."

$$\text{MLR} \quad y = b_0 + b_1 x_1 + b_2 x_2 + b_3 x_3$$

$$\text{Cubic} \quad y = a + bx + cx^2 + dx^3$$

$$\text{MLR} \quad y = b_0 + b_1 x_1 + b_2 x_2 + b_3 x_3 + b_4 x_4$$

$$\text{Spline} \quad y = a + bx + cx^2 + dx^3 + ep^3$$

### Cubic Model

Using statistical software, we get the following results:

$$y' = a + bx + cx^2 + dx^3$$

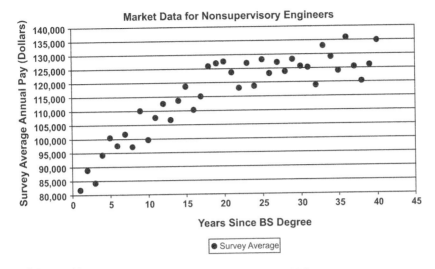

FIGURE 9.3    SURVEY AVERAGE ANNUAL PAY AND YSBS

where

$$a = 78{,}888$$

$$b = 3{,}663.9$$

$$c = -94.515$$

$$d = 0.84316$$

and

$$r^2 = 0.897$$

$$r = 0.947$$

$$\text{SEE} = 4{,}718$$

Substituting, we get

$$y' = 78{,}888 + 3{,}663.9x - 94.515x^2 + 0.84316x^3$$

Substituting for what $y'$ and $x$ represent,

predicted average survey average pay $= 78{,}888 + 3{,}663.9\,\text{YSBS} - 94.515\,\text{YSBS}^2$
$$+\, 0.84316\,\text{YSBS}^3$$

where YSBS stands for years since BS degree.

We use this equation to calculate the predicted average survey average pay for the various values of YSBS. For example, to calculate the predicted average survey average pay for a YSBS of 10, substitute 10 for YSBS and get

$$y' = 78{,}888 + (3{,}663.9)(10) - (94.515)(10^2) + (0.84316)(10^3)$$
$$y' = 78{,}888 + 36{,}639 - 9{,}451.5 + 843.16 = 106{,}919$$

**TABLE 9.2**    **CUBIC MODEL**

| YSBS | Predicted Average Survey Average Annual Pay Cubic Model | YSBS | Predicted Average Survey Average Annual Pay Cubic Model | YSBS | Predicted Average Survey Average Annual Pay Cubic Model |
|---|---|---|---|---|---|
| 1 | 82,458 | 14 | 113,971 | 27 | 125,508 |
| 2 | 85,844 | 15 | 115,426 | 28 | 125,886 |
| 3 | 89,052 | 16 | 116,768 | 29 | 126,218 |
| 4 | 92,085 | 17 | 118,002 | 30 | 126,507 |
| 5 | 94,950 | 18 | 119,133 | 31 | 126,759 |
| 6 | 97,651 | 19 | 120,165 | 32 | 126,978 |
| 7 | 100,193 | 20 | 121,105 | 33 | 127,171 |
| 8 | 102,582 | 21 | 121,957 | 34 | 127,341 |
| 9 | 104,822 | 22 | 122,727 | 35 | 127,494 |
| 10 | 106,919 | 23 | 123,418 | 36 | 127,635 |
| 11 | 108,877 | 24 | 124,037 | 37 | 127,770 |
| 12 | 110,702 | 25 | 124,588 | 38 | 127,902 |
| 13 | 112,398 | 26 | 125,077 | 39 | 128,038 |
| | | | | 40 | 128,182 |

We get the results shown in Table 9.2, which we plot on the chart as a (curved) line in Figure 9.4.

## Cubic Model Evaluation

**Appearance**    Visually, this is a very satisfactory model. The (curved) line goes through the data and satisfactorily describes the observed trend.

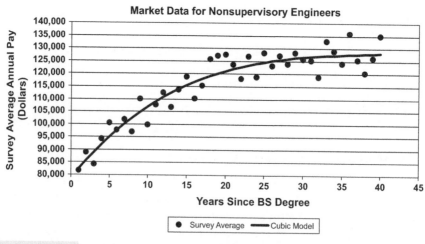

**FIGURE 9.4**    **CUBIC MODEL**

**Coefficient of Determination**   Mathematically, $r^2$ is 0.897, which means that we have explained 89.7% of the variability of survey average pay with years since BS degree.

**Correlation**   The correlation has a positive sign (since the trend is positive) and has a value of +0.947. As discussed previously, this is a qualitative measure of the association and its value indicates a very strong association.

**Standard Error of Estimate**   The SEE is 4,718, meaning that "on average" the actual survey average pay varies from the predicted average survey average pay by about 4,718. Given the magnitude of the average pay levels, this is a very small variation indeed.

**Common Sense**   In comparing the model with our experience, the rapid increase in the earlier years and the slower increase in later years are what we would expect, along with the actual pay levels.

Overall this is a good model, and it very nicely represents the predicted average survey average pay for the given years since BS degree. We decide to use this to identify our organization's market position for pay, develop a salary increase budget, and establish market-based pay ranges.

In reality, we would stop here, but to illustrate the spline model, we will continue.

## Spline Model

The spline model is a cubic model with an extra cubic term, which helps "stretch out" the flattened horizontal portion of the curve.

$$y = a + bx + cx^2 + dx^3 + ep^3$$

where

$$p = 0, \quad \text{if} \quad x < k$$
$$p = (x - k), \quad \text{if} \quad x \geq k$$

The value $k$ is known as the "knot" and based on our experience, we set the knot at YSBS = 20.

Using statistical software, we get the following results:

$$y' = a + bx + cx^2 + dx^3 + ep^3$$

where

$$a = 81,385$$
$$b = 2,665.9$$

$$c = -7.094$$
$$d = -1.2240$$
$$e = 3.9197$$

and

$$r^2 = 0.900$$
$$r = 0.949$$
$$\text{SEE} = 4,720$$

Substituting, we get

$$y' = 81,385 + 2,665.9x - 7.094x^2 - 1.2240x^3 + 3.9197p^3 \text{ or}$$

predicted average survey average pay $= 81,385 + 2,665.9\,\text{YSBS} - 7.094\,\text{YSBS}^2$
$$- 1.2240\,\text{YSBS}^3 + 3.9197p^3$$

where $p = 0$, if YSBS $< 20$, and YSBS $- 20$, if YSBS $\geq 20$

We use this equation to calculate the predicted average survey average pay for the various values of YSBS. For example, to calculate the predicted average survey average pay for a YSBS of 10, substitute 10 for YSBS and get

$$y' = 81,385 + (2,665.9)(10) - (7.094)(10^2) - (1.2240)(10^3) + (3.9197)(0^3)$$
$$y' = 81,385 + 26,659 - 709.4 - 1,224.0 + 0 = 106,111$$

For a YSBS of 30, we get

$$y' = 81,385 + (2,665.9)(30) - (7.094)(30^2) - (1.2240)(30^3) + (3.9197)((30 - 20)^3)$$
$$y' = 81,385 + 79,977 - 6,384.6 - 33,048 + 3,919.7 = 125,849$$

We get the following results in Table 9.3, which we plot on the chart as a (curved) line in Figure 9.5.

## Spline Model Evaluation

**Appearance**   Visually, this is a very satisfactory model. The (curved) line goes through the data and satisfactorily describes the observed trend.

**Coefficient of Determination**   Mathematically, $r^2$ is 0.900, which means that we have explained 90.0% of the variability of survey average pay with years since BS degree.

**Correlation**   The correlation has a positive sign (since the trend is positive) and has a value of +0.949. As discussed previously, this is a qualitative measure of the association and its value indicates a very strong association.

**TABLE 9.3    SPLINE MODEL**

| YSBS | Predicted Average Survey Average Annual Pay Spline Model | YSBS | Predicted Average Survey Average Annual Pay Spline Model | YSBS | Predicted Average Survey Average Annual Pay Spline Model |
|---|---|---|---|---|---|
| 1 | 84,043 | 14 | 113,959 | 27 | 125,445 |
| 2 | 86,679 | 15 | 115,646 | 28 | 125,606 |
| 3 | 89,286 | 16 | 117,210 | 29 | 125,735 |
| 4 | 91,857 | 17 | 118,642 | 30 | 125,849 |
| 5 | 94,384 | 18 | 119,934 | 31 | 125,964 |
| 6 | 96,861 | 19 | 121,081 | 32 | 126,095 |
| 7 | 99,279 | 20 | 122,073 | 33 | 126,259 |
| 8 | 101,631 | 21 | 122,909 | 34 | 126,472 |
| 9 | 103,911 | 22 | 123,600 | 35 | 126,751 |
| 10 | 106,111 | 23 | 124,161 | 36 | 127,112 |
| 11 | 108,222 | 24 | 124,611 | 37 | 127,570 |
| 12 | 110,239 | 25 | 124,964 | 38 | 128,142 |
| 13 | 112,154 | 26 | 125,236 | 39 | 128,844 |
|  |  |  |  | 40 | 129,692 |

**Standard Error of Estimate**    The SEE is 4,720, meaning that "on average" the actual survey average pay varies from the predicted average survey average pay by about 4,720. Given the magnitude of the average pay levels, this is a very small variation.

**Common Sense**    In comparing the model with our experience, the rapid increase in the earlier years and the slower increase in later years are what we would expect, along with the actual pay levels.

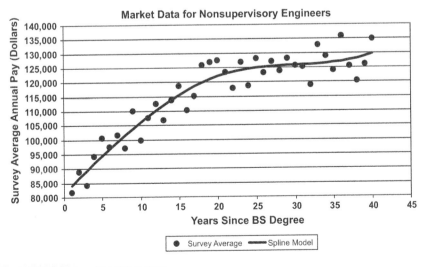

**FIGURE 9.5    SPLINE MODEL**

| TABLE 9.4 | | EVALUATION OF CUBIC AND SPLINE MODELS | |
| --- | --- | --- | --- |
| Evaluation Criteria | Symbol | Cubic Model | Spline Model |
| Appearance | | Good through middle of data and describes the trend nicely | Goes through middle of data and describes the trend nicely |
| Coefficient of determination | $r^2$ | 89.7% of variability of survey average pay associated with YSBS | 90.0% of variability of survey average pay associated with YSBS |
| Correlation | $r$ | +0.947—very strong association | +0.949—very strong association |
| Standard error of estimate | SEE | 4,718 "average" deviation of survey average pay from predicted average survey average pay—a very small variation | 4,720 "average" deviation of survey average pay from predicted average survey average pay—a very small variation |
| Common sense | | Makes sense with respect to pay levels and general shape of the relationship | Makes sense with respect to pay levels and general shape of the relationship |

Overall this is also a good model, and it very nicely represents the predicted average survey average pay for given years since BS degree. We could also use this to identify our organization's market position for pay, develop a salary increase budget, and establish market-based pay ranges.

## 9.3 COMPARISON OF MODELS

A summary evaluation of the two models is shown in Table 9.4.

The two models are nearly identical, with the spline model only better at the third significant figure. Hence, we would use the cubic model for its simplicity. Actually, with the strong evaluation of the cubic model, we would probably stop there and just use it without even developing a spline model.

With the cubic model, we can use it with confidence as the starting point to identify the organization's market position for pay, develop a salary increase budget, and establish market-based pay ranges.

## PRACTICE PROBLEMS

**9.1** You are the compensation manager for a large think tank and want to develop a model for nonsupervisory BS analysts that relates market pay to professional experience, as measured by years since BS degree. You have a very aggressive compensation policy and want to target the salaries toward the 75th percentile. You obtain the survey data shown in Table 9.5. Note that not all years are represented, due to the lack of sufficient data points for those years to calculate that percentile. The data are also in Data Set 20.

**TABLE 9.5    SURVEY 75TH PERCENTILES AND YSBS**

| YSBS | Survey 75th Percentile | YSBS | Survey 75th Percentile | YSBS | Survey 75th Percentile |
|------|------------------------|------|------------------------|------|------------------------|
| 1 | 69,693 | 12 | 113,387 | 27 | 139,427 |
| 2 | 77,027 | 13 | 125,057 | 28 | 134,235 |
| 3 | 89,497 | 14 | 136,487 | 29 | 145,677 |
| 4 | 82,647 | 15 | 130,562 | 30 | 125,507 |
| 5 | 91,739 | 16 | 115,654 | 32 | 139,061 |
| 6 | 104,842 | 17 | 134,019 | 33 | 146,148 |
| 7 | 108,819 | 21 | 128,758 | 34 | 128,187 |
| 8 | 99,996 | 22 | 121,442 | 37 | 135,779 |
| 9 | 120,950 | 23 | 143,990 | 39 | 127,104 |
| 10 | 113,840 | 25 | 141,585 | 40 | 143,221 |
| 11 | 105,277 | 26 | 126,334 | | |

Determine the type of relationship by plotting the data. State your rationale for deciding which variable is a dependent variable and which is an independent one.

9.2   If the relationship is a maturity curve, start with a cubic model and calculate the equation of the least squares line, the coefficient of determination, the correlation, and the standard error of estimate.

9.3   Draw the least squares line on the plot.

9.4   Provide a complete evaluation of the model.

9.5   Calculate the predicted average survey median annual pay for various levels of YSBS.

9.6   You decide to go ahead and fit a spline model to the data, with a knot at YSBS = 20. Calculate the equation of the least squares line, the coefficient of determination, the correlation, and the standard error of estimate.

9.7   Draw the least squares line on the plot.

9.8   Provide a complete evaluation of the model.

9.9   Calculate the predicted average survey median annual pay for various levels of YSBS.

9.10   Which model would you use and why?

CHAPTER 10

# Power Model

## 10.1 BUILDING THE MODEL

If the linear plot of two variables looks like the head of a comet arcing down toward the origin with the tail scattered behind it in the sky of the chart, that is, with a cluster of points on the lower left-hand corner of the plot with a few scattered up toward the upper right-hand corner of the plot, then we fit a power model to the data. This model is most often used when the data vary by orders of magnitude, for example, as is often the case with executive pay.

It is usually presented on a log–log plot, which is a plot where both the $x$-variable and the $y$-variable have been transformed to logs. Figure 10.1 is the final result of an example we will use to demonstrate the model-building process.

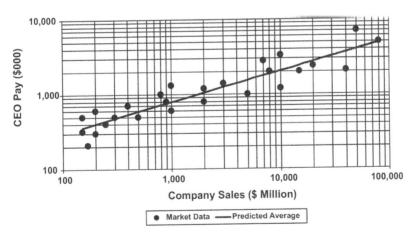

**FIGURE 10.1** POWER MODEL OF CEO PAY AND COMPANY SALES ON LOG–LOG PLOT

Power models are most often used to develop a market position and establish a market-based pay target for single incumbent executives, such as the CEO, COO, president, CFO, CIO, and so on. Usually, they relate pay to company size as measured by sales or revenue. For nonprofit organizations, often the organization size is measured by budget.

## CASE STUDY 8

The revenue of BPD is $5 billion and the board has decided that a set of 26 companies is an appropriate group to use for executive compensation comparison purposes for the CEO position, and wants you to do a market analysis for it. You gather the data shown in Table 10.1, verify it, and then plot it as shown in Figure 10.2. The data are also in Data Set 21. The reader may want to enter this data into a computer with statistical software and follow along.

**TABLE 10.1    SURVEY OF CEO PAY AND COMPANY SALES**

| Sales ($Million) | CEO Pay ($000) | Sales ($Million) | CEO Pay ($000) |
|---|---|---|---|
| 150 | 320 | 2,000 | 800 |
| 150 | 500 | 2,000 | 1,200 |
| 170 | 210 | 3,000 | 1,400 |
| 200 | 300 | 5,000 | 1,000 |
| 200 | 600 | 7,000 | 2,800 |
| 250 | 400 | 8,000 | 2,000 |
| 300 | 500 | 10,000 | 1,200 |
| 400 | 700 | 10,000 | 3,300 |
| 500 | 500 | 15,000 | 2,000 |
| 800 | 1,000 | 20,000 | 2,400 |
| 900 | 800 | 40,000 | 2,100 |
| 1,000 | 600 | 50,000 | 7,000 |
| 1,000 | 1,300 | 80,000 | 5,000 |

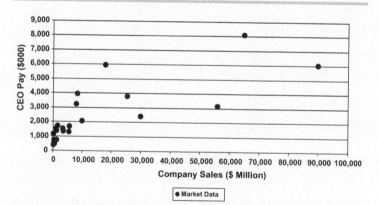

**FIGURE 10.2    CEO PAY AND COMPANY SALES ON LINEAR PLOT**

Note that sales (on the x-axis) are in millions of dollars and pay (on the y-axis) is in thousands of dollars. The pay is base pay.

We first note that the display of data points has a "comet cluster" look. We further note that sales varies from $150 million to $80,000 million ($80 billion), which spans three orders of magnitude, and that pay varies from $210 thousand to $7,000 thousand ($7 million), which spans more than one order of magnitude. These are two clues that a power model is appropriate.

To verify that fitting a power model would be appropriate, we convert both axes to logarithmic scales, shown in Figure 10.3.

While comparing Figure 10.3 with Figure 10.2, it can be seen that a logarithmic transformation visually pulls in the high values and spreads out the low values. This is a characteristic feature of a logarithmic transformation.

Here we see that the relationship between log sales and log pay is indeed linear, so we will continue and develop the equation.

Mathematically, the power model has the equation

$$\log y = a + b \log x$$

To calculate the coefficients of the equation, first transform sales to log sales and pay to log pay as shown in Table 10.2, and then regress log pay as the $y$-variable against log sales as the $x$-variable.

Using software to conduct the regression, the resulting equation is

$$\log y' = a + b \log x$$

where

$$a = 1.657$$

$$b = 0.4140$$

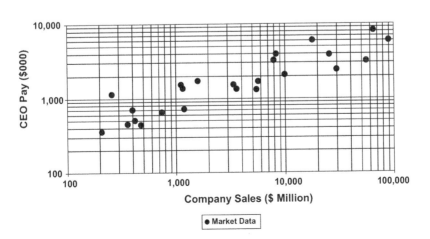

FIGURE 10.3    CEO PAY AND COMPANY SALES ON LOG–LOG PLOT

| TABLE 10.2 | LOG CEO PAY AND LOG COMPANY SALES | | |
|---|---|---|---|
| Sales ($Million) | CEO Pay ($000) | Log Sales | Log Pay |
| 150 | 320 | 2.1761 | 2.5051 |
| 150 | 500 | 2.1761 | 2.6990 |
| 170 | 210 | 2.2304 | 2.3222 |
| 200 | 300 | 2.3010 | 2.4771 |
| ⋮ | ⋮ | ⋮ | ⋮ |
| ⋮ | ⋮ | ⋮ | ⋮ |
| ⋮ | ⋮ | ⋮ | ⋮ |
| 50,000 | 7,000 | 4.6990 | 3.8451 |
| 80,000 | 5,000 | 4.9031 | 3.6990 |

and

$$r^2 = 0.849$$

$$r = 0.921$$

$$\text{SEE} = 0.152$$

Substituting, we get

$$\log y' = 1.657 + 0.4140 \log x$$
Predicted average log CEO pay $= 1.657 + 0.4140 \log \text{sales}$

Keep in mind that the CEO pay is in thousands of dollars and sales are in millions of dollars.

We want to calculate the predicted values and plot the line on our log–log chart. To do that, we take the antilogs and use the following formula to calculate the predicted value of the average pay for various values of sales. Recall from Chapter 6 that an antilog consists of raising 10 to the power of the log.

$$y' = 10^{a+b\log x} = 10^a 10^{b\log x} = 10^a 10^{(\log x)(b)} = 10^a (10^{\log x})^b$$
$$y' = 10^a x^b$$
$$y' = 10^{1.657} x^{0.4140}$$

This may look complicated, but spreadsheets and statistical software perform these calculations with ease. The predicted values are shown in Table 10.3 and the plot is shown in Figure 10.4.

You would not want to plot the line on linear scales because it is almost impossible to interpret and use, especially for low values of the variables. Figure 10.5 shows what it would look like.

**TABLE 10.3    PREDICTED AVERAGE CEO PAY AND COMPANY SALES**

| Sales ($Million) | Predicted Average CEO Pay ($000) | Sales ($Million) | Predicted Average CEO Pay ($000) |
|---|---|---|---|
| 150 | 361 | 2,000 | 1,056 |
| 150 | 361 | 2,000 | 1,056 |
| 170 | 381 | 3,000 | 1,249 |
| 200 | 407 | 5,000 | 1,543 |
| 200 | 407 | 7,000 | 1,774 |
| 250 | 446 | 8,000 | 1,874 |
| 300 | 481 | 10,000 | 2,056 |
| 400 | 542 | 10,000 | 2,056 |
| 500 | 595 | 15,000 | 2,432 |
| 800 | 723 | 20,000 | 2,739 |
| 900 | 759 | 40,000 | 3,650 |
| 1,000 | 793 | 50,000 | 4,003 |
| 1,000 | 793 | 80,000 | 4,863 |

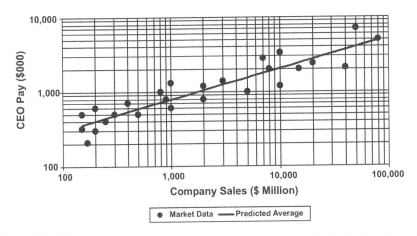

**FIGURE 10.4    PREDICTED AVERAGE CEO PAY AND COMPANY SALES ON LOG–LOG PLOT**

So, we will use the model as displayed on the log–log plot to develop a market-based pay target for our CEO, but first we will evaluate the model.

## 10.2  MODEL EVALUATION

### Appearance

Visually, this is a good model. The line goes through the data and describes the general linear trend of the logarithmic transformations.

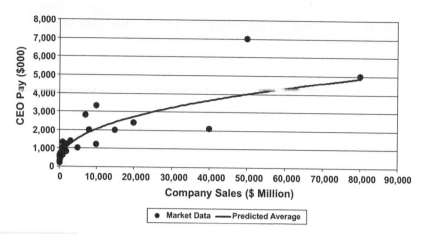

**FIGURE 10.5**   PREDICTED AVERAGE CEO PAY AND COMPANY SALES ON LINEAR PLOT

## Coefficient of Determination

Mathematically, $r^2$ is 0.849, which means the model has explained 84.9% of the variability in log CEO pay with log company sales. If we calculate $r^2$ in terms of the original metric, we would get a value of 0.741, which means the model has explained 74.1% of the variability in CEO pay with company sales.

Using either one, you conclude the model is a good one on the basis of your experience.

## Correlation

The correlation between log CEO pay and log company sales has a value of 0.921. The correlation between CEO pay and company sales has a value of 0.861. There is a strong association between the two variables.

## Standard Error of Estimate

The standard error of estimate (SEE) is 0.152. This means that "on average" the actual log of CEO pay varies from the predicted average log CEO pay by about 0.152. Using logarithmic identities shown in Chapter 8, we get

$$\text{On "average"} \quad y/y' = 10^{0.152} = 1.42$$

After MOTHing (Minus One Times Hundred), we get

$$\text{On "average"} \quad y/y' = (1.42 - 1)(100) = (0.42)(100) = 42\%$$

This means that on "average" the actual CEO pay values vary by about 42% from the predicted average CEO pay. Our experience is that there is a wide variation in CEO pay as measured by base pay, even with similar company sales.

**TABLE 10.4    SUMMARY OF MODEL EVALUATION**

| Evaluation Criteria | Symbol | In Terms of Log CEO Pay | In Terms of CEO Pay |
|---|---|---|---|
| Appearance | | Goes through middle of the data and describes the trend nicely | |
| Coefficient of determination | $r^2$ | 84.9% of variability in log CEO pay is associated with log company sales | 74.1% of variability in CEO pay is associated with company sales |
| Correlation | $r$ | +0.921, strong association | +0.861, strong association |
| Standard error of estimate | SEE | 0.152 is "average" deviation of log CEO pay from predicted average log CEO pay | 42% is "average" deviation of CEO pay from predicted average CEO pay |
| Common sense | | | Makes sense based on experience |

## Common Sense

In comparing the model with our experience, it makes sense that the CEO pay increases with company sales as shown.

## Summary of Evaluation

A summary of the evaluation is shown in Table 10.4.

Overall, we are very satisfied with this model and will use it with confidence as a basis for developing a market based pay range for the CEO.

BPD has sales of $5 billion or $5,000 million. To calculate the predicted average pay for a company of that size, we use the formula developed for the model.

$$y' = 10^{1.657} x^{0.4140}$$
$$y' = (10^{1.657})(5,000^{0.4140}) = 1,543$$

The predicted average CEO pay for a company of this size is $1,543 thousand or $1,543,000.

This will be the starting point to identify the organization's market position and establish a market-based pay target for the pay of the CEO. This is continued in Chapter 16 as Case Study 9, Part 6 of 6.

## RELATED TOPICS IN THE APPENDIX

A.12   Logarithmic Conversion

| TABLE 10.5 | | SURVEY OF CEO PAY AND COMPANY SALES | |
|---|---|---|---|
| Sales ($Million) | CEO Pay ($000) | Sales ($Million) | CEO Pay ($000) |
| 210 | 360 | 3,600 | 1,370 |
| 260 | 1,140 | 5,500 | 1,300 |
| 360 | 450 | 5,700 | 1,680 |
| 400 | 700 | 8,000 | 3,200 |
| 420 | 500 | 8,400 | 3,920 |
| 480 | 440 | 10,000 | 2,040 |
| 750 | 650 | 18,000 | 5,940 |
| 1,120 | 1,500 | 25,500 | 3,800 |
| 1,170 | 1,360 | 30,000 | 2,400 |
| 1,200 | 720 | 56,000 | 4,500 |
| 1,600 | 1,690 | 65,000 | 8,600 |
| 3,400 | 1,520 | 90,000 | 5,500 |

# PRACTICE PROBLEMS

**10.1**  You are the compensation manager for a company with $13 billion sales and have been given the task of developing a pay target for base salary for your CEO that is competitive with that of your chosen competition of 24 other companies. You collect the data in Table 10.5. The data are also in Data Set 22.

Determine the type of relationship by plotting the data. State your rationale for deciding which variable is the dependent variable and which variable is the independent variable.

**10.2**  If the relationship is a power model, calculate the equation of the least squares line, the coefficient of determination, the correlation, and the standard error of estimate. Also, report the average CEO pay and the average company sales.

**10.3**  Draw the least squares line on the plot.

**10.4**  Provide a complete evaluation of the model.

**10.5**  Calculate the predicted average CEO pay for a company of your size.

**10.6**  Give two reasons why one might consider using a power model.

**10.7**  What are the two visual characteristics of a logarithmic transformation of data?

**10.8**  What jobs are typically modeled by a power model?

**10.9**  What are the typical variables in a power model of salary survey data?

**10.10**  Why would you not want to plot a power model on a chart with linear scales?

# CHAPTER 11

# Market Models and Salary Survey Analysis

## 11.1 INTRODUCTION

This chapter and the next five chapters focus on the application of the statistics presented in the previous chapters to analyze salary surveys to

- determine the market position for a group of employees (department or entire company)
- develop a market-based pay structure for them
- develop an initial market-based salary increase budget

all effective from the beginning of the new salary administration year (aka plan year). Then, there are two additional steps:

- Develop a final market-based salary increase budget.
- Recommend a final salary increase budget.

Indeed, this process is perhaps the most important annual task done by compensation professionals.

There are two general methods used for the initial analysis: the *integrated market model* and the *job pricing market model*. There are other approaches, but these two are the ones the author has found work quite well.

The integrated market model is appropriate when the company has a grading system, where all the jobs are placed in a grade hierarchy of jobs, through a job

**TABLE 11.1    TWO MAJOR MARKET MODELS**

| | Integrated Market Model | Job Pricing Market Model |
|---|---|---|
| When to use | When there is a hierarchy of jobs, such as a grading system in which jobs are placed in a grade hierarchy, using a job evaluation system (such as point factor), or whole job ranking, or a maturity curve system with a hierarchy based on professional experience | When either there is no grading system or there is a group of jobs or skills that do not fit in with a grading system |
| Main underlying assumption | The market model is a good predictor of the average pay for all jobs in each grade or each year of experience, including both surveyed jobs and those not surveyed | The market position for jobs/skills not surveyed is the same as the overall market position for the surveyed jobs/skills |
| Characteristics | With the grading system or maturity curve system, it is not necessary to get survey data on "all" the jobs. You only need those data that cover major populations, major job families, the span of grades or experience, and jobs critical to the company. This will give credibility to the model | There is pressure to collect data on as many jobs/skills as possible, so each may be "market priced" |

evaluation system such as point factor or through whole job ranking. This model links market pay for matched benchmark jobs to the internal grades for those jobs, resulting in a model that estimates the range of market pay for all jobs in those grades. It is called an integrated market model because it integrates external value as represented by market pay with internal value as represented by grade.

The integrated market model approach is also appropriate for jobs administered in a maturity curve system, where the hierarchy is by years of professional experience rather than by grade.

The job pricing market model is appropriate when the company has no grading system or when there is a group of jobs or skills that do not fit in with a grading system. An example of the latter case would be when the market values for certain skills such as software engineers and data architects exceed the pay structure maximums for the grades in which those skills reside. The job pricing market model applies to both individuals in jobs and individuals with skills. We will call it "job pricing" even though we can apply it to either jobs or skills.

The job pricing market model approach is also appropriate for individual executive jobs where pay is related to organization size as typically measured by revenue or budget.

This is summarized in Table 11.1.

The starting point for building market models is salary surveys. For a thorough discussion of salary surveys, see Davis and Koechel [4].

---

**CASE STUDY 9, PART 1 OF 6**

It is that time of year when you analyze salary surveys and internal employee data to identify your market position, develop salary structures, and develop salary increase budgets. If your organization is working on a calendar year basis for salary administration, you are typically doing this analysis around late September or early October.

There are five general methods of salary survey analysis. We will use example employee populations from BPD to illustrate them in the next five chapters, as shown in Table 11.2.

**TABLE 11.2    FIVE METHODS OF SALARY SURVEY ANALYSIS**

| Compensation Situation | Example Employee Population | Type of Analysis | Variables for Market Model |
|---|---|---|---|
| Grades spanning a narrow range of pay | Hourly | Integrated market model: linear | External pay versus internal grades |
| Grades spanning a wide range of pay | Salaried nonexempts and most exempts | Integrated market model: exponential | External pay versus internal grades |
| Maturity curve | R&D professionals | Integrated market model: maturity curve | External pay versus external experience |
| Market pricing a group of jobs in a similar function | IT professionals | Job pricing market model: group of jobs | Internal pay versus external pay |
| Market pricing individual executive jobs | CEO | Job pricing market model: power model | External pay versus external sales |

It should be understood that the employee populations listed in this table are only for illustrative purposes, to demonstrate the five types of analyses. You may use a particular analysis for a different employee population from the one shown here.

BPD gives salary increases to its employees throughout the year, on anniversary dates or promotion dates. The company's pay policy is that it would like to target the average pay of its employees to be 2.0% above the market at the end of the plan year.

---

## 11.2  COMMONALITIES OF APPROACHES

All approaches used for analyzing salary surveys share certain commonalities. In each one, you

1. gather market data,
2. age data to a common date (usually to the start of your salary administration year),

3. create a market model effective at the start of the new plan year,

4. compare employee pay with market model.

We will use these steps in the next five chapters.

In addition, after the initial market model is developed, there are three more analyses that should be done as part of a complete model development.

- Identify outliers. Typically, this is done visually and the outliers will be the ones above the top line or below the bottom line of your model. Often, these raise job match issues. You have to investigate each one to confirm or delete the data.

   Do *not* discard a data point just because it is high or low or because it is so many SEEs from the predicted market average. There is nothing wrong with widely varying data if you are comfortable with the job matches. If you do discard data for mathematical reasons, you must report what you did and why you did it to maintain the ethical foundation of your analysis.

- If you are using data from more than one survey, identify if any particular survey is predominantly on the high side or low side. This can be done by plotting each survey (perhaps one at a time) in a different color or symbol with the rest of the data. If the survey tends to be high or low, then investigation should be done to find out why. The data may be perfectly valid, but you just need to understand the difference.

- Identify if any particular family of jobs is predominantly on the high side or low side. This can be done by plotting each family (perhaps one at a time) in a different color or symbol with the rest of the data. If a family tends to be high or low, then investigation should be done to find out why. In addition, this might be a rationale for creating a separate market model and subsequent structure for that family.

Similarly, when comparing employee pay to the market model, it is important to know which, if any, employees are outside the proposed structure and why because when you present the results, one of the first questions asked is "Who are those people?"

And you have to have the answer because . . .

| |
|---|
| Behind every data point, there is a story. |

Here are two examples:

1. "That is Jane Smith. She used to be a supervisor and didn't want to supervise anymore, and she is now a senior accountant. We didn't cut her pay, but she is above the maximum of her lower grade."

2. "That is the president's nephew." And of course everyone understands that one.

> The result of this analysis is the identification of the market position, the creation of a market-based salary structure, and the creation of an initial market-based salary increase budget, all as of the beginning of the new salary administration year.

These approaches are usually applied with base pay, but they can also be applied with total cash compensation.

In all five examples, we will develop a market position, and for the nonexecutive jobs, a market-based salary structure and salary increase budget, all effective as of the beginning of the new salary administration plan year. For the individual executive jobs, there is no salary structure involved. In the examples presented in the next five chapters, these analyses are as of January 1, 2012, the start of the new plan year.

There are two additional steps that need to be followed before the analyses are complete.

- Develop a final market-based salary increase budget.
- Recommend a final salary increase budget that takes into account other factors.

## 11.3  FINAL MARKET-BASED SALARY INCREASE BUDGET

### Initial Market-Based Salary Increase Budget and Market Position

Sometimes, you know the market position and want to quickly determine the salary increase budget percentage needed to meet the market. There is a relation between the market position and the initial market-based salary increase budget, given by the following formula:

Let    $MP\%$ = the market position, the % from the market

$SIB\%$ = the % salary increase budget required to meet the market

Then, $SIB\% = (-MP\%)/(100\% + MP\%)$

This is illustrated in Table 11.3.

**TABLE 11.3  SALARY INCREASE BUDGET FROM MARKET POSITION**

|  | Example Below the Market | Example Above the Market | Example at the Market |
|---|---|---|---|
| Market position, % from market | −4.0% | 3.0% | 0.0% |
| Numerator | 4.0% | −3.0% | 0.0% |
| Denominator | 96.0% | 103.0% | 100.0% |
| Salary increase budget, % to meet the market | 4.2% | −2.9% | 0.0% |

## Final Market-Based Salary Increase Budget

Determining the final market-based salary increase budget is relatively straight-forward once the initial market-based salary increase budget required to match the market at the beginning of the new plan year is known.

The final market-based salary increase budget has three components:

- Catch-up (or fallback) to the market at the beginning of the salary administration year. This is the initial market-based salary increase budget.
- Anticipated market movement in the future.
- Pay policy. This is the desired pay position with respect to the market at a given point in time.

The sum of these three items produces a final market-based salary increase budget. All we need to do is to fill in Table 11.4.

There are two scenarios that must be considered:

- The company gives raises throughout the year (e.g., on anniversary or promotion dates).
- The company gives raises to everyone on one common date.

## Raises Given Throughout the Year

Figure 11.1 illustrates the scenario of giving raises throughout the year. The solid line indicates the market movement throughout the year. The dashed line

**TABLE 11.4  TEMPLATE FOR FINAL MARKET-BASED SALARY INCREASE BUDGET**

| | |
|---|---|
| Catch-up (or fallback) at the beginning of the plan year | ___% |
| Anticipated market movement | ___% |
| Pay policy | ___% |
| Total | ___% |

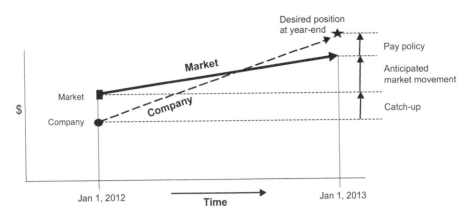

**FIGURE 11.1    RAISES GIVEN THROUGHOUT THE YEAR**

indicates the progression of the overall average company pay throughout the year as raises are given toward a desired position at year-end.

Suppose for BPD, we calculate that we need a salary increase budget of 1.0% to meet the market as of January 1, 2012. (How to calculate this is given in the next five chapters.) This is the catch-up that is the first entry in Table 11.4.

The second entry is the anticipated market movement for the forthcoming year. This value is a judgment call. Many consulting firms and professional associations are good source for this information as to what salary increases companies plan in the future. For this example, we will use a value of 2.0% annual movement.

The pay policy is the third entry, and is where the company wants its pay levels to be with respect to the market at a given point in time. For the situation where raises are given throughout the year, the point in time is usually the end of the plan year. For this example, BPD's pay policy is that it wants to be 2.0% above the market at the end of the plan year.

**A Note on Pay Policy**    More often than not, a company does not know what its pay policy should be. *All other things being equal, a good pay policy is one that states that the pay levels will be "competitive."* Translating this into numbers means that the desired market position is to be equal to the market pay at the end of the plan year. Numerically, this is 0.0% above the market.

In any case, the assumptions and context for the pay policy should be made explicit rather than a pay policy "adopted" by default. That way, if any of the assumptions is violated or if the context changes, the policy can be addressed openly and in an objective manner, and decided in accordance with supporting the strategic plans of the organization.

| TABLE 11.5 | FINAL MARKET-BASED SALARY INCREASE BUDGET EXAMPLE 1 |
|---|---|
| Catch-up (or fallback) | 1.0% |
| Anticipated market movement | 2.0% |
| Pay policy | 2.0% |
| Total | 5.0% |

In addition, a company may have a stated pay policy that sounds very nice, including one of being competitive, but many companies really do not decide until they see how much it costs.

For our example, then, we have completed the table, shown in Table 11.5 to get the final market-based salary increase budget of 5.0%.

## Raises Given on a Common Date

Figure 11.2 illustrates a "competitive" pay policy for a company giving all raises on January 1. Its pay levels match the market in the middle of the plan year, leading the market during the first 6 months of each year, and lagging the market during the last 6 months of each year. The pay level "stair-steps" the market pay over time.

With this picture in mind, Figure 11.3 illustrates the scenario for our market-based salary increase budget.

We have already determined the "catch-up" from our market position analysis. We need a salary increase budget of 1.0% to meet the market as of January 1, 2012.

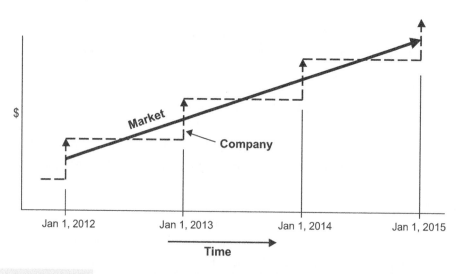

**FIGURE 11.2    RAISES GIVEN ON A COMMON DATE**

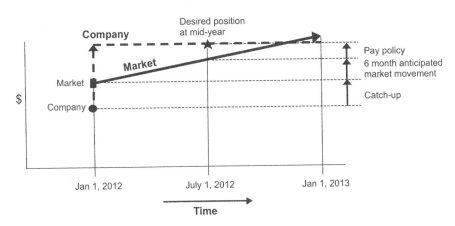

**FIGURE 11.3**    MARKET-BASED SALARY INCREASE BUDGET FOR RAISES GIVEN ON A
COMMON DATE

The value for the anticipated market movement for the forthcoming year
is a judgment call. As before, we will use a value of 2.0% annual movement, which in this scenario translates into a value of 1.0% for a 6 month
movement.

Using the same pay policy, the company would like its pay levels to be 2.0%
above the market at the middle of the plan year.

For our example, then, we have completed Table 11.6 to get the final market-
based salary increase budget of 4.0%.

If it turns out that your situation is different from these two scenarios,
then simply draw a picture that describes the situation, and fill out the table
accordingly.

To continue, we will use the scenario where raises are given throughout
the year. We now have a final market-based salary increase budget of 5.0%
for BPD.

*A final market-based salary increase budget is an IF–THEN proposition that is
driven by the pay policy.* IF you want the pay levels of BPD to be 2% above the
market at year-end, THEN it will cost 5.0%.

**TABLE 11.6    FINAL MARKET-BASED SALARY
INCREASE BUDGET EXAMPLE 2**

| | |
|---|---|
| Catch-up (or fallback) | 1.0% |
| Anticipated market movement | 1.0% |
| Pay policy | 2.0% |
| Total | 4.0% |

## 11.4 OTHER FACTORS INFLUENCING THE FINAL SALARY INCREASE BUDGET RECOMMENDATION

We have developed a final market-based salary increase budget. This is the budget required to be at a certain position with respect to the chosen market at a given point in time, in this case, at the end of the new plan year. However, there are many other factors that are considered when making the final budget decision.

We must always keep the option open to value jobs and people by our own standards and exercise our independent judgment rather than going along with the crowd (the market). Certainly, market pay is a major and important input, but it is not the only thing.

In addition to maintaining our company independence, in deciding what to recommend for a salary increase budget, other factors need to be taken into account in addition to our pay policy. Some of these are as follows:

- Balance with benefits
- Business plans of the company
- Affordability
- Management style
- Maturity of the company
- Industry practices
- Acceptability and marketability to a board
- Union contracts
- Internal equity
  - One division versus another
  - Compression
  - Critical skills
- Government factors and regulations
- Past practices
- Turnover
- Inflation
- State of the economy
- Comfort with whole numbers
- What was read in *The Wall Street Journal*
- What was heard at the country club

The compensation professional should be aware that the "hard" numbers based on salary survey data are, indeed, the major input to such a decision, but the "soft" numbers from the listed factors are equally important. So, when making a final recommendation, you should take both types of numbers into account to come up with a final budget recommendation, which may or may not be the same as the original market-based salary increase budget.

For example, after taking into account these other factors, your final recommendation may be to have a salary increase budget of 3.5% for BPD.

As mentioned in Chapter 1, many times our decisions are done in the face of uncertainty. When we make our final recommendations, we are making a host of assumptions in predicting the future. Here are some typical ones. You should come up with your own list.

### Assumptions

- The salary market moves as anticipated.
- The employees get raises as anticipated.
- The labor supply and demand situation goes as anticipated.
- The business plans are implemented according to plan.
- The external competitive environment goes as anticipated.
- The political and regulatory environment is stable.
- There are no disasters, natural or otherwise.

It would be appropriate and prudent to state all your assumptions when making your final recommendations, so that the executives have the proper context when making their approval decision.

## 11.5  SALARY STRUCTURE

Note that when we do our market analyses in the next five chapters, we will ignore current salary structures in these analyses and instead create structures based on the latest market analyses. This method automatically corrects for market movements that are not uniform (e.g., market salaries in lower grades moving faster than those in higher grades). This is the recommended method rather than taking the current structure and simply adjusting it uniformly.

Once a final budget decision is made, the market-based salary structure you developed (and which you will be developing in the following chapters) that was effective at the beginning of the new plan year should be adjusted appropriately

**TABLE 11.7**   ADJUSTMENTS TO MARKET-BASED SALARY STRUCTURE

| Situation | Market Position, % from Market Midpoint | Salary Increase Budget, % | Adjustment to Market-Based Salary Structure, % |
|---|---|---|---|
| 1 | −5% | 3% | −2% |
| 2 | −3% | 3% | 0% |
| 3 | 0% | 3% | 3% |
| 4 | +2% | 3% | 5% |

within which to administer the salaries for that new year. Different organizations have different views on setting a structure.

A salary structure is simply an administrative control device used to achieve consistency and ease in administering pay within company policies. It is not the be-all and end-all of a compensation program. With this perspective, one approach is as follows. Start with the market-based salary structure you developed, and adjust it so that after the approved salary increase budget is administered, the company's total salaries will be at 100% of the total midpoints of the adjusted structure (a) at the end of the plan year if raises are given throughout the year or (b) in the middle of the plan year if raises are all given at once at the beginning of the plan year. In other words, the total salaries will be at 100% of the total midpoints after all raises are given.

Obviously, there are many variations on a theme in doing this adjustment, but this is a suggestion that will help keep the salaries in the middle of the structure.

Here is the mathematics for such an adjustment. This formula applies whether you give raises throughout the year or all at once.

Let $P =$ market position, % you are from the market

$B =$ salary budget increase %

Then, $A =$ adjustment to the market-based salary structure, %

Theoretically, $A = (1 + B)(1 + P) - 1$

Practically, $A \approx B + P$

Here are four examples given in Table 11.7, using the practical approximation formula.

For example, in Situation 1, based on your analysis, you are 5% below the market at the beginning of the plan year. You have an approved 3% salary increase budget. The adjustment to the market-based salary structure, which was created to describe the market at the beginning of the plan year, will be downward 2%. Similar calculations are done for the other situations.

# PRACTICE PROBLEMS

**11.1** You are working on a calendar year basis and give raises throughout the year. Your market position at the beginning of the new plan year is 4.0% below the market. That is, your overall salaries are 4.0% below the midpoints of the market-based salary structure. The anticipated annual market movement during the new year is 3.0%. Your pay policy is to match the market at the end of the plan year. What is the market-based salary increase budget recommendation?

**11.2** Continuing, after taking into account many factors in addition to the market-based salary increase budget recommendation, your executives approve a salary increase budget of 5.0%. By how much do you adjust the market-based salary structure?

**11.3** You are working on a calendar year basis and give raises throughout the year. Your market position at the beginning of the new plan year is 3.0% above the market. That is, your overall salaries are 3.0% above the midpoints of the market-based salary structure. The anticipated annual market movement during the new year is 3.0%. Your pay policy is to match the market at the end of the plan year. What is the market-based salary increase budget recommendation?

**11.4** Continuing, after taking into account many factors in addition to the market-based salary increase budget recommendation, your executives approve a salary increase budget of 2.0%. By how much do you adjust the market-based salary structure?

**11.5** You are working on a calendar year basis and give all raises on January 1. Your market position at the beginning of the new plan year is 4.0% below the market. That is, your overall salaries are 4.0% below the midpoints of the market-based salary structure. The anticipated annual market movement during the new year is 3.0%. Your pay policy is to match the market at the middle of the plan year. What is the market-based salary increase budget recommendation?

**11.6** Continuing, after taking into account many factors in addition to the market-based salary increase budget recommendation, your executives approve a salary increase budget of 3.0%. By how much do you adjust the market-based salary structure?

**11.7** You are working on a calendar year basis and give all raises on January 1. Your market position at the beginning of the new plan year is at the market level. That is, your overall salaries equal the midpoints of the market-based salary structure. The anticipated annual market movement in the coming year is 3.0%. Your pay policy is to match the market at the middle of the plan year. What is the market-based salary increase budget recommendation?

**11.8** Continuing, after taking into account many factors in addition to the market-based salary increase budget recommendation, your executives approve a salary increase budget of 3.0%. By how much do you adjust the market-based salary structure?

**11.9** In addition to the market position, anticipated market movement, and pay policy, what are some of the additional factors that are typically taken into account when recommending a salary increase budget?

**11.10** What are some typical assumptions concerning a recommended salary increase budget?

# CHAPTER 12

# Integrated Market Model: Linear

---

**CASE STUDY 9, PART 2 OF 6**

For the BPD hourly population in the two Texas plants, as a result of a previous analysis showing a wide disparity in the grades and subsequent salaries, all jobs were re-evaluated with a new job evaluation system that used a new grade numbering system. Several salary adjustments were made along with appropriate management and policy changes. You now have to create a new salary structure and develop a salary increase budget.

---

For purposes of illustration, we will use data for just a few jobs and for the employees in those jobs.

We will proceed through the four steps listed in Chapter 11.

## 12.1  GATHER MARKET DATA

You belong to a compensation survey group among local manufacturers and have the survey data in Table 12.1. The data are also in Data Set 23. The reader might want to enter the data in a computer that has statistical software and follow along.

The Title is the job title of the BPD job that was matched to the survey. Grade is the internal grade of the matched job. Survey is a letter designating which survey was matched. In this case, there was just one survey.

The Hourly Base Salary is straight time pay and does not include overtime or shift differential. The Market-Weighted Average is the measure that was

*Statistics for Compensation: A Practical Guide to Compensation Analysis,* By John H. Davis

**TABLE 12.1    SURVEY DATA ON HOURLY MANUFACTURING JOBS**

| Job | Title | Grade | Survey | Hourly Base Salary, MarketWtd Avg |
|-----|-------|-------|--------|-----------------------------------|
| 1 | Manufacturing Technician 1 | 46 | A | 14.82 |
| 2 | Welder 1 | 46 | A | 17.94 |
| 3 | Machinist 1 | 46 | A | 19.24 |
| 4 | Manufacturing Technician 2 | 47 | A | 19.11 |
| 5 | Electrical Engineering Technician 1 | 47 | A | 20.67 |
| 6 | Welder 2 | 47 | A | 21.84 |
| 7 | Tool and Die Maker 1 | 47 | A | 22.23 |
| 8 | Machinist 2 | 47 | A | 22.75 |
| 9 | Mechanical Engineering Technician 1 | 47 | A | 24.05 |
| 10 | Manufacturing Technician 3 | 48 | A | 23.79 |
| 11 | Welder 3 | 48 | A | 24.05 |
| 12 | Mechanical Engineering Technician 2 | 48 | A | 25.74 |
| 13 | Electrical Engineering Technician 2 | 48 | A | 26.00 |
| 14 | Tool and Die Maker 2 | 48 | A | 26.13 |
| 15 | Machinist 3 | 48 | A | 26.78 |
| 16 | Tool and Die Maker 3 | 49 | A | 28.60 |
| 17 | Electrical Engineering Technician 3 | 49 | A | 30.81 |
| 18 | Mechanical Engineering Technician 3 | 49 | A | 34.58 |

extracted from the survey. You may have decided to use another measure, such as unweighted average or median. In this example, we are using the weighted average.

In addition, you may have other data on your spreadsheet that are not shown here, such as the survey job title, the survey job number, the page numbers in the survey of the job description and of the data extracted, the internal company job number, and so on. It is useful to have these "trails" so you can easily find the source data in case of questions and for use in subsequent years.

For the survey itself, you have the date of data and annual movement, as shown in Table 12.2. Typically, these two items are reported in surveys.

If the survey does not specify the date of the data, you need to contact the survey provider for that date.

If the survey does not specify the general annual movement of the jobs in the survey, there are typically three ways to get it. The first approach is to contact the survey provider for it. They may have collected general salary increase budget information and would be able to summarize it for you. The second approach

**TABLE 12.2    AGING TABLE: INITIAL ENTRIES**

| Survey | Date of Data | Annual Movement | Date Aged To | Months Aged | Aging Factor |
|--------|-------------|-----------------|--------------|-------------|--------------|
| A | 01 Jan 11 | 4.0% | | | |

is to call the companies in the survey and ask them what their salary increase budgets are, and summarize it yourself. The third approach is to conduct or participate in a survey focusing on gathering and summarizing salary increase budget information.

## 12.2 AGE DATA TO A COMMON DATE

The data are aged to a common date so they are comparable in time. It would not make sense to use data, for example, from one survey that is 1 year old and from another survey that is 6 months old without any adjustments between the two surveys, because salaries change over time. The common date aged to is usually the start of your salary administration year. We will age this survey data to January 1, 2012.

The data from the survey will be aged according to the annual movement for that survey during the time from the date of the data to the beginning of your plan year. For survey A, we will age the data from January 1, 2011 to January 1, 2012, a period of 12 months. At an annual rate of 4.0%, we will age the data 4.0% by multiplying all the data from Survey A by an aging factor of 1.040. Some of this may seem redundant or oversimplified, but it is done this way to allow ease of expansion when there is more than one survey.

The completed aging table should look like Table 12.3.

We are now ready to use this table. What we will do is multiply the survey data from Survey A by its aging factor (1.040). We get Table 12.4.

For example for Job 1, $(14.82)(1.04) = 15.41$.

## 12.3 CREATE AN INTEGRATED MARKET MODEL

We will be repeating the process we used in Chapter 7. First, we plot the aged market weighted average $(y)$ versus grade $(x)$ for the matched jobs, shown in Figure 12.1.

There is a relationship, it is positive, and comparing this relationship with the charts of functional forms in Figures 6.8 and 6.9, we will use a linear model to describe the relationship. We create the following linear model using the techniques in Chapter 7.

**TABLE 12.3    AGING TABLE: COMPLETED**

| Survey | Date of Data | Annual Movement | Date Aged To | Months Aged | Aging Factor |
|--------|--------------|-----------------|--------------|-------------|--------------|
| A | 01 Jan 11 | 4.0% | 01 Jan 12 | 12.0 | 1.040 |

**TABLE 12.4**   **AGED SURVEY DATA ON HOURLY MANUFACTURING JOBS**

| Job | Title | Grade | Survey | Hourly Base Salary, Market Wtd Avg | Aging Factor | Aged Hourly Base Salary, Market Wtd Avg |
|---|---|---|---|---|---|---|
| 1 | Manufacturing Technician 1 | 46 | A | 14.82 | 1.04 | 15.41 |
| 2 | Welder 1 | 46 | A | 17.94 | 1.04 | 18.66 |
| 3 | Machinist 1 | 46 | A | 19.24 | 1.04 | 20.01 |
| 4 | Manufacturing Technician 2 | 47 | A | 19.11 | 1.04 | 19.87 |
| 5 | Electrical Engineering Technician 1 | 47 | A | 20.67 | 1.04 | 21.50 |
| 6 | Welder 2 | 47 | A | 21.84 | 1.04 | 22.71 |
| 7 | Tool and Die Maker 1 | 47 | A | 22.23 | 1.04 | 23.12 |
| 8 | Machinist 2 | 47 | A | 22.75 | 1.04 | 23.66 |
| 9 | Mechanical Engineering Technician 1 | 47 | A | 24.05 | 1.04 | 25.01 |
| 10 | Manufacturing Technician 3 | 48 | A | 23.79 | 1.04 | 24.74 |
| 11 | Welder 3 | 48 | A | 24.05 | 1.04 | 25.01 |
| 12 | Mechanical Engineering Technician 2 | 48 | A | 25.74 | 1.04 | 26.77 |
| 13 | Electrical Engineering Technician 2 | 48 | A | 26.00 | 1.04 | 27.04 |
| 14 | Tool and Die Maker 2 | 48 | A | 26.13 | 1.04 | 27.18 |
| 15 | Machinist 3 | 48 | A | 26.78 | 1.04 | 27.85 |
| 16 | Tool and Die Maker 3 | 49 | A | 28.60 | 1.04 | 29.74 |
| 17 | Electrical Engineering Technician 3 | 49 | A | 30.81 | 1.04 | 32.04 |
| 18 | Mechanical Engineering Technician 3 | 49 | A | 34.58 | 1.04 | 35.96 |

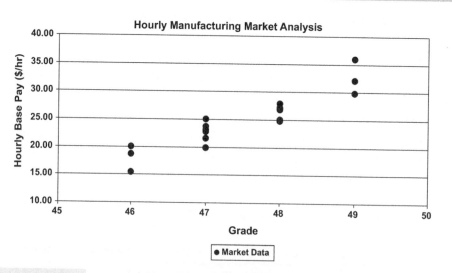

**FIGURE 12.1**   **AGED HOURLY MARKET PAY AND GRADE**

Recall that the equation for a linear model is $y = a + bx$ and that the prediction equation (or model), $y' = a + bx$, is derived from the data points using the method of least squares.

The resulting prediction equation and coefficient of determination are

$$y' = -196.47 + 4.6583x$$

$$r^2 = 85.7\%$$

## Interpretations

Notice the negative intercept of $-196.47$. Mathematically, that is the value of $y'$ when $x$ is zero. Problem-wise it is the value of the predicted average pay for grade zero. But there is no grade zero. Our grades start at grade 46. Hence, there is no problem interpretation of the intercept.

Regarding the slope of 4.6583, mathematically it is the change in $y'$ when $x$ increases by 1. Problem-wise, there is an increase of \$4.66/hr (rounded) in the predicted average pay for each increase in grade.

The mathematical interpretation of $r^2$ is that 85.7% of the variability in $y$ can be attributed to the variability in $x$. Problem-wise, 85.7% of the variability in pay can be attributed to the variability in grade.

Applying the equation, we calculate the predicted average market pay for each grade, shown in Table 12.5.

For example for grade 46,

$$y' = -196.47 + (4.6583)(46) = -196.47 + 214.28 = 17.81$$

We note that for a linear market model, the progression from one grade to the next is a constant dollar amount. Namely, the difference from one grade to the next is the value of the slope, \$4.66/hr (rounded).

We now plot the predicted average market pay on the chart as a line, shown in Figure 12.2.

The line goes through the data and $r^2$ is very good. The values of the numbers make sense. We decide that this is a good model.

**TABLE 12.5   PREDICTED AVERAGE HOURLY PAY AND GRADE**

| Grade | Predicted Average Market Pay |
| --- | --- |
| 46 | 17.81 |
| 47 | 22.47 |
| 48 | 27.13 |
| 49 | 31.79 |

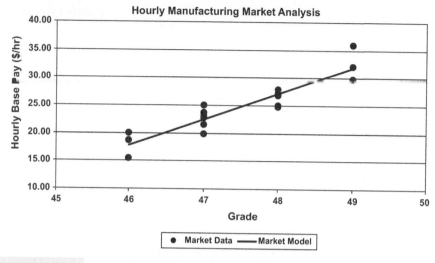

**FIGURE 12.2    PREDICTED AVERAGE HOURLY MARKET PAY AND GRADE**

Our model nicely *describes the relationship between external value (market pay) and internal value (grade).* This graphical display gives us confidence in the model.

> As an organization, we want our pay programs to be "internally equitable and externally competitive," or "internally fair and externally fair." This chart and the associated equation are where these two concepts are integrated together. In this case, external value is represented by market pay, the *y*-variable (dependent variable), and internal value is represented by grade, the *x*-variable (independent variable).

We will use this model to create a market-based salary structure. We first designate the predicted average market pay as the market-based midpoint.

The next step is to add minimum and maximum lines. We will select a range spread of 40% for this example. This means that the maximum is 40% more than the minimum. This amount both provides ample pay growth opportunities within a grade and delimits what we will pay for jobs in a grade. The most predominant range spread is 50%, although they range from 30% to over 70% in practice. Here is how to calculate the minimum and maximum starting with a midpoint and range spread, where the range spread is expressed as a decimal (e.g., 40% = 0.40).

$$\text{Minimum} = \frac{\text{midpoint}}{(1 + (\text{range spread}/2))}$$

$$\text{Maximum} = (\text{minimum})(1 + \text{range spread})$$

**TABLE 12.6    MARKET-BASED SALARY STRUCTURE**

**Market-Based Salary Structure**

| Grade | Minimum | Midpoint | Maximum |
|-------|---------|----------|---------|
| 46 | 14.84 | 17.81 | 20.78 |
| 47 | 18.73 | 22.47 | 26.22 |
| 48 | 22.61 | 27.13 | 31.65 |
| 49 | 26.49 | 31.79 | 37.09 |

For example, for grade 46,

$$\text{Minimum} = \frac{17.81}{(1 + (0.40/2))} = \frac{17.81}{1.20} = 14.84$$

$$\text{Maximum} = (14.84)(1 + 0.40) = (14.84)(1.40) = 20.78$$

We calculate the structure in Table 12.6 and plot it in Figure 12.3.

What this structure represents is a description of the span of average market pay as it relates to internal grades. The middle line is the predicted average pay for a given grade, which we have designated now as the midpoint, and the top and bottom lines form the basis for a market-based salary structure (assuming a desired range spread of 40%).

So, now we have a market-based salary structure as of January 1, 2012. You examine the data for any outliers or to see if any family needs to be segregated, and none does. At this stage, there is no further analysis of the survey data.

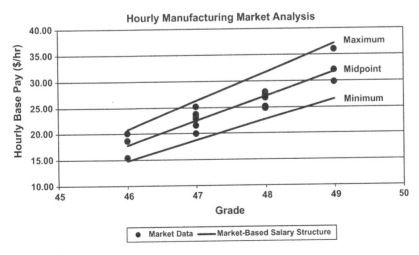

**FIGURE 12.3    MARKET-BASED SALARY STRUCTURE**

## 12.4  COMPARE EMPLOYEE PAY WITH MARKET MODEL

This step involves comparison of salaries of all employees in these grades, whether or not their jobs were surveyed, with the predicted average market pay (the midpoint line). We will have both a tabular display and a graphical one. At this point, we make explicit the assumption underlying this approach.

> The integrated market model approach assumes that for each grade, the predicted average market pay from the model is a good estimator for the average market pay for all jobs in that grade, whether or not they were surveyed.

This assumption is based on the internal evaluation system (grades) that internally evaluates all jobs in a given grade equivalently. In most cases, the market and a company evaluate jobs similarly, as evident by a typically strong relationship (i.e., high $r^2$) between survey pay and grades for benchmark jobs.

It is under this assumption that we are able to compare the salaries of all employees with the predicted average market pay for their jobs to determine an overall market position.

Since the market model is based on data aged to January 1, 2012, the employee data should also be as of that date, before any raise is given in the new plan year. If you know the projected or planned raises for individual employees between the date of the employee data you have at hand and the date you have aged the market data to, then it would be appropriate to take those into account. Otherwise, just note that there are differences in the dates and any implication these may lead to.

To determine the market position, along with other useful information, we need to compare the employee data with the market-based salary structure. The employee data are shown in Table 12.7 and are also in Data Set 24.

This comparison is shown in Table 12.8.

For purposes of illustration, we have just indicated an employee number. When doing this analysis for your company, you would probably have additional fields of data such as employee name, job title, job code, department, location, department and/or location code, and so on to allow additional analyses to be done.

Please note that we often interchange the terms "salary structure" and "pay structure," whether the employees are salaried, hourly, exempt, or nonexempt.

After the employee identification, we have the grade and the pay of the employee. The next three columns contain the market-based salary structure you have just created.

**TABLE 12.7    EMPLOYEE SALARIES AND GRADES**

| Employee No. | Grade | Pay | Employee No. | Grade | Pay | Employee No. | Grade | Pay |
|---|---|---|---|---|---|---|---|---|
| 1 | 46 | 14.50 | 51 | 47 | 22.24 | 101 | 48 | 28.19 |
| 2 | 46 | 14.75 | 52 | 47 | 22.24 | 102 | 48 | 28.19 |
| 3 | 46 | 15.94 | 53 | 47 | 22.60 | 103 | 48 | 28.41 |
| 4 | 46 | 16.07 | 54 | 47 | 22.60 | 104 | 48 | 28.63 |
| 5 | 46 | 16.33 | 55 | 47 | 22.95 | 105 | 48 | 28.63 |
| 6 | 46 | 16.72 | 56 | 47 | 22.95 | 106 | 48 | 28.85 |
| 7 | 46 | 16.72 | 57 | 47 | 22.95 | 107 | 48 | 29.52 |
| 8 | 46 | 16.72 | 58 | 47 | 23.30 | 108 | 48 | 29.97 |
| 9 | 46 | 16.98 | 59 | 47 | 23.30 | 109 | 48 | 32.00 |
| 10 | 46 | 16.98 | 60 | 47 | 23.48 | 110 | 48 | 32.50 |
| 11 | 46 | 17.11 | 61 | 47 | 23.48 | 111 | 49 | 26.09 |
| 12 | 46 | 17.24 | 62 | 47 | 23.83 | 112 | 49 | 26.63 |
| 13 | 46 | 17.24 | 63 | 47 | 23.83 | 113 | 49 | 27.70 |
| 14 | 46 | 17.37 | 64 | 47 | 24.54 | 114 | 49 | 27.97 |
| 15 | 46 | 17.37 | 65 | 47 | 24.71 | 115 | 49 | 28.24 |
| 16 | 46 | 17.50 | 66 | 47 | 26.50 | 116 | 49 | 28.51 |
| 17 | 46 | 17.63 | 67 | 48 | 22.40 | 117 | 49 | 28.78 |
| 18 | 46 | 17.76 | 68 | 48 | 23.07 | 118 | 49 | 28.78 |
| 19 | 46 | 17.76 | 69 | 48 | 23.74 | 119 | 49 | 29.31 |
| 20 | 46 | 17.89 | 70 | 48 | 23.96 | 120 | 49 | 29.31 |
| 21 | 46 | 18.02 | 71 | 48 | 24.18 | 121 | 49 | 29.85 |
| 22 | 46 | 18.02 | 72 | 48 | 24.41 | 122 | 49 | 29.85 |
| 23 | 46 | 18.15 | 73 | 48 | 24.63 | 123 | 49 | 30.12 |
| 24 | 46 | 18.41 | 74 | 48 | 24.63 | 124 | 49 | 30.39 |
| 25 | 46 | 18.54 | 75 | 48 | 24.63 | 125 | 49 | 30.39 |
| 26 | 46 | 18.67 | 76 | 48 | 25.30 | 126 | 49 | 30.66 |
| 27 | 46 | 18.80 | 77 | 48 | 25.30 | 127 | 49 | 30.66 |
| 28 | 46 | 19.33 | 78 | 48 | 25.30 | 128 | 49 | 30.93 |
| 29 | 46 | 19.98 | 79 | 48 | 25.52 | 129 | 49 | 30.93 |
| 30 | 47 | 18.72 | 80 | 48 | 25.52 | 130 | 49 | 31.19 |
| 31 | 47 | 19.07 | 81 | 48 | 25.52 | 131 | 49 | 31.19 |
| 32 | 47 | 19.77 | 82 | 48 | 25.96 | 132 | 49 | 31.46 |
| 33 | 47 | 19.95 | 83 | 48 | 25.96 | 133 | 49 | 31.46 |
| 34 | 47 | 20.13 | 84 | 48 | 25.96 | 134 | 49 | 32.00 |
| 36 | 47 | 20.48 | 86 | 48 | 26.19 | 136 | 49 | 32.54 |
| 37 | 47 | 20.48 | 87 | 48 | 26.19 | 137 | 49 | 32.54 |
| 38 | 47 | 20.83 | 88 | 48 | 26.41 | 138 | 49 | 32.54 |
| 39 | 47 | 20.83 | 89 | 48 | 26.41 | 139 | 49 | 33.07 |
| 40 | 47 | 21.18 | 90 | 48 | 26.63 | 140 | 49 | 33.07 |
| 41 | 47 | 21.18 | 91 | 48 | 26.85 | 141 | 49 | 33.34 |
| 42 | 47 | 21.36 | 92 | 48 | 26.85 | 142 | 49 | 33.34 |
| 43 | 47 | 21.54 | 93 | 48 | 26.85 | 143 | 49 | 33.88 |
| 44 | 47 | 21.54 | 94 | 48 | 27.52 | 144 | 49 | 33.88 |
| 45 | 47 | 21.71 | 95 | 48 | 27.52 | 145 | 49 | 34.95 |
| 46 | 47 | 21.71 | 96 | 48 | 27.52 | 146 | 49 | 35.22 |
| 47 | 47 | 21.89 | 97 | 48 | 27.74 | 147 | 49 | 36.30 |
| 48 | 47 | 21.89 | 98 | 48 | 27.74 | | | |
| 49 | 47 | 22.07 | 99 | 48 | 27.74 | | | |
| 50 | 47 | 22.07 | 100 | 48 | 27.97 | | | |

**TABLE 12.8**    **EMPLOYEE SALARIES AND MARKET-BASED SALARY STRUCTURE**

| Emp. No. | Grade | Pay | Market-Based Pay Structure | | | Pay as % of Midpoint | $/hr Under Min. | $/hr Over Max. |
|---|---|---|---|---|---|---|---|---|
| | | | Minimum | Midpoint | Maximum | | | |
| 1 | 46 | 14.50 | 14.84 | 17.81 | 20.78 | 81% | 0.34 | – |
| 2 | 46 | 14.75 | 14.84 | 17.81 | 20.78 | 83% | 0.09 | – |
| 3 | 46 | 15.94 | 14.84 | 17.81 | 20.78 | 90% | – | – |
| 4 | 46 | 16.07 | 14.84 | 17.81 | 20.78 | 90% | – | – |
| ⋮ | ⋮ | ⋮ | ⋮ | ⋮ | ⋮ | ⋮ | ⋮ | ⋮ |
| ⋮ | ⋮ | ⋮ | ⋮ | ⋮ | ⋮ | ⋮ | ⋮ | ⋮ |
| 145 | 49 | 34.95 | 26.49 | 31.79 | 37.09 | 110% | – | – |
| 146 | 49 | 35.22 | 26.49 | 31.79 | 37.09 | 111% | – | – |
| 147 | 49 | 36.30 | 26.49 | 31.79 | 37.09 | 114% | – | – |

Next is an indication of the employee's pay as it relates to the midpoint. Finally, if the employee's pay is under the minimum, we calculate by how much. Similarly, if the employee's pay is over the maximum, we calculate by how much.

This table furnishes the raw data to identify the overall market position. We use this data to create the summary table in Table 12.9, calculating subtotals by grade and annualizing the hourly pay to annual pay. We annualize the pay for budgeting purposes.

The Total Pay (Annual Dollars) is the total annual pay of the 147 BPD employees as of January 1, 2012. The Total Mkt Midpoint (Annual Dollars) is the total annual midpoint of the grades of the market-based salary structure we created for all 147 employees.

We see that in terms of annual dollars, the BPD total pay is 160,330 below the total market midpoint. The Difference is the Total Mkt Midpoint minus the Total Pay.

We now want to calculate two percentages. First, there are two definitions:

1. "Market position" is the % we are from the market midpoint. The market midpoint is the reference. In Table 12.9, this is in the column headed % from Mkt Mid.

2. "Budget needed to meet the market" is the % increase (or decrease) in our pay needed to meet the market. Our pay is the reference. In Table 12.9, this is in the column headed % to Meet Mkt Mid.

Recall from Chapter 2, how to calculate the percentage difference between two terms.

$$\text{Percent difference} = \frac{(\text{data} - \text{reference})}{(\text{reference})}(100)$$

**TABLE 12.9    MARKET ANALYSIS**

Hourly Manufacturing Market Analysis

Comparison of Employee Annual Pay with the Market-Based Pay Structure

| Grade | No. Emp. | Total Pay (Annual Dollars) | Total Mkt Midpoint (Annual Dollars) | Difference | % From Mkt Mid. | % to Meet Mkt Mid. | No. Emp. Under Min. | No. Emp. Over Max. | Dollars Under Min. | Dollars Over Max. |
|---|---|---|---|---|---|---|---|---|---|---|
| 46 | 29 | 1,052,387 | 1,077,398 | 25,011 | −2.3% | 2.4% | 2 | 0 | 897 | 0 |
| 47 | 37 | 1,698,421 | 1,734,280 | 35,858 | −2.1% | 2.1% | 1 | 1 | 21 | 584 |
| 48 | 44 | 2,447,295 | 2,490,100 | 42,805 | −1.7% | 1.7% | 1 | 2 | 438 | 2,503 |
| 49 | 37 | 2,396,960 | 2,453,616 | 56,656 | −2.3% | 2.4% | 1 | 0 | 834 | 0 |
| Total | 147 | 7,595,063 | 7,755,393 | 160,330 | −2.1% | 2.1% | 5 | 3 | 2,190 | 3,087 |

For the market position, % from Mkt Mid, the data is our total pay and the reference is the total market midpoint.

$$\% \text{ from Mkt Mid} = \frac{(7{,}595{,}063 - 7{,}755{,}393)}{(7{,}755{,}393)}(100) = \frac{-160{,}330}{7{,}755{,}393}(100)$$
$$= (-0.021)(100) = -2.1\%$$

Our market position is that we are 2.1% below the market as of January 1, 2012. Overall, the company's hourly employees are 2.1% below the market midpoint, which represents the market-weighted average.

For the budget needed to meet the market, % to Meet Mkt Mid, the data is the total market midpoint and the reference is our total pay.

$$\% \text{ to Meet Mkt Mid} = \frac{(7{,}755{,}393 - 7{,}595{,}063)}{(7{,}595{,}063)}(100) = \frac{160{,}330}{7{,}595{,}063}(100)$$
$$= (0.021)(100) = 2.1\%$$

It would take a 2.1% budget to equal the market midpoint. Table 12.9 both indicates that some grades have different market positions from others and summarizes how many employees have pay outside the pay structure and by how much.

If desired, you could have subtotals by job family, location, and so on.

The next step is to present the results in a graphical form, shown in Figure 12.4. In this chart, you compare the employee pay with the market-based salary structure you have just created.

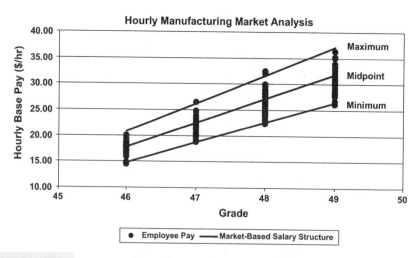

**FIGURE 12.4**    EMPLOYEE SALARIES AND MARKET-BASED SALARY STRUCTURE

**TABLE 12.10   TEMPLATE FOR FINAL MARKET-BASED SALARY INCREASE BUDGET**

| | |
|---|---|
| Catch-up (or fallback) | —— |
| Anticipated market movement | —— |
| Pay policy | —— |
| Total | —— |

This chart shows that the bulk of the employees' pay is within the market-based pay structure, and it highlights the pay that is outside the structure.

You identify the outliers and find out why they are outside the salary range.

In summary, we have completed the following four steps to conduct a market analysis where we have grades, resulting in an identification of the market position, the creation of a market-based salary structure, and the development of an initial market-based salary increase budget, all as of the beginning of the new plan year.

1. Gather market data

2. Age data to a common date

3. Create an integrated market model

4. Compare employee pay with the market model

For the final market-based salary increase budget, we complete Table 12.10.

The catch-up is 2.1%, which is our initial market-based salary increase budget as of the beginning of the new plan year. For this population, from survey providers we ascertain that the anticipated market movement is 1.5% for the coming year. Our pay policy is to be 2.0% above the market at the end of the plan year. So our results are in Table 12.11.

The final market-based salary increase budget is 5.6%. For the final recommendations, you now take into account the "soft" information described in Section 11.4.

**TABLE 12.11   FINAL MARKET-BASED SALARY INCREASE BUDGET**

| | |
|---|---|
| Catch-up (or fallback) | 2.1% |
| Anticipated market movement | 1.5% |
| Pay policy | 2.0% |
| Total | 5.6% |

## RELATED TOPICS IN THE APPENDIX

A.13  Range Spread Relationships

# PRACTICE PROBLEMS

**12.1**  BPD has a building construction department in Huntsville, Alabama. It is October 2012 and you are preparing to recommend a salary increase budget for 2013. Your salary administration year is a calendar year. You give all raises on January 1. Your compensation philosophy is that you want your average pay to equal the market pay at the middle of the plan year. You have recently obtained two surveys for some of the skilled trades. The data you extract from the surveys are the unweighted averages. These are shown in Table 12.12, and also in Data Set 25. The date of the data for Survey A is July 1, 2012 and for Survey B is April 1, 2012. From the surveys, the general market movement for 2012 for Survey A is 5.0% annually and for Survey B is 4% annually.

**TABLE 12.12    SURVEY DATA ON HOURLY CONSTRUCTION JOBS**

| Job | Title | Grade | Survey | Hourly Base Salary, Market Unwtd Avg |
|---|---|---|---|---|
| 1 | Carpenter 1 | 32 | A | 21.09 |
| 2 | Carpenter 2 | 33 | A | 22.75 |
| 3 | Carpenter 3 | 33 | A | 25.63 |
| 4 | Electrician 1 | 32 | B | 23.29 |
| 5 | Electrician 2 | 33 | B | 24.91 |
| 6 | Electrician 3 | 34 | B | 27.78 |
| 7 | HVAC Mechanic 1 | 32 | B | 19.90 |
| 8 | HVAC Mechanic 3 | 34 | B | 29.23 |
| 9 | Painter 1 | 31 | A | 19.28 |
| 10 | Painter 2 | 32 | A | 21.64 |
| 11 | Painter 3 | 33 | A | 26.68 |
| 12 | Plumber 1 | 32 | A | 21.39 |
| 13 | Plumber 2 | 33 | A | 23.34 |
| 14 | Plumber 3 | 34 | A | 26.05 |

How much will you age the data from Survey A? How much will you age the data from Survey B?

**12.2**  Age the survey data appropriately.

**12.3**  Plot the data. What kind of relationship is shown by the plot?

**12.4**  Conduct a regression. State the market model equation, the coefficient of determination, and the standard error of estimate (SEE).

**12.5**  Plot, interpret, and evaluate the model.

**12.6**  You decide to use a range spread of 40%. Setting the market model as the midpoint, create a market-based salary structure and plot it. What is the midpoint progression?

**12.7** The employee data for the 48 skilled trade employees are in Table 12.13 and also in Data Set 26. Complete Table 12.14. Create a plot that compares employee pay with the market-based salary structure.

**TABLE 12.13    EMPLOYEE SALARIES AND GRADES**

| Emp. No. | Grade | Pay | Emp. No. | Grade | Pay |
|---|---|---|---|---|---|
| 1 | 31 | 14.56 | 25 | 32 | 20.30 |
| 2 | 31 | 15.92 | 26 | 32 | 20.52 |
| 3 | 31 | 16.67 | 27 | 32 | 22.55 |
| 4 | 31 | 17.11 | 28 | 32 | 23.34 |
| 5 | 31 | 18.71 | 29 | 32 | 23.40 |
| 6 | 31 | 18.79 | 30 | 32 | 24.19 |
| 7 | 31 | 19.26 | 31 | 33 | 20.76 |
| 8 | 31 | 19.87 | 32 | 33 | 23.38 |
| 9 | 31 | 20.77 | 33 | 33 | 24.21 |
| 10 | 31 | 21.71 | 34 | 33 | 24.77 |
| 11 | 31 | 22.97 | 35 | 33 | 24.85 |
| 12 | 32 | 18.03 | 36 | 33 | 25.90 |
| 13 | 32 | 18.03 | 37 | 33 | 27.46 |
| 14 | 32 | 18.41 | 38 | 33 | 30.19 |
| 15 | 32 | 18.49 | 39 | 34 | 23.85 |
| 16 | 32 | 18.55 | 40 | 34 | 24.19 |
| 17 | 32 | 19.05 | 41 | 34 | 25.32 |
| 18 | 32 | 19.44 | 42 | 34 | 25.75 |
| 19 | 32 | 19.53 | 43 | 34 | 27.09 |
| 20 | 32 | 19.84 | 44 | 34 | 27.39 |
| 21 | 32 | 20.00 | 45 | 34 | 27.61 |
| 22 | 32 | 20.02 | 46 | 34 | 28.21 |
| 23 | 32 | 20.04 | 47 | 34 | 29.01 |
| 24 | 32 | 20.05 | 48 | 34 | 31.09 |

**TABLE 12.14    TEMPLATE FOR EMPLOYEE SALARIES AND MARKET-BASED SALARY STRUCTURE**

| Emp. No. | Grade | Pay | Minimum | Midpoint | Maximum | Pay as % of Midpoint | $ Under Min. | $ Over Max. |
|---|---|---|---|---|---|---|---|---|
| 1 | 31 | 14.56 | | | | | | |
| 2 | 31 | 15.92 | | | | | | |
| 3 | 31 | 16.67 | | | | | | |
| : | : | : | | | | | | |
| : | : | : | | | | | | |

**12.8** Complete the summary table shown in Table 12.15. Convert hourly pay to annual pay, using 2,086 hours per year.

**TABLE 12.15  TEMPLATE FOR MARKET ANALYSIS**

**Hourly Manufacturing Market Analysis**
**Comparison of Employee Pay with the Market-Based Pay Structure**

| Grade | No. Emp. | Total Pay Annual Dollars | Total Market Midpoint Annual Dollars | Difference | % From Mkt Mid. | % To Meet Mkt Mid. | No. Emp. Under Min. | No. Emp. Over Max. | Dollars Under Min. | Dollars Over Max. |
|-------|----------|--------------------------|--------------------------------------|------------|-----------------|--------------------|---------------------|--------------------|--------------------|--------------------|
| 31 | | | | | | | | | | |
| 32 | | | | | | | | | | |
| 33 | | | | | | | | | | |
| 34 | | | | | | | | | | |
| Total | | | | | | | | | | |

**12.9** For both surveys, you estimate that the market will move 4% annually during 2013. What is the market-based salary increase budget recommendation?

**12.10** Your executives approve a salary increase budget of 4.5%. By how much do you adjust the market-based salary structure to have the average pay at the midpoint after raises are given?

# Integrated Market Model: Exponential

We will use the four steps listed in Chapter 11.

1. *Gather Market Data.* We have data from three surveys: A, B, and C, for jobs in the BPD finance department, as shown in Table 13.1. The data are also in Data Set 27. The reader might wish to enter the data in a computer and follow along.

    The company job title is the title of the company job that was matched to the surveys. Grade is the internal grade of the matched job. Survey is a letter designating which survey was matched.

    The market weighted average annual base pay is the measure that was extracted from the surveys. You may have decided to use another measure, such as unweighted average or median. However, in this example, we are using the weighted average.

*Statistics for Compensation: A Practical Guide to Compensation Analysis,* By John H. Davis
Copyright © 2011 John Wiley & Sons Inc.

**TABLE 13.1    SURVEY DATA ON FINANCE DEPARTMENT JOBS**

| Job | Company Job Title | Grade | Survey | Market Weighted Average Annual Base Pay |
|---|---|---|---|---|
| 1 | Office Assistant General | 21 | A | 34,286 |
| 2 | Office Services Assistant | 21 | A | 36,190 |
| 3 | Receptionist | 21 | B | 39,664 |
| 4 | Accts Payable Acctg Assistant | 22 | A | 37,714 |
| 5 | Accts Receivable Acctg Assistant | 22 | A | 40,857 |
| 6 | Admin Assistant | 22 | A | 45,571 |
| 7 | General Ledger Acctg Assistant | 22 | B | 52,356 |
| 8 | Sr. Accts Payable Acctg Assistant | 23 | A | 40,857 |
| 9 | Sr. Admin Assistant | 23 | A | 47,142 |
| 10 | Sr. Accts Receivable Acctg Assistant | 23 | B | 49,183 |
| 11 | Accountant—Entry | 24 | A | 47,142 |
| 12 | Financial Analyst—Entry | 24 | A | 50,000 |
| 13 | Sr. Admin Assistant | 24 | A | 51,429 |
| 14 | Sr. Admin Assistant | 24 | B | 53,366 |
| 15 | Data Coordinator | 24 | B | 54,808 |
| 16 | Executive Assistant | 24 | B | 64,904 |
| 17 | Financial Analyst | 25 | A | 54,286 |
| 18 | Accountant | 25 | A | 61,429 |
| 19 | Sr. Financial Analyst | 26 | A | 61,429 |
| 20 | Sr. Accountant | 26 | B | 64,904 |
| 21 | Sr. Financial Analyst | 26 | B | 70,673 |
| 22 | Sr. Financial Analyst | 26 | C | 88,995 |
| 23 | Sr. Accounting Specialist | 27 | B | 76,442 |
| 24 | Sr. Financial Specialist | 27 | B | 86,538 |
| 25 | Supv. Financial Reports | 28 | A | 68,571 |
| 26 | Supv. Accts Payable | 28 | B | 92,308 |
| 27 | Supv. Accts Receivable | 28 | C | 99,044 |
| 28 | Manager Info Sys Interface | 29 | B | 93,750 |
| 29 | Manager Financial Analysis | 29 | C | 107,656 |
| 30 | Manager Accounting | 29 | C | 123,445 |

In addition, you may have other data on your spreadsheet that are not shown here, such as the survey job title, the survey job number, the page numbers in the survey of the job description and of the data extracted, the internal company job number, and so on. It is useful to have these "trails" so that you can easily find the source data in case of questions and for use in subsequent years.

For the surveys themselves, you have the dates of data and annual movements, as shown in Table 13.2. Typically, these items are reported in the surveys.

If the survey does not specify the date of the data, you need to contact the survey provider for that date.

**TABLE 13.2    AGING TABLE: INITIAL ENTRIES**

| Survey | Date of Data | Annual Movement (%) | Date Aged To | Months Aged | Aging Factor |
|--------|--------------|---------------------|--------------|-------------|--------------|
| A | 01-1-11 | 5.0 | | | |
| B | 01-7-11 | 8.0 | | | |
| C | 01-4-11 | 6.0 | | | |

If the survey does not specify the general annual movement of the jobs in the survey, there are mainly three ways to get it. First, contact the survey provider for it. They may have collected general salary increase budget information and would be able to summarize it for you. Second, call the companies in the survey, ask them what their salary increase budgets are, and summarize it on your own. Third, conduct or participate in a survey focusing on gathering and summarizing salary increase budget information.

2. *Age Data to a Common Date.* The data are aged to a common date so that they are comparable in time. It would not make sense to use data, for example, from one survey that is a year old and from another survey that is 6 months old without any adjustments between the two surveys, because salaries change over time. The common date aged to is usually the start of your salary administration year (salary plan year). We will age these survey data to January 1, 2012.

The data from each survey will be aged according to the annual movement for that survey during the time from the date of the data to the beginning of your plan year. For example, for survey B, we will age the data from July 1, 2011 to January 1, 2012, a period of 6 months. At an annual rate of 8.0%, we will age the data 4.0% by multiplying all the data from survey B by an aging factor of 1.040.

The completed Table 13.2 should look like Table 13.3.

We are now ready to use this table. We will multiply the survey data from survey A, B, and C by their corresponding aging factors: 1.050, 1.040, and 1.045, respectively. For example, for job 1, we multiply 34,286 by 1.050 to get 36,000. The results are shown in Table 13.4.

**TABLE 13.3    AGING TABLE: COMPLETED**

| Survey | Date of Data | Annual Movement | Date Aged To | Months Aged | Aging Factor |
|--------|--------------|-----------------|--------------|-------------|--------------|
| A | 01-1-11 | 5.0% | 01-1-12 | 12.0 | 1.050 |
| B | 01-7-11 | 8.0% | 01-1-12 | 6.0 | 1.040 |
| C | 01-4-11 | 6.0% | 01-1-12 | 9.0 | 1.045 |

**TABLE 13.4    AGED SURVEY DATA ON FINANCE DEPARTMENT JOBS**

| Job | Company Job Title | Grade | Survey | Market Weighted Average Annual Base Pay | Aging Factor | Aged Market Weighted Average Annual Base Pay |
|-----|-------------------|-------|--------|------|------|------|
| 1 | Office Assistant General | 21 | A | 34,286 | 1.050 | 36,000 |
| 2 | Office Services Assistant | 21 | A | 36,190 | 1.050 | 38,000 |
| 3 | Receptionist | 21 | B | 39,664 | 1.040 | 41,250 |
| 4 | Accts Payable Acctg Asst | 22 | A | 37,714 | 1.050 | 39,600 |
| 5 | Accts Receivable Acctg Asst | 22 | A | 40,857 | 1.050 | 42,900 |
| 6 | Admin Asst | 22 | A | 45,571 | 1.050 | 47,850 |
| 7 | General Ledger Acctg Asst | 22 | B | 52,356 | 1.040 | 54,450 |
| 8 | Sr. Accts Payable Acctg Asst | 23 | A | 40,857 | 1.050 | 42,900 |
| 9 | Sr. Admin Asst | 23 | A | 47,142 | 1.050 | 49,500 |
| 10 | Sr. Accts Receivable Acctg Asst | 23 | B | 49,183 | 1.040 | 51,151 |
| 11 | Accountant—Entry | 24 | A | 47,142 | 1.050 | 49,500 |
| 12 | Financial Analyst—Entry | 24 | A | 50,000 | 1.050 | 52,500 |
| 13 | Sr. Admin Assistant | 24 | A | 51,429 | 1.050 | 54,000 |
| 14 | Sr. Admin Assistant | 24 | B | 53,366 | 1.040 | 55,500 |
| 15 | Data Coordinator | 24 | B | 54,808 | 1.040 | 57,000 |
| 16 | Executive Assistant | 24 | B | 64,904 | 1.040 | 67,500 |
| 17 | Financial Analyst | 25 | A | 54,286 | 1.050 | 57,000 |
| 18 | Accountant | 25 | A | 61,429 | 1.050 | 64,500 |
| 19 | Sr. Financial Analyst | 26 | A | 61,429 | 1.050 | 64,500 |
| 20 | Sr. Accountant | 26 | B | 64,904 | 1.040 | 67,500 |
| 21 | Sr. Financial Analyst | 26 | B | 70,673 | 1.040 | 73,500 |
| 22 | Sr. Financial Analyst | 26 | C | 88,995 | 1.045 | 93,000 |
| 23 | Sr. Accounting Specialist | 27 | B | 76,442 | 1.040 | 79,500 |
| 24 | Sr. Financial Specialist | 27 | B | 86,538 | 1.040 | 90,000 |
| 25 | Supv. Financial Reports | 28 | A | 68,571 | 1.050 | 72,000 |
| 26 | Supv. Accts Payable | 28 | B | 92,308 | 1.040 | 96,000 |
| 27 | Supv. Accts Receivable | 28 | C | 99,044 | 1.045 | 103,500 |
| 28 | Manager Info Sys Interface | 29 | B | 93,750 | 1.040 | 97,500 |
| 29 | Manager Financial Analysis | 29 | C | 107,656 | 1.045 | 112,501 |
| 30 | Manager Accounting | 29 | C | 123,445 | 1.045 | 129,000 |

3. *Create an Integrated Market Model.* We will now repeat the process we used in Chapter 8. First, we will plot the aged market weighted average ($y$) versus grade ($x$) for the matched jobs, as shown in Figure 13.1.

There is a relationship, which is positive, and comparing this relationship with the charts of functional forms in Figures 6.8 and 6.9, we will use an exponential model to describe the relationship. We will thus create the following exponential model using the techniques in Chapter 8.

Recall that the equation for an exponential model is $\log y = a + bx$ and that the prediction equation (or model) $\log y' = a + bx$ is derived from the data points using the method of least squares.

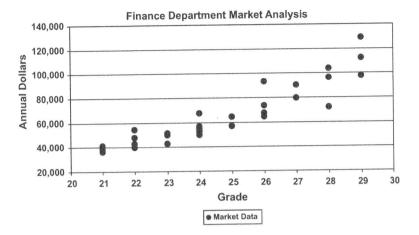

**FIGURE 13.1     AGED MARKET PAY AND GRADE**

The resulting prediction equation and coefficient of determination are

$$\log y' = 3.417 + 0.0556x$$

or

$$y' = 10^{(3.417+0.0556x)}$$

and

$$r^2 = 88.8\% \text{ (between log pay and grade)}$$

A linear model could fit the data. Indeed, if you fit a straight line, you would get an $r^2$ of 84.0%. However, since we will be using the model to create a salary structure, the exponential model will result in a constant *percentage* midpoint progression. (A linear model results in a constant *dollar* midpoint progression.)

The constant percentage progression from one grade to the next is

$$10^{\text{Slope}} = 10^{0.0556} = 1.137$$

On MOTHing (minus one times hundred) we get a 13.7% progression from one grade to the next.

$$(1.137 - 1) \times 100 = 0.137 \times 100 = 13.7\%$$

Applying the prediction equation, we calculate the predicted average market pay for each grade, as shown in Table 13.5, and plot them on the chart as a (curved) line, as shown in Figure 13.2.

| TABLE 13.5 | PREDICTED AVERAGE MARKET PAY AND GRADE |
| --- | --- |

| Grade | Predicted Average Market Pay |
| --- | --- |
| 21 | 38,424 |
| 22 | 43,672 |
| 23 | 49,636 |
| 24 | 56,416 |
| 25 | 64,121 |
| 26 | 72,879 |
| 27 | 82,832 |
| 28 | 94,146 |
| 29 | 107,004 |

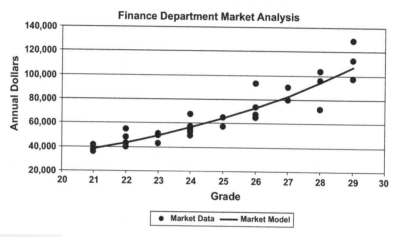

**FIGURE 13.2    PREDICTED AVERAGE MARKET PAY AND GRADE**

The line goes through the data and $r^2$ is very good. We decide that this is a good model.

Our model nicely *describes the relationship between external value (market pay) and internal value (grade)*. The graphical display gives us confidence in the model.

As an organization we want our pay programs to be "internally equitable and externally competitive," or "internally fair and externally fair." This chart and the associated equation are where these two concepts are integrated together. In this case, external value is represented by market pay, the $y$-variable, and internal value is represented by grade, the $x$-variable.

We will use this model to create a market-based salary structure. We will first designate the predicted average market pay as the market-based midpoint.

The next step is to add minimum and maximum lines. We will select a range spread of 60% for this example. This means that the maximum is 60% more than the minimum. This amount provides ample pay growth opportunities within a grade as well as delimits what we will pay for jobs in a grade. The most predominant range spread is 50%, although range spreads go from 30% to over 70% in practice. Here is how to calculate the minimum and maximum starting with a midpoint and range spread, where the range spread is expressed as a decimal (e.g., 60% = 0.60).

$$\text{Minimum} = \text{midpoint}/ \left(1 + \frac{\text{range spread}}{2}\right)$$

$$\text{Maximum} = \text{minimum} \times (1 + \text{range spread})$$

For example, for grade 21,

$$\text{Minimum} = 38{,}424/ \left(1 + \frac{0.60}{2}\right) = \frac{38{,}424}{1.30} = 29{,}557$$

$$\text{Maximum} = (29{,}557)(1 + 0.60) = (29{,}557)(1.60) = 47{,}291$$

We calculate the structure in Table 13.6 and plot it as (curved) lines in Figure 13.3.

What this market-based salary structure represents is a description of the span of average market pay as it relates to internal grades. The middle line is the predicted average pay for a given grade, and the top and bottom lines form a basis for a market-based salary structure (assuming a desired range spread of 60%).

So, now we have a market-based salary structure as of January 1, 2012.

**TABLE 13.6    MARKET-BASED SALARY STRUCTURE**

| Grade | Minimum | Midpoint | Maximum |
|-------|---------|----------|---------|
| 21 | 29,557 | 38,424 | 47,291 |
| 22 | 33,594 | 43,672 | 53,750 |
| 23 | 38,182 | 49,636 | 61,091 |
| 24 | 43,397 | 56,416 | 69,435 |
| 25 | 49,324 | 64,121 | 78,918 |
| 26 | 56,061 | 72,879 | 89,698 |
| 27 | 63,717 | 82,832 | 101,947 |
| 28 | 72,420 | 94,146 | 115,872 |
| 29 | 82,311 | 107,004 | 131,698 |

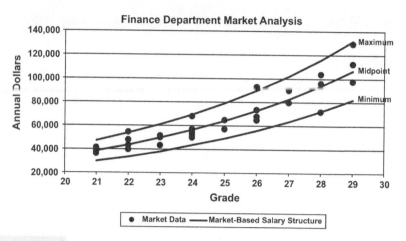

**FIGURE 13.3** **MARKET-BASED SALARY STRUCTURE**

Before we go on, we conduct three additional analyses as needed that should be done as part of a complete structure development.

- Identify outliers. If present, verify their validity.
- Identify if any particular survey is predominantly on the high side or low side. If so, verify the appropriateness of the market that survey represents.
- Identify if any particular family of jobs is predominantly on the high side or low side. If so, determine if a separate structure is needed for that family.

We will assume this has been done and there are no changes.

4. *Compare Employee Pay with the Market Model.* This step involves comparing the salaries of all employees in these grades, irrespective of whether their jobs were surveyed or not, with the predicted average market pay (the midpoint line). We will have a tabular as well as a graphical display. At this point, we repeat the assumption underlying this approach.

> The integrated market model approach assumes that for each grade, the predicted average market pay from the model is a good estimator for the average market pay for all jobs in that grade, whether or not they were surveyed.

This assumption is based on the internal valuing system (grades) that internally values all jobs in a given grade equivalently. In most cases, the market and a company value jobs similarly, as evidenced by a typically strong relationship (i.e., high coefficient of determination, $r^2$) between survey pay and grades for benchmark jobs.

It is under this assumption that we are able to compare the salaries of *all* employees with the predicted average market pay for their jobs to determine an overall market position.

Since the market model is based on data aged to January 1, 2012, the employee data should also be as of that date, before any raises are given in the new plan year. If you know the projected or planned raises for individual employees between the date of the employee data you have at hand and the date you have aged the market data to, then it would be appropriate to take those into account. Otherwise just note that there are differences in the dates of the data and any implications these may lead to.

The employee data are in Table 13.7, and also in Data Set 28.

**TABLE 13.7    EMPLOYEE SALARIES AND GRADE**

| Emp. No. | Grade | Pay | Emp. No. | Grade | Pay |
|---|---|---|---|---|---|
| 1 | 21 | 26,265 | 35 | 25 | 66,435 |
| 2 | 21 | 27,810 | 36 | 25 | 66,435 |
| 3 | 21 | 37,080 | 37 | 25 | 67,980 |
| 4 | 21 | 48,805 | 38 | 25 | 69,525 |
| 5 | 22 | 30,900 | 39 | 25 | 71,070 |
| 6 | 22 | 35,535 | 40 | 25 | 71,070 |
| 7 | 22 | 40,170 | 41 | 25 | 72,615 |
| 8 | 22 | 41,715 | 42 | 25 | 72,615 |
| 9 | 22 | 43,260 | 43 | 25 | 74,160 |
| 10 | 22 | 44,805 | 44 | 26 | 49,440 |
| 11 | 23 | 35,535 | 45 | 26 | 57,165 |
| 12 | 23 | 40,170 | 46 | 26 | 58,710 |
| 13 | 23 | 41,715 | 47 | 26 | 60,255 |
| 14 | 23 | 43,260 | 48 | 26 | 61,800 |
| 15 | 23 | 46,350 | 49 | 26 | 69,525 |
| 16 | 23 | 50,985 | 50 | 26 | 77,250 |
| 17 | 23 | 52,530 | 51 | 26 | 80,340 |
| 18 | 23 | 55,620 | 52 | 26 | 81,885 |
| 19 | 24 | 37,080 | 53 | 26 | 106,605 |
| 20 | 24 | 41,715 | 54 | 27 | 71,070 |
| 21 | 24 | 44,805 | 55 | 27 | 74,160 |
| 22 | 24 | 47,895 | 56 | 27 | 77,250 |
| 23 | 24 | 52,530 | 57 | 27 | 80,340 |
| 24 | 24 | 58,710 | 58 | 27 | 83,430 |
| 25 | 24 | 60,255 | 59 | 27 | 89,610 |
| 26 | 24 | 63,345 | 60 | 28 | 74,160 |
| 27 | 24 | 66,435 | 61 | 28 | 86,520 |
| 28 | 24 | 75,705 | 62 | 28 | 100,425 |
| 29 | 25 | 43,260 | 63 | 28 | 101,970 |
| 30 | 25 | 46,350 | 64 | 28 | 111,240 |
| 31 | 25 | 49,440 | 65 | 29 | 94,245 |
| 32 | 25 | 50,985 | 66 | 29 | 108,150 |
| 33 | 25 | 52,530 | 67 | 29 | 122,055 |
| 34 | 25 | 54,075 | 68 | 29 | 138,070 |

**TABLE 13.8    EMPLOYEE SALARIES AND MARKET-BASED SALARY STRUCTURE**

| Emp. No. | Grade | Pay | Market-Based Salary Structure | | | Pay as % of Mid | Dollars Under Min | Dollars Over Max |
|---|---|---|---|---|---|---|---|---|
| | | | Minimum | Midpoint | Maximum | | | |
| 1 | 21 | 26,265 | 29,557 | 38,424 | 47,291 | 68 | 3,292 | – |
| 2 | 21 | 27,810 | 29,557 | 38,424 | 47,291 | 72 | 1,747 | – |
| 3 | 21 | 37,080 | 29,557 | 38,424 | 47,291 | 97 | – | – |
| 4 | 21 | 48,805 | 29,557 | 38,424 | 47,291 | 127 | – | 1,514 |
| 5 | 22 | 30,900 | 33,594 | 43,672 | 53,750 | 71 | 2,694 | – |
| 6 | 22 | 35,535 | 33,594 | 43,672 | 53,750 | 81 | – | – |
| : | : | : | : | : | : | : | : | : |
| : | : | : | : | : | : | : | : | : |
| 67 | 29 | 122,055 | 82,311 | 107,004 | 131,698 | 114 | – | – |
| 68 | 29 | 138,070 | 82,311 | 107,004 | 131,698 | 129 | – | 6,372 |

To determine the market position along with other useful information, we need to create Table 13.8.

For purposes of illustration, we have just indicated an employee number. When doing this analysis for your company, you would probably have additional fields of data such as employee name, job title, job code, department, location, department and/or location code, and so on that allow additional analyses.

After the employee identification, we have the grade and pay of the employee. The next three columns contain the market-based salary structure we have just created.

Next is an indication of the employee's pay as it relates to the midpoint. Finally, if the employee's pay is under the minimum, we calculate by how much. Similarly, if the employee's pay is over the maximum, we calculate by how much.

This table furnishes the raw data to identify the overall market position. We use these data to create a summary table, Table 13.9, calculating subtotals by grade.

In the total row, the total pay of 4,335,200 is the total annual pay of the 68 BPD employees as of January 1, 2012. The total market midpoint of 4,463,319 is the total annual midpoint of the grades of the market-based salary structure we created for all 68 employees.

We see that in terms of annual dollars, the BPD total pay is 128,119 below the total market midpoint. The difference is the total market midpoint minus the total pay.

We now want to calculate two percentages. First are the two definitions:

"Market position" is the % we are from the market midpoint. The market midpoint is the reference. In Table 13.8, this is in the column headed % from Market Mid.

**TABLE 13.9    MARKET ANALYSIS**

Finance Department Market Analysis: Comparison of Employee Pay with the Market-Based Salary Structure

| Grade | No. of Employees | Total Pay | Total Market Midpoint | Difference | % from Market Mid | % to Meet Market Mid | No. of Employees Under Min | No. of Employees Over Max | Dollars Under Min | Dollars Over Max |
|---|---|---|---|---|---|---|---|---|---|---|
| 21 | 4 | 139,960 | 153,696 | 13,736 | −8.9 | 9.8 | 2 | 1 | 5,039 | 1,514 |
| 22 | 6 | 236,385 | 262,032 | 25,647 | −9.8 | 10.8 | 1 | 0 | 2,694 | 0 |
| 23 | 8 | 366,165 | 397,088 | 30,923 | −7.8 | 8.4 | 1 | 0 | 2,647 | 0 |
| 24 | 10 | 548,475 | 564,160 | 15,685 | −2.8 | 2.9 | 2 | 1 | 7,999 | 6,270 |
| 25 | 15 | 928,545 | 961,815 | 33,270 | −3.5 | 3.6 | 2 | 0 | 9,038 | 0 |
| 26 | 10 | 702,975 | 728,790 | 25,815 | −3.5 | 3.7 | 1 | 1 | 6,621 | 16,907 |
| 27 | 6 | 475,860 | 496,992 | 21,132 | −4.3 | 4.4 | 0 | 0 | 0 | 0 |
| 28 | 5 | 474,315 | 470,730 | −3,585 | 0.8 | −0.8 | 0 | 0 | 0 | 0 |
| 29 | 4 | 462,520 | 428,016 | −34,504 | 8.1 | −7.5 | 0 | 1 | 0 | 6,372 |
| **Total** | **68** | **4,335,200** | **4,463,319** | **128,119** | **−2.9** | **3.0** | **9** | **4** | **34,038** | **31,063** |

"Budget needed to meet the market" is the % increase (decrease) in our pay needed to meet the market. Our pay is the reference. In Table 13.8, this is in the column headed % to Meet Market Mid.

Recall from Chapter 2 how to calculate the percent difference between the two terms.

$$\text{Percent difference} = \frac{\text{data} - \text{reference}}{\text{reference}} \times 100$$

For the market position (% from Mkt Mid), the data are our total pay and the reference is the total market midpoint.

$$\% \text{ from Mkt Mid} = \frac{4,335,200 - 4,463,319}{4,463,319} \times 100$$

$$= \frac{-128,119}{4,463,319} \times 100 = -0.029 \times 100 = -2.9\%$$

For the budget needed to meet the market (% to meet Mkt Mid), the data are the total market midpoint and the reference is our total pay.

$$\% \text{ to meet Mkt Mid} = \frac{4,463,319 - 4,335,200}{4,335,200} \times 100$$

$$= \frac{128,119}{4,335,200} \times 100 = 0.030 \times 100 = 3.0\%$$

Overall, the company's finance department employees are 2.9% below the market midpoint, which represents the market weighted average. It would take a 3.0% budget to equal the market midpoint. This table also indicates that some grades have different market positions than others, as well as summarizes how many employees have pay outside the pay structure and by how much.

If desired, you can have subtotals by job family, location, and so on.

The next step is to present the results in a graphical form. In Figure 13.4, you compare the employee pay with the market-based salary structure you have just created.

This chart shows that the bulk of the employees' pay is within the market-based salary structure, and it also highlights the pay that is outside the structure.

As some employees are outside the structure, before presenting these results you research those data points and identify who they are and why they are like they are.

In summary, we have completed the following four steps to conduct a market analysis where we have grades, resulting in an identification of the market position, the creation of a market-based salary structure, and the development of

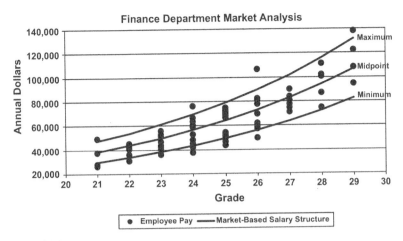

**FIGURE 13.4    EMPLOYEE SALARIES AND MARKET-BASED SALARY STRUCTURE**

an initial market-based salary increase budget, all as of the beginning of the new plan year.

1. Gather market data
2. Age data to a common date
3. Create an integrated market model
4. Compare employee pay with the market model

For the final market-based salary increase budget, we complete Table 13.10.

The catch up is 3.0%, which is our initial market-based salary increase budget as of the beginning of the new plan year. For this group of finance jobs, you obtain from the survey providers the anticipated market movement for 2012. This is what the survey participants think (guess) will happen in the future. The number you get from the providers is 1.0%. BPD's pay policy is to be 2.0% above the market at the end of the plan year. Table 13.11 is the completed table.

The final market-based salary increase budget is 6.0%. For the final recommendations, you now take into account the "soft" information described in Section 11.4.

**TABLE 13.10    TEMPLATE FOR FINAL MARKET-BASED SALARY INCREASE BUDGET**

| | |
|---|---|
| Catch-up (or fall back) | —— |
| Anticipated market movement | —— |
| Pay policy | —— |
| **Total** | —— |

| TABLE 13.11 | FINAL MARKET-BASED SALARY INCREASE BUDGET |
|---|---|

| Catch-up (or fall back) | 3.0% |
| Anticipated market movement | 1.0% |
| Pay policy | 2.0% |
| **Total** | **6.0 %** |

## RELATED TOPICS IN THE APPENDIX

A.13   Range Spread Relationships

## PRACTICE PROBLEMS

**13.1**   BPD has a mine in Gillette, Wyoming. It is October 2012 and you are preparing to recommend a salary increase budget for 2013. Your salary administration year is a calendar year. You give raises throughout the year on anniversary dates or promotion dates. Your compensation philosophy is that you want your average pay to be 3.0% below the market pay at the end of the plan year, because you have a generous benefits plan and you are managing on total employee costs. You have recently obtained three surveys for some of the professional and managerial jobs in the mine. The data you extract from the surveys are the weighted averages. These are shown in Table 13.12, and also in Data Set 29. The date of the data for Survey A is July 1, 2012, for Survey B is April 1, 2012, and for Survey C is January 1, 2012. From the surveys, the general market movement for 2012 for Survey A, B, and C annually is 3.6%, 4%, and 6.0%, respectively.

How much will you age the data from Survey A? From Survey B? From Survey C?

| TABLE 13.12 | SURVEY DATA ON MINING JOBS |
|---|---|

| Job | Company Job Title | Grade | Survey | Market Weighted Average |
|---|---|---|---|---|
| 1 | Mining Engineer 1 | 41 | A | 44,399 |
| 2 | Production Scheduler | 41 | A | 48,586 |
| 3 | Accountant 2 | 42 | B | 38,678 |
| 4 | Chemist 2 | 42 | C | 42,142 |
| 5 | Geologist 2 | 42 | A | 50,011 |
| 6 | Cost Accountant 2 | 42 | B | 52,469 |
| 7 | Mining Engineer 2 | 42 | A | 61,699 |
| 8 | Accountant 3 | 43 | B | 46,411 |
| 9 | Geologist 3 | 43 | A | 54,842 |
| 10 | Network Engineer 2 | 43 | C | 57,194 |
| 11 | HR Generalist 3 | 43 | C | 57,331 |
| 12 | Mining Engineer 3 | 43 | A | 65,315 |
| 13 | Billing Supervisor | 44 | B | 51,762 |
| 14 | Accountant 4 | 44 | B | 60,367 |

**TABLE 13.12    (*CONTINUED*)**

| Job | Company Job Title | Grade | Survey | Market Weighted Average |
|-----|-------------------|-------|--------|-------------------------|
| 15 | Geologist 4 | 44 | A | 69,293 |
| 16 | Accounting Supervisor | 45 | B | 58,597 |
| 17 | Mining Engineer 4 | 45 | A | 86,716 |
| 18 | Lead Mining Engineer | 46 | A | 61,211 |
| 19 | Mine Foreman | 46 | A | 75,912 |
| 20 | Operations Supervisor | 46 | A | 84,648 |
| 21 | Project Engr Manager | 47 | A | 81,160 |
| 22 | Health and Safety Manager | 47 | A | 92,537 |
| 23 | Accounting Manager | 48 | B | 110,577 |
| 24 | Engineering Supervisor | 49 | A | 94,642 |
| 25 | Sr. Engineering Manager | 50 | A | 138,444 |
| 26 | Mine Manager | 51 | A | 145,205 |

**13.2**  Age the survey data appropriately.

**13.3**  Plot the data. What kind of relationship is shown by the plot?

**13.4**  Conduct a regression. State the market model equation, the coefficient of determination, and the standard error of estimate.

**13.5**  Plot, interpret, and evaluate the model.

**13.6**  You decide to use a range spread of 50%. Setting the market model as the midpoint, create a market-based salary structure and plot it.

**13.7**  The employee data for the 85 professional and managerial employees are in Table 13.13 and also in Data Set 30. Complete Table 13.14. Create a plot that compares employee pay with the market-based salary structure.

**TABLE 13.13    EMPLOYEE SALARIES AND GRADES**

| Emp. No. | Grade | Pay | Emp. No. | Grade | Pay |
|----------|-------|-----|----------|-------|-----|
| 1 | 41 | 49,589 | 44 | 44 | 57,091 |
| 2 | 41 | 32,966 | 45 | 44 | 65,198 |
| 3 | 41 | 50,926 | 46 | 45 | 61,503 |
| 4 | 41 | 35,737 | 47 | 45 | 69,445 |
| 5 | 41 | 48,079 | 48 | 45 | 64,933 |
| 6 | 41 | 33,131 | 49 | 45 | 62,706 |
| 7 | 41 | 40,944 | 50 | 45 | 74,966 |
| 8 | 41 | 37,412 | 51 | 45 | 71,783 |
| 9 | 41 | 50,785 | 52 | 45 | 66,652 |
| 10 | 42 | 53,315 | 53 | 45 | 69,482 |
| 11 | 42 | 44,145 | 54 | 45 | 59,057 |
| 12 | 42 | 43,293 | 55 | 45 | 75,953 |
| 13 | 42 | 41,579 | 56 | 46 | 81,003 |
| 14 | 42 | 55,839 | 57 | 46 | 80,001 |

(*continued*)

**TABLE 13.13** (*CONTINUED*)

| Emp. No. | Grade | Pay | Emp. No. | Grade | Pay |
|---|---|---|---|---|---|
| 15 | 42 | 51,452 | 58 | 46 | 81,396 |
| 16 | 42 | 52,710 | 59 | 46 | 77,193 |
| 17 | 42 | 46,133 | 60 | 46 | 73,153 |
| 18 | 42 | 40,377 | 61 | 46 | 68,169 |
| 19 | 42 | 41,164 | 62 | 46 | 72,018 |
| 20 | 42 | 55,621 | 63 | 46 | 72,993 |
| 21 | 42 | 49,554 | 64 | 47 | 88,756 |
| 22 | 42 | 44,230 | 65 | 47 | 96,272 |
| 23 | 42 | 58,548 | 66 | 47 | 96,463 |
| 24 | 42 | 39,699 | 67 | 47 | 85,399 |
| 25 | 43 | 45,137 | 68 | 47 | 79,192 |
| 26 | 43 | 45,294 | 69 | 47 | 93,717 |
| 27 | 43 | 63,014 | 70 | 47 | 84,452 |
| 28 | 43 | 61,366 | 71 | 47 | 91,343 |
| 29 | 43 | 46,962 | 72 | 47 | 92,344 |
| 30 | 43 | 56,621 | 73 | 48 | 101,977 |
| 31 | 43 | 52,560 | 74 | 48 | 102,711 |
| 32 | 43 | 51,940 | 75 | 48 | 105,316 |
| 33 | 43 | 62,143 | 76 | 48 | 107,232 |
| 34 | 43 | 62,517 | 77 | 48 | 96,502 |
| 35 | 43 | 54,481 | 78 | 48 | 89,574 |
| 36 | 44 | 69,606 | 79 | 48 | 96,360 |
| 37 | 44 | 69,485 | 80 | 49 | 105,383 |
| 38 | 44 | 65,384 | 81 | 49 | 109,437 |
| 39 | 44 | 58,981 | 82 | 49 | 119,611 |
| 40 | 44 | 61,168 | 83 | 50 | 127,978 |
| 41 | 44 | 59,550 | 84 | 50 | 125,578 |
| 42 | 44 | 68,277 | 85 | 51 | 130,733 |
| 43 | 44 | 52,525 | | | |

**TABLE 13.14** TEMPLATE FOR EMPLOYEE SALARIES AND MARKET-BASED SALARY STRUCTURE

| Emp. No. | Grade | Pay | Minimum | Midpoint | Maximum | Pay as % of Midpoint | Dollars Under Min | Dollars Over Max |
|---|---|---|---|---|---|---|---|---|
| 1 | 41 | 49,589 | | | | | | |
| 2 | 41 | 32,966 | | | | | | |
| 3 | 41 | 50,926 | | | | | | |
| : | : | : | | | | | | |
| : | : | : | | | | | | |

**13.8** Complete the summary table shown in Table 13.15.

**13.9** For all three surveys, you estimate that the market will move 4% annually during 2013. What is the market-based salary increase budget recommendation?

**TABLE 13.15    TEMPLATE FOR MARKET ANALYSIS**

Market Analysis for Gillette Mine: Comparison of Employee Pay with the Market-Based Salary Structure

| Grade | No. of Employees | Total Pay | Total Market Midpoint | Difference | % from Mkt Mid | % to Meet Mkt Mid | No. of Employees Under Min | No. of Employees Over Max | Dollars Under Min | Dollars Over Max |
|---|---|---|---|---|---|---|---|---|---|---|
| 41 | | | | | | | | | | |
| 42 | | | | | | | | | | |
| 43 | | | | | | | | | | |
| 44 | | | | | | | | | | |
| 45 | | | | | | | | | | |
| 46 | | | | | | | | | | |
| 47 | | | | | | | | | | |
| 48 | | | | | | | | | | |
| 49 | | | | | | | | | | |
| 50 | | | | | | | | | | |
| 51 | | | | | | | | | | |
| Total | | | | | | | | | | |

**13.10**   Your executives approve a salary increase budget of 3.5%. By how much do you adjust the market-based salary structure to have the average pay be at the midpoint after raises are given?

# Integrated Market Model: Maturity Curve

We will use the four steps listed in Chapter 11.

1. *Gather Market Data.*
2. *Age Data to a Common Date.*
   We have data shown in Table 14.1, also shown in Data Set 31, from a major maturity curve survey that reports the five standard percentiles for each year since bachelor of science degree (YSBS). We have already aged the data to January 1, 2012.
3. *Create an Integrated Market Model.* We will be repeating the process we used in Chapter 9. First, we plot the aged percentiles ($y$) versus YSBS ($x$) in Figure 14.1.
   There is a relationship for each percentile, each one is positive, and comparing these relationships with the charts of functional forms in Figures 6.8

*Statistics for Compensation: A Practical Guide to Compensation Analysis,* By John H. Davis
Copyright © 2011 John Wiley & Sons Inc.

**TABLE 14.1**   SURVEY PERCENTILES AND YSBS

| YSBS | Actual 10th Percentile | Actual 25th Percentile | Actual 50th Percentile | Actual 75th Percentile | Actual 90th Percentile |
|---|---|---|---|---|---|
| 1 | 56,829 | 60,349 | 64,174 | 69,601 | 87,385 |
| 2 | 57,000 | 61,800 | 66,360 | 72,431 | 82,944 |
| 3 | 60,348 | 63,604 | 68,688 | 75,904 | 85,764 |
| 4 | 62,400 | 67,200 | 72,925 | 79,800 | 90,727 |
| 5 | 63,408 | 69,906 | 78,000 | 86,400 | 98,400 |
| 6 | 64,200 | 71,842 | 81,600 | 91,519 | 103,200 |
| 7 | 66,720 | 75,600 | 86,402 | 99,480 | 109,368 |
| 8 | 66,600 | 77,203 | 89,899 | 102,000 | 112,800 |
| 9 | 70,376 | 80,496 | 92,407 | 104,022 | 116,015 |
| 10 | 72,735 | 83,291 | 95,760 | 106,800 | 117,792 |
| 11 | 71,580 | 82,811 | 96,000 | 107,400 | 118,500 |
| 12 | 73,188 | 84,874 | 97,200 | 108,000 | 119,795 |
| 13 | 75,600 | 87,470 | 99,360 | 110,444 | 121,920 |
| 14 | 75,600 | 87,600 | 100,585 | 112,669 | 124,834 |
| 15 | 78,103 | 89,520 | 102,240 | 114,360 | 127,550 |
| 16 | 79,488 | 91,200 | 104,582 | 116,160 | 127,248 |
| 17 | 81,578 | 93,059 | 106,200 | 118,058 | 131,421 |
| 18 | 81,120 | 93,359 | 107,182 | 120,000 | 133,378 |
| 19 | 83,400 | 96,000 | 108,120 | 121,440 | 135,964 |
| 20 | 83,849 | 96,000 | 109,200 | 122,370 | 137,933 |
| 21 | 86,043 | 97,800 | 110,401 | 123,660 | 136,776 |
| 22 | 84,972 | 98,789 | 110,760 | 123,840 | 138,120 |
| 23 | 85,970 | 99,403 | 112,032 | 127,136 | 141,739 |
| 24 | 86,514 | 99,840 | 112,873 | 126,794 | 142,705 |
| 25 | 84,144 | 99,066 | 112,680 | 127,200 | 141,681 |
| 26 | 86,192 | 100,080 | 111,601 | 125,676 | 141,073 |
| 27 | 90,000 | 102,593 | 115,555 | 130,804 | 146,653 |
| 28 | 87,280 | 100,320 | 114,000 | 129,382 | 146,076 |
| 29 | 89,712 | 102,083 | 116,400 | 130,061 | 147,600 |
| 30 | 88,801 | 103,200 | 116,179 | 130,192 | 146,752 |
| 31 | 89,970 | 103,440 | 116,784 | 132,060 | 146,616 |
| 32 | 91,149 | 103,801 | 117,420 | 133,160 | 149,638 |
| 33 | 94,215 | 105,536 | 119,040 | 134,965 | 150,234 |
| 34 | 91,634 | 104,400 | 117,869 | 133,096 | 147,929 |
| 35 | 97,603 | 107,170 | 120,240 | 135,608 | 149,323 |
| 36 | 95,387 | 108,000 | 121,440 | 138,528 | 153,935 |
| 37 | 95,645 | 108,374 | 122,063 | 137,066 | 151,008 |
| 38 | 96,324 | 110,196 | 121,723 | 137,741 | 157,386 |
| 39 | 96,531 | 108,982 | 123,602 | 139,926 | 154,786 |
| 40 | 96,072 | 108,240 | 123,288 | 136,999 | 153,683 |

and 6.9, we will use cubic models to describe the relationships. We create the following cubic models using the techniques described in Chapter 9.

$$y' = a + bx + cx^2 + dx^3$$

$y'$ = predicted average percentile of pay

$x$ = YSBS (years since BS degree)

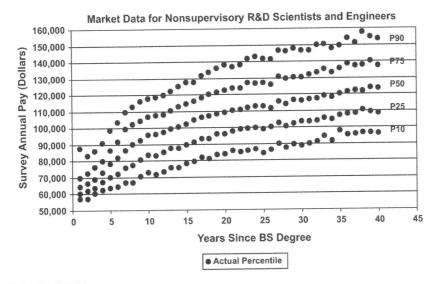

**FIGURE 14.1    AGED PERCENTILES AND YSBS**

The coefficients for the five percentile prediction equations and the coefficients of determination are shown in Table 14.2.

Applying the equations, we calculate the predicted average market pay for each YSBS and plot them on the chart as lines. For example, for YSBS = 2 for the 10th percentile,

$$y' = 54,544 + (2,216 \times 2) - (46.86 \times 2^2) + (0.4713 \times 2^3)$$

$$= 54,544 + 4,432 - 187.44 + 3.77 = 57,792$$

We also calculate the ratio of the 90th percentile to the 10th percentile to see if those two percentiles would be appropriate for a salary range. These results are in Table 14.3 and Figure 14.2.

The lines go through the data very nicely and the plot verifies the extremely high values of $r^2$. It should be noted that the coefficient of determination will always be higher for calculated values (in this case, percentiles) than for the raw data.

**TABLE 14.2    REGRESSION COEFFICIENTS AND COEFFICIENT OF DETERMINATION**

| Model | a | b | c | d | $r^2$ |
|-------|------|------|--------|--------|-------|
| P10 | 53,544 | 2,216 | −46.86 | 0.4713 | 0.988 |
| P25 | 55,249 | 3,447 | −89.45 | 0.9343 | 0.996 |
| P50 | 57,410 | 4,965 | −154.76 | 1.8167 | 0.996 |
| P75 | 62,963 | 5,677 | −174.55 | 2.0278 | 0.992 |
| P90 | 76,348 | 5,144 | −134.32 | 1.3808 | 0.989 |

**TABLE 14.3**   PREDICTED AVERAGE PERCENTILES AND RATIOS

| YSBS | Actual 10th Percentile | Actual 25th Percentile | Actual 50th Percentile | Actual 75th Percentile | Actual 90th Percentile | Predicted 10th Percentile | ... | Predicted 50th Percentile | ... | Predicted 90th Percentile | Ratio of P90/P10 |
|---|---|---|---|---|---|---|---|---|---|---|---|
| 1 | 56,829 | 60,349 | 64,174 | 69,601 | 87,385 | 55,714 | ... | 62,222 | ... | 81,359 | 1.46 |
| 2 | 57,000 | 61,800 | 66,360 | 72,431 | 82,944 | 57,792 | ... | 66,735 | ... | 86,110 | 1.49 |
| 3 | 60,348 | 63,604 | 68,688 | 75,904 | 85,764 | 59,783 | ... | 70,961 | ... | 90,608 | 1.52 |
| 4 | 62,400 | 67,200 | 72,925 | 79,800 | 90,727 | 61,688 | ... | 74,910 | ... | 94,863 | 1.54 |
| 5 | 63,408 | 69,906 | 78,000 | 86,400 | 98,400 | 63,511 | ... | 78,593 | ... | 98,883 | 1.56 |
| 6 | 64,200 | 71,842 | 81,600 | 91,519 | 103,200 | 65,255 | ... | 82,021 | ... | 102,675 | 1.57 |
| 7 | 66,720 | 75,600 | 86,402 | 99,480 | 109,368 | 66,922 | ... | 85,205 | ... | 106,248 | 1.59 |
| 8 | 66,600 | 77,203 | 89,899 | 102,000 | 112,800 | 68,514 | ... | 88,156 | ... | 109,610 | 1.60 |
| 9 | 70,376 | 80,496 | 92,407 | 104,022 | 116,015 | 70,036 | ... | 90,884 | ... | 112,771 | 1.61 |
| 10 | 72,735 | 83,291 | 95,760 | 106,800 | 117,792 | 71,489 | ... | 93,401 | ... | 115,737 | 1.62 |
| ... | ... | ... | ... | ... | ... | ... | ... | ... | ... | ... | ... |
| 35 | 97,603 | 107,170 | 120,240 | 135,608 | 149,323 | 93,907 | ... | 119,495 | ... | 151,048 | 1.61 |
| 36 | 95,387 | 108,000 | 121,440 | 138,528 | 153,935 | 94,578 | ... | 120,341 | ... | 151,876 | 1.61 |
| 37 | 95,645 | 108,374 | 122,063 | 137,066 | 151,008 | 95,257 | ... | 121,270 | ... | 152,734 | 1.60 |
| 38 | 96,324 | 110,196 | 121,723 | 137,741 | 157,386 | 95,947 | ... | 122,293 | ... | 153,629 | 1.60 |
| 39 | 96,531 | 108,982 | 123,602 | 139,926 | 154,786 | 96,651 | ... | 123,420 | ... | 154,571 | 1.60 |
| 40 | 96,072 | 108,240 | 123,288 | 136,999 | 153,683 | 97,371 | ... | 124,663 | ... | 155,567 | 1.60 |

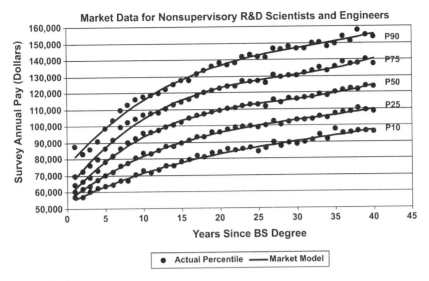

**FIGURE 14.2    PREDICTED AVERAGE PERCENTILES AND YSBS**

With these results, we do not need to fit spline models to the data. The cubic models are more than sufficient.

The values of P90/P10 range from 1.46 to 1.65. (All data are not shown here.) If these two percentiles were used to define the minimum and maximum of a salary structure, the range spreads would be from 46% to 65%. (These are obtained after MOTHing 1.46 and 1.65.)

As our compensation structures for exempt personnel typically have 50–60% ranges, these values would fit right in and, hence, are satisfactory and are what we will recommend.

Our model nicely *describes the relationship between external value (market pay) and internal value (YSBS)*. This graphical display gives us confidence in the model.

As an organization we want our pay programs to be "internally equitable and externally competitive," or "internally fair and externally fair." This chart and the associated equations are where these two concepts are integrated together. In this case, external value is represented by market pay, the *y*-variable, and internal value is represented by YSBS, the *x*-variable.

We will use this model to create a market–based salary structure by designating the minimum as P10, the midpoint at P50, and the maximum as P90. Theoretically, the midpoint is halfway between the minimum and the maximum. Here it is not, as P10, P50, and P90 are calculated independent of each other, but we will use the term "midpoint" nonetheless.

4. *Compare Employee Pay with the Market Model.* This step involves comparing the salaries of all employees with the predicted median market pay (the midpoint line). We will have a tabular display as well as a graphical one.

> The integrated market model approach assumes that for each YSBS, the predicted average market pay from the model is a good estimator for the average market pay for all employees with that YSBS, whether or not their jobs were surveyed.

Since the market model is based on data aged to January 1, 2012, the employee data should also be as of that date, before any raises are given in the new plan year. If you know the projected or planned raises for individual employees between the date of the employee data you have at hand and the date you have aged the market data to, then it would be appropriate to take those into account. Otherwise, just note that there are differences in the dates and any implications these may lead to.

To determine the market position, along with other useful information, we compare employee data with the five curves we just created. Employee data are shown in Table 14.4 and also in Data Set 32. The comparison will be the completed Table 14.5.

For purposes of illustration, we have just indicated an employee number. When doing this analysis for your company, you would probably have additional fields of data such as employee name, job title, job code, department, location, department and/or location code, and so on to allow additional analyses to be done.

After the employee identification, we have the YSBS and the pay of the employee. The next three columns contain the market-based salary structure we have just created.

Next is an indication of the employee's pay as it relates to the midpoint. Finally, if the employee's pay is under the minimum, we calculate by how much. Similarly, if the employee's pay is over the maximum, we calculate by how much.

This table furnishes the raw data to identify the overall market position. We use these data to create the following summary table, Table 14.6, calculating subtotals by YSBS.

In Chapter 12 with a linear model for hourly manufacturing employees and in Chapter 13 with an exponential model for financial department employees, the market position for each was that BPD was *below* the market as of January 1, 2012. But here, for the research and development scientists and engineers, BPD is *above* the market. Let us review the calculations.

The total pay is the total annual pay of the 120 BPD employees as of January 1, 2012. The total market midpoint is the total annual midpoint of the YSBSs of the market-based salary structure we created for all 120 employees.

**TABLE 14.4    EMPLOYEE SALARIES AND YSBS**

| Emp. No. | YSBS | Salary | Emp. No. | YSBS | Salary | Emp. No. | YSBS | Salary |
|---|---|---|---|---|---|---|---|---|
| 1 | 1 | 59,000 | 41 | 12 | 81,000 | 81 | 26 | 98,000 |
| 2 | 1 | 64,000 | 42 | 12 | 93,000 | 82 | 26 | 113,000 |
| 3 | 1 | 75,000 | 43 | 12 | 118,000 | 83 | 26 | 135,000 |
| 4 | 1 | 79,000 | 44 | 13 | 85,000 | 84 | 27 | 110,000 |
| 5 | 2 | 60,000 | 45 | 13 | 100,000 | 85 | 27 | 126,000 |
| 6 | 2 | 65,000 | 46 | 13 | 128,000 | 86 | 28 | 113,000 |
| 7 | 2 | 70,000 | 47 | 14 | 88,000 | 87 | 29 | 100,000 |
| 8 | 2 | 78,000 | 48 | 14 | 95,000 | 88 | 29 | 117,000 |
| 9 | 3 | 64,000 | 49 | 14 | 97,000 | 89 | 29 | 129,000 |
| 10 | 3 | 67,000 | 50 | 14 | 99,000 | 90 | 29 | 142,000 |
| 11 | 3 | 81,000 | 51 | 15 | 106,000 | 91 | 30 | 110,000 |
| 12 | 3 | 92,000 | 52 | 16 | 82,000 | 92 | 30 | 115,000 |
| 13 | 4 | 69,000 | 53 | 16 | 95,000 | 93 | 31 | 101,000 |
| 14 | 4 | 72,000 | 54 | 16 | 110,000 | 94 | 31 | 138,000 |
| 15 | 4 | 78,000 | 55 | 16 | 120,000 | 95 | 32 | 100,000 |
| 16 | 4 | 85,000 | 56 | 17 | 100,000 | 96 | 32 | 112,000 |
| 17 | 4 | 90,000 | 57 | 17 | 132,000 | 97 | 32 | 117,000 |
| 18 | 5 | 70,000 | 58 | 18 | 95,000 | 98 | 33 | 129,000 |
| 19 | 5 | 85,000 | 59 | 18 | 105,000 | 99 | 34 | 101,000 |
| 20 | 5 | 90,000 | 60 | 18 | 113,000 | 100 | 34 | 112,000 |
| 21 | 5 | 95,000 | 61 | 19 | 110,000 | 101 | 34 | 124,000 |
| 22 | 6 | 85,000 | 62 | 19 | 130,000 | 102 | 34 | 128,000 |
| 23 | 6 | 100,000 | 63 | 20 | 100,000 | 103 | 35 | 115,000 |
| 24 | 7 | 81,000 | 64 | 20 | 118,000 | 104 | 35 | 140,000 |
| 25 | 7 | 88,000 | 65 | 20 | 140,000 | 105 | 36 | 103,000 |
| 26 | 7 | 101,000 | 66 | 2 | 188,000 | 106 | 36 | 125,000 |
| 27 | 8 | 74,000 | 67 | 21 | 110,000 | 107 | 36 | 145,000 |
| 28 | 8 | 83,000 | 68 | 21 | 123,000 | 108 | 37 | 114,000 |
| 29 | 8 | 87,000 | 69 | 22 | 100,000 | 109 | 37 | 119,000 |
| 30 | 8 | 104,000 | 70 | 22 | 116,000 | 110 | 37 | 126,000 |
| 31 | 8 | 112,000 | 71 | 22 | 118,000 | 111 | 38 | 116,000 |
| 32 | 9 | 85,000 | 72 | 23 | 100,000 | 112 | 38 | 119,000 |
| 33 | 9 | 95,000 | 73 | 23 | 106,000 | 113 | 38 | 125,000 |
| 34 | 9 | 115,000 | 74 | 23 | 109,000 | 114 | 38 | 138,000 |
| 35 | 10 | 80,000 | 75 | 23 | 113,000 | 115 | 39 | 107,000 |
| 36 | 10 | 86,000 | 76 | 24 | 95,000 | 116 | 39 | 120,000 |
| 37 | 10 | 94,000 | 77 | 24 | 116,000 | 117 | 39 | 125,000 |
| 38 | 10 | 113,000 | 78 | 25 | 92,000 | 118 | 40 | 102,000 |
| 39 | 11 | 85,000 | 79 | 25 | 111,000 | 119 | 40 | 115,000 |
| 40 | 11 | 96,000 | 80 | 25 | 134,000 | 120 | 40 | 137,000 |

We see that in terms of annual dollars, the BPD total pay is 166,459 above the total market midpoint. The difference is the total market midpoint minus the total pay.

We now want to calculate two percentages. First, recall two definitions:

"Market position" is the % we are from the market midpoint. The market midpoint is the reference. In Table 14.6, this is in the column headed % from Market Mid.

**TABLE 14.5**  **EMPLOYEE SALARIES AND MARKET-BASED SALARY STRUCTURE**

| Emp. No. | YSBS | Pay | Market-Based Pay Structure | | | Pay as % of Mid | Dollars Under Min | Dollars Over Max |
|---|---|---|---|---|---|---|---|---|
| | | | Minimum | Midpoint | Maximum | | | |
| 1 | 1 | 59,000 | 55,714 | 62,222 | 81,359 | 95 | – | – |
| 2 | 1 | 64,000 | 55,714 | 62,222 | 81,359 | 103 | – | – |
| 3 | 1 | 75,000 | 55,714 | 62,222 | 81,359 | 121 | – | – |
| 4 | 1 | 79,000 | 55,714 | 62,222 | 81,359 | 127 | – | – |
| 5 | 2 | 60,000 | 57,792 | 66,735 | 86,110 | 90 | – | – |
| 6 | 2 | 65,000 | 57,792 | 66,735 | 86,110 | 97 | – | – |
| 7 | 2 | 70,000 | 57,792 | 66,735 | 86,110 | 105 | – | – |
| 8 | 2 | 78,000 | 57,792 | 66,735 | 86,110 | 117 | – | – |
| 9 | 3 | 64,000 | 59,783 | 70,961 | 90,608 | 90 | – | – |
| 10 | 3 | 67,000 | 59,783 | 70,961 | 90,608 | 94 | – | – |
| 11 | 3 | 81,000 | 59,783 | 70,961 | 90,608 | 114 | – | – |
| 12 | 3 | 92,000 | 59,783 | 70,961 | 90,608 | 130 | – | 1,392 |
| 13 | 4 | 69,000 | 61,688 | 74,910 | 94,863 | 92 | – | – |
| ⋮ | ⋮ | ⋮ | ⋮ | ⋮ | ⋮ | ⋮ | ⋮ | ⋮ |
| 117 | 39 | 125,000 | 96,651 | 123,420 | 154,571 | 101 | – | – |
| 118 | 40 | 102,000 | 97,371 | 124,663 | 155,567 | 82 | – | – |
| 119 | 40 | 115,000 | 97,371 | 124,663 | 155,567 | 92 | – | – |
| 120 | 40 | 137,000 | 97,371 | 124,663 | 155,567 | 110 | – | – |

"Budget needed to meet the market" is the % increase (decrease) in our pay to meet the market. Our pay is the reference. In Table 14.6, this is in the column headed % to Meet Market Mid.

Recall from Chapter 2 how to calculate the percent difference between the two terms.

$$\text{Percent difference} = \frac{\text{data} - \text{reference}}{\text{reference}} \times 100$$

For the market position, % from Mkt Mid, the data are our total pay and the reference is the total market midpoint.

$$\% \text{ from Mkt Mid} = \frac{12{,}384{,}000 - 12{,}217{,}541}{12{,}217{,}541} \times 100$$

$$= \frac{166{,}459}{12{,}217{,}444} \times 100 = 0.014 \times 100 = 1.4\%$$

For the budget needed to meet the market, % to Meet Mkt Mid, the data are the total market midpoint and the reference is our total pay.

$$\% \text{ to meet Mkt Mid} = \frac{12{,}217{,}541 - 12{,}384{,}000}{12{,}384{,}000} \times 100$$

$$= \frac{-166{,}459}{12{,}384{,}000} \times 100 = -0.013 \times 100 = -1.3\%$$

**TABLE 14.6   MARKET ANALYSIS**

Market Analysis of Nonsupervisory Scientists and Engineers: Comparison of Employee Pay with the Market-Based Pay Structure

| YSBS | No. of Employees | Total Pay | Total Market Midpoint | Difference | % from Market Mid | % to Meet Market Mid | No. of Employees Under Min | No. of Employees Over Max | Dollars Under Min | Dollars Over Max |
|------|------------------|-----------|-----------------------|------------|-------------------|----------------------|----------------------------|----------------------------|-------------------|------------------|
| 1 | 4 | 277,000 | 248,888 | −28,112 | 11.3% | −10.1% | 0 | 0 | 0 | 0 |
| 2 | 4 | 273,000 | 266,940 | −6,060 | 2.3% | −2.2% | 0 | 0 | 0 | 0 |
| 3 | 4 | 304,000 | 283,844 | −20,156 | 7.1% | −6.6% | 0 | 1 | 0 | 1,392 |
| 4 | 5 | 394,000 | 374,550 | −19,450 | 5.2% | −4.9% | 0 | 0 | 0 | 0 |
| ... | ... | ... | ... | ... | ... | ... | ... | ... | ... | ... |
| 38 | 4 | 498,000 | 489,172 | −8,828 | 1.8% | −1.8% | 0 | 0 | 0 | 0 |
| 39 | 3 | 352,000 | 370,260 | 18,260 | −4.9% | 5.2% | 0 | 0 | 0 | 0 |
| 40 | 3 | 354,000 | 373,989 | 19,989 | −5.3% | 5.6% | 0 | 0 | 0 | 0 |
| **Total** | **120** | **12,384,000** | **12,217,541** | **−166,459** | **1.4%** | **−1.3%** | **0** | **6** | **0** | **14,150** |

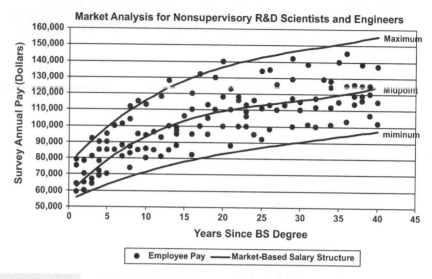

**FIGURE 14.3**   EMPLOYEE SALARIES AND MARKET-BASED SALARY STRUCTURE

Overall, the pay of the company's R&D professionals is 1.4% above the market midpoint, which represents the market median. It would take a 1.3% reduction to equal the market midpoint. This table also indicates that some YSBSs have different market positions than others, as well as summarizes how many employees have pay outside the pay structure and by how much. If desired you could have subtotals by discipline, location, and so on.

The next step is to present the results in a graphical form, as shown in Figure 14.3. In this chart, you compare the employee pay with the market-based salary structure you have just created.

This chart shows that the bulk of the employees' pay is within the market-based pay structure and it highlights the pay that is outside the structure.

As some employees are outside the structure, before presenting these results you research on those data points to identify who they are and why they are like they are.

In summary, we have conducted the following four steps to conduct a market analysis where we have maturity curve salary administration, resulting in an identification of the market position, the creation of a market-based salary structure, and the development of an initial market-based salary increase (in this case, decrease) budget, all as of the beginning of the new plan year.

1. Gather market data

2. Age data to a common date

3. Create an integrated market model

4. Compare employee pay with the market model

**TABLE 14.7    TEMPLATE FOR FINAL MARKET-BASED SALARY INCREASE BUDGET**

| | |
|---|---|
| Catch-up (or fall back) | —— |
| Anticipated market movement | —— |
| Pay policy | —— |
| **Total** | —— |

**TABLE 14.8    FINAL MARKET-BASED SALARY INCREASE BUDGET**

| | |
|---|---|
| Catch-up (or fall back) | −1.3% |
| Anticipated market movement | 3.0% |
| Pay policy | 2.0% |
| **Total** | **3.7%** |

For the final market-based salary increase budget, we complete Table 14.7.

The fall back is −1.3%, which is our initial market-based salary increase (decrease) budget as of the beginning of the new plan year. For this population, you find from the survey provider that the anticipated market movement is 3.0%. Our pay policy is to be 2.0% above the market at the end of the plan year. So our results are in Table 14.8.

The final market-based salary increase budget is 3.7%. For the final recommendations, you now take into account the "soft" information described in Section 11.4.

## PRACTICE PROBLEMS

14.1   In the BPD research and development division, there are 22 first-level supervisors who are administered on a maturity curve basis. It is October 2012 and you are preparing to recommend a salary increase budget for 2013. Your salary administration year is a calendar year. You give raises throughout the year on anniversary dates or promotion dates. Your compensation philosophy for this division is that you want your average pay to be 2.0% above the market pay at the end of the plan year. You have recently obtained the raw data from a survey of first-level R&D supervisors in your industry. There are 99 incumbents in the survey. There are not enough data points to calculate averages or percentiles for each YSBS, so you will just calculate the predicted averages from the raw data and create a structure from that. The data are shown in Table 14.9 and also in Data Set 33. The date of the data for this survey is January 1, 2012. From the survey, the general market movement for 2012 is 4.0% annually.

How much will you age the data from this survey?

14.2   Age the survey data appropriately.

14.3   Plot the data. What kind of relationship is shown by the plot?

**TABLE 14.9**    **SURVEY DATA ON FIRST-LEVEL R&D SUPERVISORS**

| Incumbent No. | YSBS | Market Pay | Incumbent No. | YSBS | Market Pay | Incumbent No. | YSBS | Market Pay |
|---|---|---|---|---|---|---|---|---|
| 1 | 6 | 85,600 | 34 | 16 | 164,900 | 67 | 29 | 138,100 |
| 2 | 6 | 90,600 | 35 | 17 | 121,900 | 68 | 29 | 147,300 |
| 3 | 7 | 77,700 | 36 | 17 | 127,900 | 69 | 29 | 160,000 |
| 4 | 7 | 101,300 | 37 | 18 | 127,500 | 70 | 30 | 153,700 |
| 5 | 7 | 115,000 | 38 | 18 | 135,800 | 71 | 30 | 155,800 |
| 6 | 8 | 91,900 | 39 | 18 | 154,200 | 72 | 31 | 138,700 |
| 7 | 8 | 92,300 | 40 | 19 | 124,000 | 73 | 31 | 144,600 |
| 8 | 8 | 95,000 | 41 | 19 | 138,300 | 74 | 32 | 134,600 |
| 9 | 8 | 97,300 | 42 | 20 | 128,500 | 75 | 32 | 147,700 |
| 10 | 8 | 105,200 | 43 | 20 | 135,800 | 76 | 32 | 151,700 |
| 11 | 9 | 97,500 | 44 | 20 | 179,200 | 77 | 33 | 155,000 |
| 12 | 9 | 108,700 | 45 | 21 | 122,300 | 78 | 34 | 132,100 |
| 13 | 9 | 119,400 | 46 | 21 | 133,500 | 79 | 34 | 138,500 |
| 14 | 10 | 110,200 | 47 | 21 | 160,600 | 80 | 34 | 150,600 |
| 15 | 10 | 116,200 | 48 | 22 | 129,200 | 81 | 34 | 180,600 |
| 16 | 10 | 123,100 | 49 | 22 | 131,300 | 82 | 35 | 153,100 |
| 17 | 10 | 140,600 | 50 | 22 | 133,500 | 83 | 35 | 164,800 |
| 18 | 11 | 96,600 | 51 | 23 | 110,800 | 84 | 36 | 141,000 |
| 19 | 11 | 123,700 | 52 | 23 | 130,200 | 85 | 36 | 144,200 |
| 20 | 12 | 124,200 | 53 | 23 | 140,400 | 86 | 36 | 156,500 |
| 21 | 12 | 128,700 | 54 | 23 | 144,800 | 87 | 37 | 131,900 |
| 22 | 12 | 133,700 | 55 | 24 | 135,600 | 88 | 37 | 148,500 |
| 23 | 13 | 120,800 | 56 | 24 | 154,600 | 89 | 37 | 160,200 |
| 24 | 13 | 127,700 | 57 | 25 | 131,200 | 90 | 38 | 134,000 |
| 25 | 13 | 138,800 | 58 | 25 | 141,000 | 91 | 38 | 142,300 |
| 26 | 14 | 138,800 | 59 | 25 | 166,500 | 92 | 38 | 147,300 |
| 27 | 14 | 123,500 | 60 | 26 | 146,200 | 93 | 38 | 172,900 |
| 28 | 14 | 127,500 | 61 | 26 | 148,800 | 94 | 39 | 133,500 |
| 29 | 14 | 140,400 | 62 | 26 | 165,600 | 95 | 39 | 145,200 |
| 30 | 15 | 144,000 | 63 | 27 | 136,900 | 96 | 39 | 160,200 |
| 31 | 16 | 126,200 | 64 | 27 | 154,400 | 97 | 40 | 135,600 |
| 32 | 16 | 136,000 | 65 | 28 | 175,800 | 98 | 40 | 150,200 |
| 33 | 16 | 151,900 | 66 | 29 | 126,000 | 99 | 40 | 152,300 |

**14.4**  Conduct a regression. State the market model equation, the coefficient of determination, and the standard error of estimate (SEE).

**14.5**  Plot, interpret, and evaluate the model.

**14.6**  You decide to use a range spread of 55%. Setting the market model as the midpoint, create a market-based salary structure and plot it.

**14.7**  The employee data for the 22 first-level R&D supervisors are in Table 14.10 and also in Data Set 34. Complete Table 14.11. Create a plot that compares employee pay with the market-based salary structure.

**14.8**  Complete the summary table shown in Table 14.12.

**14.9**  For this survey, you estimate that the market will move 3% annually during 2013. What is the market-based salary increase budget recommendation?

**TABLE 14.10    EMPLOYEE SALARIES AND YSBS**

| Emp. No. | YSBS | Pay |
|---|---|---|
| 1 | 7 | 100,000 |
| 2 | 8 | 120,000 |
| 3 | 11 | 110,000 |
| 4 | 11 | 123,000 |
| 5 | 14 | 150,000 |
| 6 | 16 | 160,000 |
| 7 | 18 | 140,000 |
| 8 | 20 | 150,000 |
| 9 | 21 | 135,000 |
| 10 | 21 | 150,000 |
| 11 | 21 | 163,000 |
| 12 | 23 | 155,000 |
| 13 | 24 | 145,000 |
| 14 | 27 | 165,000 |
| 15 | 28 | 145,000 |
| 16 | 29 | 135,000 |
| 17 | 30 | 162,000 |
| 18 | 30 | 173,000 |
| 19 | 31 | 150,000 |
| 20 | 34 | 139,000 |
| 21 | 35 | 155,000 |
| 22 | 35 | 160,000 |

**TABLE 14.11    TEMPLATE FOR EMPLOYEE SALARIES AND MARKET-BASED SALARY STRUCTURE**

| Emp. No. | YSBS | Pay | Minimum | Midpoint | Maximum | Pay as % of Midpoint | Dollars Under Min | Dollars Over Max |
|---|---|---|---|---|---|---|---|---|
| 1 | 7 | 100,000 | | | | | | |
| 2 | 8 | 120,000 | | | | | | |
| 3 | 11 | 110,000 | | | | | | |
| ⋮ | ⋮ | ⋮ | | | | | | |

**TABLE 14.12    TEMPLATE FOR MARKET ANALYSIS**

Market Analysis of First-Level R&D Supervisors: Comparison of Employee Pay with the Market-Based Salary Structure

| YSBS | No. of Employees | Total Pay | Total Market Midpoint | Difference | % from Mkt Mid | % to Meet Mkt Mid | No. of Employees Under Min | No. of Employees Over Max | Dollars Under Min | Dollars Over Max |
|---|---|---|---|---|---|---|---|---|---|---|
| 7 | | | | | | | | | | |
| 8 | | | | | | | | | | |
| 11 | | | | | | | | | | |

*(continued)*

**TABLE 14.12** *(CONTINUED)*

Market Analysis of First-Level R&D Supervisors: Comparison of Employee Pay with the Market-Based Salary Structure

| YSBS | No. of Employees | Total Pay | Total Market Midpoint | Difference | % from Mkt Mid | % to Meet Mkt Mid | No. of Employees Under Min | No. of Employees Over Max | Dollars Under Min | Dollars Over Max |
|---|---|---|---|---|---|---|---|---|---|---|
| 14 | | | | | | | | | | |
| 16 | | | | | | | | | | |
| 18 | | | | | | | | | | |
| 20 | | | | | | | | | | |
| 21 | | | | | | | | | | |
| 23 | | | | | | | | | | |
| 24 | | | | | | | | | | |
| 27 | | | | | | | | | | |
| 28 | | | | | | | | | | |
| 29 | | | | | | | | | | |
| 30 | | | | | | | | | | |
| 31 | | | | | | | | | | |
| 34 | | | | | | | | | | |
| 35 | | | | | | | | | | |
| Total | | | | | | | | | | |

**14.10** Your executives approve a salary increase budget of 4.0%. By how much do you adjust the market-based salary structure to have the average pay be at the midpoint after raises are given?

# CHAPTER 15

# Job Pricing Market Model: Group of Jobs

This approach is briefly in four steps and can be done with base pay or total cash compensation.

We reverse the last two steps listed in Chapter 11.

---

**CASE STUDY 9, PART 5 OF 6**

To illustrate this approach, we will conduct a market analysis of a group of 100 BPD employees with certain IT skills at different levels. For this example, we will be using skills rather than jobs. You have to create a salary structure and develop a salary increase budget for these employees.

---

1. *Gather Market Data.*
2. *Age Data to a Common Date.* We will assume that we have chosen the median as our measure of the market and have already aged the data to January 1, 2012, as shown in Table 15.1. The data are also in Data Set 35. Shown are the employee number, the skill and level, the aged survey median for that skill and level, and the employee pay. The reader might wish to enter the data in a computer that has spreadsheet or statistical software and follow along.

   The last two steps are done simultaneously.
3. *Compare Employee Pay with the Market Pay.*

*Statistics for Compensation: A Practical Guide to Compensation Analysis,* By John H. Davis
Copyright © 2011 John Wiley & Sons Inc.

| TABLE 15.1 | | SURVEY DATA AND EMPLOYEE SALARIES FOR IT SKILLS | | | | | |
|---|---|---|---|---|---|---|---|

| Emp. No. | Skill/Level | Aged Survey Median | Employee Pay | Emp. No. | Skill/Level | Aged Survey Median | Employee Pay |
|---|---|---|---|---|---|---|---|
| 1 | Data Min 1 | 62,000 | 56,000 | 51 | Int/Int 3 | 111,000 | 128,000 |
| 2 | Data Min 1 | 62,000 | 60,000 | 52 | Int/Int 4 | 150,000 | 138,000 |
| 3 | Data Min 1 | 62,000 | 63,000 | 53 | Int/Int 4 | 150,000 | 150,000 |
| 4 | Data Min 1 | 62,000 | 71,000 | 54 | Int/Int 4 | 150,000 | 168,000 |
| 5 | Data Min 2 | 84,000 | 66,000 | 55 | Net Eng 1 | 66,000 | 60,000 |
| 6 | Data Min 2 | 84,000 | 78,000 | 56 | Net Eng 1 | 66,000 | 62,000 |
| 7 | Data Min 2 | 84,000 | 83,000 | 57 | Net Eng 1 | 66,000 | 63,000 |
| 8 | Data Min 2 | 84,000 | 87,000 | 58 | Net Eng 1 | 66,000 | 63,000 |
| 9 | Data Min 2 | 84,000 | 101,000 | 59 | Net Eng 1 | 66,000 | 71,000 |
| 10 | Data Min 3 | 99,000 | 78,000 | 60 | Net Eng 1 | 66,000 | 71,000 |
| 11 | Data Min 3 | 99,000 | 93,000 | 61 | Net Eng 1 | 66,000 | 86,000 |
| 12 | Data Min 3 | 99,000 | 99,000 | 62 | Net Eng 2 | 90,000 | 71,000 |
| 13 | Data Min 3 | 99,000 | 102,000 | 63 | Net Eng 2 | 90,000 | 78,000 |
| 14 | Data Min 3 | 99,000 | 104,000 | 64 | Net Eng 2 | 90,000 | 81,000 |
| 15 | Data Min 3 | 99,000 | 105,000 | 65 | Net Eng 2 | 90,000 | 84,000 |
| 16 | Data Min 4 | 123,000 | 86,000 | 66 | Net Eng 2 | 90,000 | 86,000 |
| 17 | Data Min 4 | 123,000 | 105,000 | 67 | Net Eng 2 | 90,000 | 87,000 |
| 18 | Data Min 4 | 123,000 | 105,000 | 68 | Net Eng 2 | 90,000 | 90,000 |
| 19 | Data Min 4 | 123,000 | 116,000 | 69 | Net Eng 2 | 90,000 | 93,000 |
| 20 | Data Min 4 | 123,000 | 131,000 | 70 | Net Eng 2 | 90,000 | 98,000 |
| 21 | EDI 1 | 63,000 | 51,000 | 71 | Net Eng 2 | 90,000 | 101,000 |
| 22 | EDI 1 | 63,000 | 66,000 | 72 | Net Eng 3 | 120,000 | 108,000 |
| 23 | EDI 2 | 80,000 | 74,000 | 73 | Net Eng 3 | 120,000 | 116,000 |
| 24 | EDI 2 | 80,000 | 78,000 | 74 | Net Eng 3 | 120,000 | 120,000 |
| 25 | EDI 2 | 80,000 | 86,000 | 75 | Net Eng 3 | 120,000 | 126,000 |
| 26 | EDI 2 | 80,000 | 93,000 | 76 | Net Eng 3 | 120,000 | 135,000 |
| 27 | EDI 3 | 95,000 | 78,000 | 77 | Net Eng 4 | 147,000 | 131,000 |
| 28 | EDI 3 | 95,000 | 78,000 | 78 | Net Eng 4 | 147,000 | 138,000 |
| 29 | EDI 3 | 95,000 | 83,000 | 79 | Net Eng 4 | 147,000 | 147,000 |
| 30 | EDI 3 | 95,000 | 93,000 | 80 | Net Eng 4 | 147,000 | 153,000 |
| 31 | EDI 4 | 143,000 | 105,000 | 81 | Net Eng 4 | 147,000 | 153,000 |
| 32 | EDI 4 | 143,000 | 111,000 | 82 | Net Eng 4 | 147,000 | 156,000 |
| 33 | EDI 4 | 143,000 | 116,000 | 83 | Sys Adm 1 | 51,000 | 44,000 |
| 34 | EDI 4 | 143,000 | 131,000 | 84 | Sys Adm 1 | 51,000 | 57,000 |
| 35 | EDI 4 | 143,000 | 138,000 | 85 | Sys Adm 2 | 75,000 | 60,000 |
| 36 | EDI 4 | 143,000 | 141,000 | 86 | Sys Adm 2 | 75,000 | 62,000 |
| 37 | EDI 4 | 143,000 | 150,000 | 87 | Sys Adm 2 | 75,000 | 63,000 |
| 38 | Int/Int 1 | 57,000 | 53,000 | 88 | Sys Adm 2 | 75,000 | 71,000 |
| 39 | Int/Int 1 | 57,000 | 62,000 | 89 | Sys Adm 2 | 75,000 | 74,000 |
| 40 | Int/Int 2 | 71,000 | 56,000 | 90 | Sys Adm 2 | 75,000 | 78,000 |
| 41 | Int/Int 2 | 71,000 | 59,000 | 91 | Sys Adm 2 | 75,000 | 83,000 |
| 42 | Int/Int 2 | 71,000 | 63,000 | 92 | Sys Adm 3 | 105,000 | 93,000 |
| 43 | Int/Int 2 | 71,000 | 63,000 | 93 | Sys Adm 3 | 105,000 | 116,000 |
| 44 | Int/Int 2 | 71,000 | 71,000 | 94 | Sys Adm 3 | 105,000 | 138,000 |
| 45 | Int/Int 2 | 71,000 | 78,000 | 95 | Sys Adm 4 | 132,000 | 93,000 |
| 46 | Int/Int 2 | 71,000 | 101,000 | 96 | Sys Adm 4 | 132,000 | 116,000 |
| 47 | Int/Int 3 | 111,000 | 83,000 | 97 | Sys Adm 4 | 132,000 | 123,000 |
| 48 | Int/Int 3 | 111,000 | 101,000 | 98 | Sys Adm 4 | 132,000 | 123,000 |
| 49 | Int/Int 3 | 111,000 | 105,000 | 99 | Sys Adm 4 | 132,000 | 131,000 |
| 50 | Int/Int 3 | 111,000 | 120,000 | 100 | Sys Adm 4 | 132,000 | 153,000 |

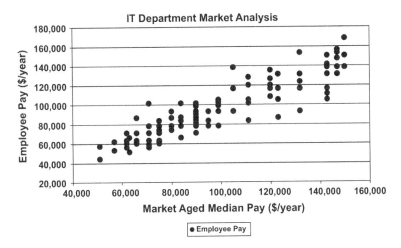

**IT Department Market Analysis**

**FIGURE 15.1    AGED MARKET PAY AND EMPLOYEE PAY**

4. *Create a Job Pricing Market Model.* We will first do a graphical comparison, and then a tabular one. In this approach, we will compare the pay of individual employees with the aged median market pay for their individual skill/level, as shown in Figure 15.1.

Recall that in the previous models (linear, exponential, and maturity curve) the market pay was the dependent or *y*-variable. For this model, the employee pay is the dependent or *y*-variable and the market pay is the independent or *x*-variable. It is done this way for ease of interpretation when we develop the market-based salary structure.

We note that there is a relationship that is positive, linear, and not perfect. This means that even though there is a lot of variability in the data, in general the skills that the market values highly, we do too.

However, unlike the situations when you have grades or experience, with this technique we do not conduct a regression. So it does not matter whether the relationship is linear or nonlinear.

Next we would like to add three lines to create our model: a midpoint line that represents the survey pay (in this case, the median) and the upper and lower lines that represent the maximum and the minimum, respectively, of our market-based salary structure. In this example we will choose a 50% range spread.

We first create a midpoint that is equal to the aged survey median. Then we create the minimum and maximum using the formulas below.

$$\text{Minimum} = \text{midpoint} / \left( 1 + \frac{\text{range spread}}{2} \right)$$

$$\text{Maximum} = \text{minimum} \times (1 + \text{range spread})$$

**TABLE 15.2**     EMPLOYEE SALARIES AND MARKET-BASED SALARY STRUCTURE

| Emp. No. | Skill/Level | Aged Survey Median | Employee Pay | Market-Based Salary Structure Minimum | Midpoint | Maximum |
|---|---|---|---|---|---|---|
| 1 | Data Min 1 | 62,000 | 56,000 | 49,600 | 62,000 | 74,400 |
| 2 | Data Min 1 | 62,000 | 60,000 | 49,600 | 62,000 | 74,400 |
| 3 | Data Min 1 | 62,000 | 63,000 | 49,600 | 62,000 | 74,400 |
| 4 | Data Min 1 | 62,000 | 71,000 | 49,600 | 62,000 | 74,400 |
| 5 | Data Min 2 | 84,000 | 66,000 | 67,200 | 84,000 | 100,800 |
| 6 | Data Min 2 | 84,000 | 78,000 | 67,200 | 84,000 | 100,800 |
| ⋮ | ⋮ | ⋮ | ⋮ | ⋮ | ⋮ | ⋮ |
| ⋮ | ⋮ | ⋮ | ⋮ | ⋮ | ⋮ | ⋮ |

For example, for employee number 1, whose survey job match had an aged survey median of 62,000, we get

$$\text{Midpoint} = \text{aged survey median of matched job} = 62,000$$
$$\text{Minimum} = 62,000/(1 + 0.50/2) = 62,000/1.25 = 49,600$$
$$\text{Maximum} = 49,600 \times (1 + 0.50) = 49,600 \times 1.50 = 74,400$$

After completing the calculations, we obtain Table 15.2.

We see that each employee has an individualized market-based salary structure, based on the particular skill and level.

The minimum, midpoint, and maximum values are all added to the first chart as dependent variables against the aged survey median as the independent variable and shown as lines. The resulting chart is Figure 15.2.

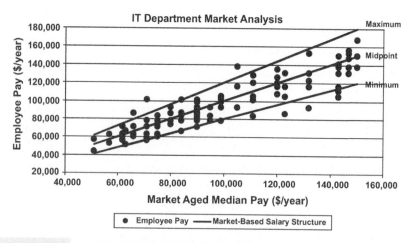

**FIGURE 15.2**     EMPLOYEE SALARIES AND MARKET-BASED SALARY STRUCTURE

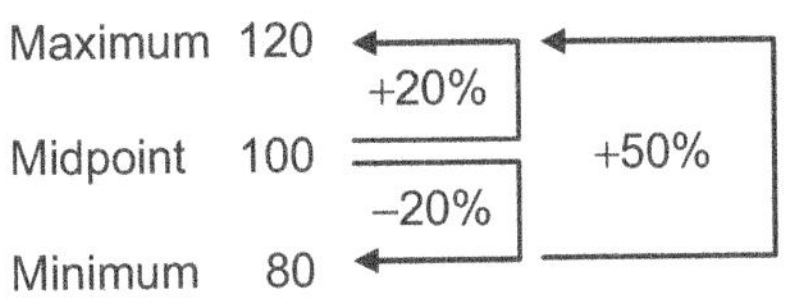

| Maximum | 120 | | |
|---|---|---|---|
| | | +20% | |
| Midpoint | 100 | | +50% |
| | | −20% | |
| Minimum | 80 | | |

**FIGURE 15.3    FIFTY PERCENT RANGE SPREAD RELATIONSHIPS**

The middle line represents the market median, which is our midpoint. The top line is 20% above the median and the bottom line is 20% below the median.

Here is where the 20% comes from, starting with a 50% range spread, illustrated in Figure 15.3. The range spread, sometimes called the range, is the percent that the maximum is more than the minimum.

In this example, with a 50% range spread, the maximum of 120 is 50% more than the minimum of 80. The reference is the minimum. But the maximum is only 20% more than the midpoint of 100. The reference here is the midpoint. There is a change in the reference point, or denominator, from 80 to 100. Similarly, the minimum is only 20% below the midpoint. See the Appendix for more discussion of the range spread.

As before, we see that our model nicely *describes the relationship between internal value (employee pay) and external value (market pay)*. In this case, internal value is represented by the *y*-variable and external value is represented by the *x*-variable. Having it this way makes it easy to interpret the chart.

A data point *above* the middle line means that the employee's pay is *above* the market median. With a 50% range spread, if the data point is above the top line it means that the employee's pay is more than 20% above the market median.

Similarly, a data point *below* the middle line means that the employee's pay is *below* the market median. With a 50% range spread, if the data point is below the bottom line it means that the employee's pay is more than 20% below the market median.

We see from the chart that there are a few points that need to be subjected to some aggressive inquisitiveness. Assume that has been done and no changes are made.

We expand the data table to Table 15.3 and calculate the position of each employee's salary as a percent of midpoint, and if the salary is outside the range, by what amount. This table corresponds to the chart and allows you to identify those data points that need further investigation.

We now create a final summary table, Table 15.4, of this analysis.

| TABLE 15.3 | | INDIVIDUAL EMPLOYEE MARKET ANALYSIS | | | | | |
|---|---|---|---|---|---|---|---|

| Emp. No. | .. | Employee Pay | Market-Based Salary Structure | | | Pay as % of Mid | Dollars Under Min | Dollars Above Max |
|---|---|---|---|---|---|---|---|---|
| | | | Minimum | Midpoint | Maximum | | | |
| 1 | .. | 56,000 | 49,600 | 62,000 | 74,400 | 90 | – | – |
| 2 | .. | 60,000 | 49,600 | 62,000 | 74,400 | 97 | – | – |
| 3 | .. | 63,000 | 49,600 | 62,000 | 74,400 | 102 | – | – |
| 4 | .. | 71,000 | 49,600 | 62,000 | 74,400 | 115 | – | – |
| 5 | .. | 66,000 | 67,200 | 84,000 | 100,800 | 79 | 1,200 | – |
| 6 | .. | 78,000 | 67,200 | 84,000 | 100,800 | 93 | – | – |
| : | .. | : | : | : | : | : | : | : |

This table shows the overall market position, namely, the IT department, is 3.4% below the market-based midpoint, and that it will take a 3.5% budget to catch up to that midpoint.

> With the job pricing market model approach, what do you do with jobs and skills that were not matched? Here, the assumption is made that the market position of the unmatched jobs/skills is the same as the market position of the matched jobs/skills. In this example then, it is assumed that the market position for the unmatched skills is also 3.4% below the market median.

For an individual employee, the salary structure is based on the market (median in this case) pay of his skill/job, and not on a grade. Everyone has an "individualized" pay range.

In summary, we have completed the following four steps to conduct a market analysis when we do not have grades, resulting in an identification of the market position, the creation of a market-based salary structure, and the development of an initial market-based salary increase budget, all as of the beginning of the new plan year.

1. Gather market data
2. Age data to a common date
3. Compare employee pay with the market pay
4. Create a job pricing market model

For the final market-based salary increase budget, we complete Table 15.5.

The catch-up is 3.5%, which is our initial market-based salary increase budget as of the beginning of the new plan year. For this population, we learn from surveys that the anticipated market movement is 1.5%. Our pay policy is to be 2.0% above the market at the end of the plan year. So we complete Table 15.6.

TABLE 15.4   MARKET ANALYSIS

## IT Department Market Analysis: Comparison of Employee Pay with the Market-Based Salary Structure

| Skill/Level | No. of Employees | Total Employee Pay | Total Market Midpoint | Difference | % from Mkt Mid | % to Meet Mkt Mid | No. of Employees < Min | No. of Employees > Max | Total $ < Min | Total $ > Max |
|---|---|---|---|---|---|---|---|---|---|---|
| Data Min 1 | 4 | 250,000 | 248,000 | -2,000 | 0.8 | -0.8 | 0 | 0 | 0 | 0 |
| Data Min 2 | 5 | 415,000 | 420,000 | 5,000 | -1.2 | 1.2 | 1 | 1 | 1,200 | 200 |
| Data Min 3 | 6 | 581,000 | 594,000 | 13,000 | -2.2 | 2.2 | 1 | 0 | 1,200 | 0 |
| Data Min 4 | 5 | 543,000 | 615,000 | 72,000 | -11.7 | 13.3 | 1 | 0 | 12,400 | 0 |
| **Data Min Total** | **20** | **1,789,000** | **1,877,000** | **88,000** | **-4.7** | **4.9** | **3** | **1** | **14,800** | **200** |
| EDI 1 | 2 | 117,000 | 126,000 | 9,000 | -7.1 | 7.7 | 0 | 0 | 0 | 0 |
| EDI 2 | 4 | 331,000 | 320,000 | -11,000 | 3.4 | -3.3 | 0 | 0 | 0 | 0 |
| EDI 3 | 4 | 332,000 | 380,000 | 48,000 | -12.6 | 14.5 | 0 | 0 | 0 | 0 |
| EDI 4 | 7 | 892,000 | 1,001,000 | 109,000 | -10.9 | 12.2 | 2 | 0 | 12,800 | 0 |
| **EDI Total** | **17** | **1,672,000** | **1,827,000** | **155,000** | **-8.5** | **9.3** | **2** | **0** | **12,800** | **0** |
| Int/Int 1 | 2 | 115,000 | 114,000 | -1,000 | 0.9 | -0.9 | 0 | 0 | 0 | 0 |
| Int/Int 2 | 7 | 491,000 | 497,000 | 6,000 | -1.2 | 1.2 | 1 | 1 | 800 | 15,800 |
| Int/Int 3 | 5 | 537,000 | 555,000 | 18,000 | -3.2 | 3.4 | 1 | 0 | 5,800 | 0 |
| Int/Int 4 | 3 | 456,000 | 450,000 | -6,000 | 1.3 | -1.3 | 0 | 0 | 0 | 0 |
| **Int/Int Total** | **17** | **1,599,000** | **1,616,000** | **17,000** | **-1.1** | **1.1** | **2** | **1** | **6,600** | **15,800** |
| Net Eng 1 | 7 | 476,000 | 462,000 | -14,000 | 3.0 | -2.9 | 0 | 1 | 0 | 6,800 |
| Net Eng 2 | 10 | 869,000 | 900,000 | 31,000 | -3.4 | 3.6 | 1 | 0 | 1,000 | 0 |
| Net Eng 3 | 5 | 605,000 | 600,000 | -5,000 | 0.8 | -0.8 | 0 | 0 | 0 | 0 |
| Net Eng 4 | 6 | 878,000 | 882,000 | 4,000 | -0.5 | 0.5 | 0 | 0 | 0 | 0 |
| **Net Eng Total** | **28** | **2,828,000** | **2,844,000** | **16,000** | **-0.6** | **0.6** | **1** | **1** | **1,000** | **6,800** |
| Sys Admin 1 | 2 | 101,000 | 102,000 | 1,000 | -1.0 | 1.0 | 0 | 0 | 0 | 0 |
| Sys Admin 2 | 7 | 491,000 | 525,000 | 34,000 | -6.5 | 6.9 | 0 | 0 | 0 | 0 |
| Sys Admin 3 | 3 | 347,000 | 315,000 | -32,000 | 10.2 | -9.2 | 0 | 1 | 0 | 12,000 |
| Sys Admin 4 | 6 | 739,000 | 792,000 | 53,000 | -6.7 | 7.2 | 1 | 0 | 12,600 | 0 |
| **Sys Adm Total** | **18** | **1,678,000** | **1,734,000** | **56,000** | **-3.2** | **3.3** | **1** | **1** | **12,600** | **12,000** |
| **Grand Total** | **100** | **9,566,000** | **9,898,000** | **332,000** | **-3.4** | **3.5** | **9** | **4** | **47,800** | **34,800** |

| TABLE 15.5 | TEMPLATE FOR FINAL MARKET-BASED SALARY INCREASE BUDGET |
|---|---|
| Catch-up (or fall back) | — |
| Anticipated market movement | — |
| Pay policy | — |
| **Total** | — |

| TABLE 15.6 | FINAL MARKET-BASED SALARY INCREASE BUDGET |
|---|---|
| Catch-up (or fall back) | 3.5% |
| Anticipated market movement | 1.5% |
| Pay policy | 2.0% |
| **Total** | **7.0 %** |

The final market-based salary increase budget is 7.0%. For the final recommendations you now take into account the "soft" information described in Section 11.4.

## RELATED TOPICS IN THE APPENDIX

A.13   Range Spread Relationships

## PRACTICE PROBLEMS

**15.1**   BPD has a financial center in its Eastern Division, with 33 nonexecutive employees. There are no grades and the jobs are market priced. It is October 2012 and you are preparing to recommend a salary increase budget for 2013. Your salary administration year is a calendar year. You give raises throughout the year on anniversary dates or promotion dates. Your compensation philosophy is that you want your average pay to match the market at the end of the plan year. You have recently obtained a salary survey for these jobs. The data you extract from the survey are the medians, and are combined with the corresponding employee data. These are shown in Table 15.7 and also in Data Set 36. The date of the data for Survey A is January 1, 2012. From the surveys, the general market movement for 2012 is 4.0% annually.

How much will you age the data from the survey?

**15.2**   Age the survey data appropriately.

**15.3**   Plot the data.

**15.4**   What kind of relationship is shown by the plot? Does it matter what kind of a relationship it is? Why or why not?

**15.5**   Create a market-based salary structure for each employee's job using a range spread of 50%. Complete Table 15.8.

**TABLE 15.7    SURVEY DATA ON FINANCE JOBS**

| Emp. No. | Family | Job Title | Survey Median | Employee Pay |
|---|---|---|---|---|
| 1 | Accounting | Accountant | 49,000 | 43,800 |
| 2 | Accounting | Accountant | 49,000 | 47,900 |
| 3 | Accounting | Accountant | 49,000 | 50,100 |
| 4 | Accounting | Accountant | 49,000 | 52,300 |
| 5 | Accounting | Sr. Accountant | 56,000 | 50,100 |
| 6 | Accounting | Sr. Accountant | 56,000 | 51,200 |
| 7 | Accounting | Sr. Accountant | 56,000 | 52,300 |
| 8 | Accounting | Sr. Accountant | 56,000 | 55,600 |
| 9 | Accounting | Sr. Accountant | 56,000 | 57,800 |
| 10 | Accounting | Sr. Accountant | 56,000 | 58,900 |
| 11 | Accounting | Sr. Accountant | 56,000 | 62,000 |
| 12 | Accounting | Staff Accountant | 75,000 | 66,500 |
| 13 | Accounting | Staff Accountant | 75,000 | 72,700 |
| 14 | Accounting | Staff Accountant | 75,000 | 77,500 |
| 15 | Accounting | Staff Accountant | 75,000 | 88,500 |
| 16 | Accounting | Staff Accountant | 75,000 | 94,000 |
| 17 | Financial Analysis | Financial Analyst | 63,000 | 60,600 |
| 18 | Financial Analysis | Financial Analyst | 63,000 | 63,800 |
| 19 | Financial Analysis | Financial Analyst | 63,000 | 58,900 |
| 20 | Financial Analysis | Financial Analyst | 63,000 | 60,000 |
| 21 | Financial Analysis | Sr. Financial Analyst | 86,000 | 76,500 |
| 22 | Financial Analysis | Sr. Financial Analyst | 86,000 | 82,000 |
| 23 | Financial Analysis | Sr. Financial Analyst | 86,000 | 87,500 |
| 24 | Financial Analysis | Sr. Financial Analyst | 86,000 | 98,500 |
| 25 | Financial Analysis | Staff Financial Analyst | 106,000 | 108,400 |
| 26 | Financial Analysis | Staff Financial Analyst | 106,000 | 113,900 |
| 27 | Mgmt/Supv | Supv Accts Payable | 81,000 | 76,500 |
| 28 | Mgmt/Supv | Supv Accts Payable | 81,000 | 82,000 |
| 29 | Mgmt/Supv | Supv Accts Receivable | 85,000 | 79,800 |
| 30 | Mgmt/Supv | Supv Accts Receivable | 85,000 | 85,300 |
| 31 | Mgmt/Supv | Supv Payroll | 89,000 | 94,100 |
| 32 | Mgmt/Supv | Manager Accounting | 132,000 | 126,000 |
| 33 | Mgmt/Supv | Manager Financial Analysis | 144,000 | 148,000 |

**TABLE 15.8    TEMPLATE FOR EMPLOYEE SALARIES AND MARKET-BASED SALARY STRUCTURE**

| Emp. No. | ... | Employee Pay | Minimum | Midpoint | Maximum | Pay as % of Midpoint | Dollars Under Min | Dollars Over Max |
|---|---|---|---|---|---|---|---|---|
| 1 | ... | 43,800 | | | | | | |
| 2 | ... | 47,900 | | | | | | |
| 3 | ... | 50,100 | | | | | | |
| ... | ... | ... | | | | | | |
| ... | ... | ... | | | | | | |

**TABLE 15.9** TEMPLATE FOR MARKET ANALYSIS

**Market Analysis for Eastern Financial Division:**
**Comparison of Employee Pay with the Market-Based Salary Structure**

| Job | No. of Employees | Total Employee Pay | Market Midpoint | Total Difference | % from Mkt Mid | % to Meet Mkt Mid | No. of Employees < Min | No. of Employees > Max | Total $ < Min | Total $ > Max |
|---|---|---|---|---|---|---|---|---|---|---|
| Acct | | | | | | | | | | |
| Sr. Acct | | | | | | | | | | |
| Staff Acct | | | | | | | | | | |
| **Acctg Total** | | | | | | | | | | |
| Financial Analyst | | | | | | | | | | |
| Sr. Financial Analyst | | | | | | | | | | |
| Staff Financial Analyst | | | | | | | | | | |
| **Financial Analyst Total** | | | | | | | | | | |
| Mgr Acctg | | | | | | | | | | |
| Mgr Financial Analyst | | | | | | | | | | |
| Supv AP | | | | | | | | | | |
| Supv AR | | | | | | | | | | |
| Supv Payroll | | | | | | | | | | |
| **Mgt/Supv Total** | | | | | | | | | | |
| **Grand Total** | | | | | | | | | | |

**15.6**   Create a plot that compares employee pay with the market-based salary structure.

**15.7**   Looking at the plot, you notice that the dot for employee number 16 is above the top line. What does this mean?

**15.8**   Complete the summary table shown in Table 15.9.

**15.9**   For this survey, you estimate that the market will move 3.0% annually during 2013. What is the market-based salary increase budget recommendation?

**15.10**   Your executives approve a salary increase budget of 4.0%. By how much do you adjust the market-based salary structure to have the average pay be at the midpoint after raises are given?

# CHAPTER 16

# Job Pricing Market Model: Power Model

Many executive jobs are priced individually using the technique of a power model, where external salaries are modeled as a function of organization size, usually measured by company revenue for profit-making companies or by budget for nonprofit organizations.

Often the analyses for executives are done separately from that of the other employees, as decisions on executive pay is done by the board, while decisions on the other employees are made by the executives themselves (CEO and/or President).

---

**CASE STUDY 9, PART 6 OF 6**

You have the task of developing a market model for your executives. We will illustrate the power model by completing the development of a model and market position for base pay for the BPD CEO started in Chapter 10, *Power Model* in Case Study 8.

---

We will follow the same steps as before.

1. *Gather Market Data.*
2. *Age Data to a Common Date.*
3. *Create a Market Model.*

---

*Statistics for Compensation: A Practical Guide to Compensation Analysis,* By John H. Davis
Copyright © 2011 John Wiley & Sons Inc.

These three steps were completed previously in Chapter 10, starting with the survey data in Table 10.1 and also in Data Set 21. The results are repeated here, where the dependent variable, CEO pay, is in thousands of dollars and the independent variable, sales, is in millions of dollars.

Recall that for a power model both the dependent and the independent variables are transformed into logs.

The general equation for a power model is

$$\log y = a + b \log x$$

Bringing forth the regression results from Chapter 10, we have the following prediction equation, which was derived from the raw data using the method of least squares.

$$\log y' = 1.657 + 0.4140 \log x$$

$$\text{Predicted average log CEO pay} = 1.657 + 0.4140 \log \text{ sales}$$

and

$$\text{Coefficient of determination, } r^2 = 0.849$$

$$\text{Standard error of estimate (SEE)} = 0.152$$

We calculated the predicted values and plotted the line on a log–log chart along with the raw data, as shown in Figure 16.1. To do that, we

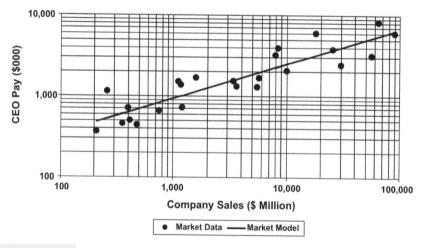

**FIGURE 16.1**   PREDICTED AVERAGE CEO PAY AND COMPANY SALES ON LOG–LOG PLOT

took the antilogs, and used the following formula to calculate the predicted values of average pay for the various values of sales.

$$y' = 10^a x^b$$
$$y' = 10^{1.657} x^{0.4140} \qquad (16.1)$$

As we stated previously, overall we are very satisfied with this model and will use it with confidence as a basis for developing our market position, a market-based pay target, and a salary increase budget for the CEO. Typically there is not a pay range for the CEO, but just a pay target.

In Chapter 10, using the SEE, we had calculated that on "average" the actual CEO pay values vary by about 42% from the predicted average CEO pay. Our experience is that there is wide variation in CEO pay, even with similar company sales.

BPD has sales of $5 billion or $5,000 million. To calculate the predicted average pay for a company of that size, we use the formula developed for the model.

$$y' = 10^{1.657} x^{0.4140}$$

$$y' = 10^{1.657} \times 5,000^{0.4140} = 1,543$$

The predicted average CEO pay for a company of our size is $1,543,000.

4. *Compare Employee Pay with the Market Model.* Keep in mind that the CEO pay we have been analyzing is base pay. The BPD board may have other compensation items, such as short-term incentives, long-term incentives, equity, and perquisites, as part of the total package. The BPD CEO base pay is $1,900,000. The comparison with the market model is shown in Table 16.1 along with a plot in Figure 16.2. In the figure, the BPD CEO pay is represented by a triangle ▲ .

**TABLE 16.1   MARKET ANALYSIS**

| | | | | | | |
|---|---|---|---|---|---|---|
| Market Analysis: Comparison of CEO Base Pay with the Market Model | | | | | | |
| BPD Sales | BPD CEO Base Pay | Market Model Predicted Average Base Pay | Difference | % BPD from Market Model | % to Meet Market Model | Average% of CEO Base Pay from Market Model |
| $5 billion | 1,900,000 | 1,543,000 | −357,000 | 23.1 | −18.8 | 42 |

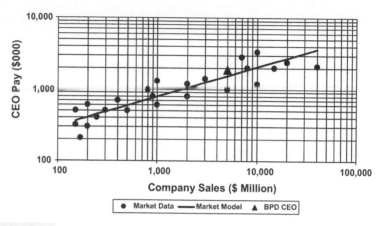

**FIGURE 16.2**  BPD CEO PAY AND MARKET MODEL

BPD's base pay is 23.1% above the market. This sounds like a lot, but in the context that the "average" percent difference from the market model is 42%, it is very reasonable. This is apparent by looking at the chart.

This table and the chart will be used by the BPD board as input for the total compensation package for the CEO for the new salary administration year. As mentioned above, at this level, there are additional factors to consider, such as short-term incentives, long-term incentives, equity, and perquisites, and the board will look at the total package and how these various factors are balanced with each other and linked with individual and company performance in making a decision on each one, including the base pay.

Typically there is not a formal salary structure with a minimum and maximum at this level. In addition, there is typically not a formal salary increase budget for the CEO's base pay, as the base pay is integrated with the total compensation package.

## PRACTICE PROBLEMS

**16.1**  You have the job of developing a market model for base salary of the president of BPD's European Division. The BPD European Division has revenues of €1 billion (€1,000 million). The president of that division has a base salary of €1.5 million (€1,500 thousand). It is October 2012 and you are preparing a base salary market model for the 2013 salary administration year. All executive raises are given on January 1. You have recently obtained an executive compensation survey, with an effective date of January 1, 2012. The data are shown in Table 16.2 and also in Data Set 37. From the survey the general market movement for executive-level jobs is 5.0% annually for 2012.

How much will you age the data from the survey?

**TABLE 16.2    SURVEY DATA FOR
PRESIDENT JOBS**

| Sales (€Million) | President Base Pay (€000) |
|---|---|
| 160 | 420 |
| 180 | 240 |
| 200 | 300 |
| 220 | 650 |
| 300 | 400 |
| 500 | 800 |
| 600 | 500 |
| 800 | 1,100 |
| 900 | 800 |
| 1,000 | 500 |
| 1,400 | 1,800 |
| 2,000 | 800 |
| 3,000 | 1,500 |
| 5,000 | 1,000 |
| 7,000 | 3,800 |
| 9,000 | 1,200 |
| 10,000 | 3,300 |
| 17,000 | 1,400 |
| 20,000 | 4,500 |
| 35,000 | 2,100 |
| 50,000 | 7,000 |
| 87,000 | 5,000 |

**16.2**  Age the survey data appropriately.

**16.3**  Plot the data. What kind of relationship is shown by the plot?

**16.4**  Convert the linear plot to a log–log plot. What kind of relationship is shown by this plot?

**16.5**  Conduct a regression. State the market model equation, the coefficient of determination, and the standard error of estimate.

**16.6**  Plot, interpret, and evaluate the model.

**16.7**  What is the predicted average pay for a president of the size of the BPD European Division? Complete Table 16.3 to identify the market position.

**TABLE 16.3    TEMPLATE FOR MARKET ANALYSIS**

| | | | | | | |
|---|---|---|---|---|---|---|
| | | Market Analysis for Base Pay: President European Division | | | | |
| BPD Sales, Euros | BPD President Base Pay, Euros | Market Model Predicted Average Base Pay | Difference | % BPD from Market Model | % to Meet Market Model | Average% from Market Model |

**16.8** Plot the BPD president's salary on the chart.

**16.9** Does the salary of the BPD president seem out of line with the other salaries on this chart? Why or why not?

**16.10** What are other parts of the direct remuneration package that the BPD board will consider when setting the base salary for 2013?

CHAPTER $17$

# Multiple Linear Regression

## 17.1 WHAT IT IS

Up to this point in modeling we have been dealing with just one $x$-variable. Sometimes we transform it into logarithms and sometimes we add powers of it, but it is still just one $x$-variable.

Often we can predict one variable quite accurately in terms of another. However, it stands to reason that predictions should improve if we consider additional relevant information. For example, we should be able to make better predictions of a person's pay if we consider not only his grade but also his experience in the field, education, time in grade, and performance. Multiple linear regression is a regression with more than one $x$-variable, and is a very powerful tool for analyzing several variables simultaneously.

There are many mathematical formulas that can express relationships between the $y$-variable and more than one $x$-variable, but those most commonly used in statistics are linear equations of the form

$$y = b_0 + b_1 x_1 + b_2 x_2 + b_3 x_3 + \cdots + b_k x_k$$

Here, $y$ is the variable of interest that is to be predicted, $x_1$, $x_2$, $x_3$, ..., $x_k$ are the $k$ known variables on which predictions are to be based, and $b_0$, $b_1$, $b_2$, $b_3$, ..., $b_k$ are numerical regression coefficients (intercept and slopes) that have to be determined on the basis of the given data.

When you conduct a multiple linear regression you are a detective trying to solve a mystery. You are on a search mission trying to discover what factors, both singly and in combination, are really important in influencing the variability of the $y$-variable. There is a lot of trial and error involved, where you try different combinations of the $x$-variables to discover meaningful results.

*Statistics for Compensation: A Practical Guide to Compensation Analysis,* By John H. Davis
Copyright © 2011 John Wiley & Sons Inc.

## 17.2 SIMILARITIES AND DIFFERENCES WITH SIMPLE LINEAR REGRESSION

There are many *similarities* between multiple linear regression and simple linear regression.

- The criterion for determining the equation is the method of least squares.

- We are predicting or estimating *averages*. We are predicting average pay, or average raise, or average bonus for given values of the x-variables.

- The interpretation of the coefficient of determination, $r^2$, is exactly the same as in simple linear regression. It is the proportion of variation of the y-variable we have explained by the particular linear combination of the x-variables.

  Indeed, a major driving force to consider multiple linear regression is to increase the value of $r^2$, the explanatory power of the model. It turns out that when you add an additional x-variable, $r^2$ never decreases. It may not increase much if the additional variable doesn't really contribute to the model, but it won't go down.

- The interpretation of the standard error of estimate, SEE, is identical. It represents an "average" difference between the actual y-values and those predicted by the model.

- The interpretation of the partitioning of the sum of squares is identical. The SS total is a measure of the total variation of the y-variable about its mean. The SS residual is a measure of the variation of the y-variable about the predicted values, and is the part of the total variation of the y-variable that the model does not explain. The SS regression is a measure of the amount of total variation of the y-variable that the model explains.

There are some *differences* between simple linear regression and multiple linear regression.

- In simple linear regression, the equation $y = a + bx$ describes a one-dimensional straight line in two dimensions. In multiple linear regression with two x-variables, the equation $y = b_0 + b_1 x_1 + b_2 x_2$ describes a two-dimensional plane in three dimensions. With $k$ x-variables, the equation $y = b_0 + b_1 x_1 + b_2 x_2 + b_3 x_3 + \cdots + b_k x_k$ describes a $k$-dimensional object in $k + 1$ dimensions.

- The interpretation of the intercept $b_0$ is slightly different. It is the predicted average value of $y$ when *all* the x-variables are zero.

- The interpretation of the slopes $b_1$, $b_2$, $b_3$, ..., $b_k$ is very different and is the *key difference between simple linear regression and multiple linear regression.*

Each slope is the impact its associated *x*-variable has (i.e., when that *x*-variable increases in value by 1) on the *y*-variable *taking into account the other x-variables*. Another way of stating this is that each slope is the impact its associated *x*-variable has on the *y*-variable *holding the other x-variables constant*.

This interpretation is what gives multiple linear regression its power. You take into account all the other *x*-variables when examining each of them individually.

- The number of data points you need to feel comfortable with your model and the conclusions depend on the number of *x*-variables.

A conservative rule of thumb is that you should have at least 10 data points for each *x*-variable plus 10 for the *y*-variable. So if your model has three *x*-variables, you will want to have at least 40 data points to have a high comfort level with the validity of the model.

A less conservative rule of thumb is that you should have at least 10 data points for each *x*-variable. So if your model has three *x*-variables, you will want to have at least 30 data points to have a good comfort level with the validity of the model.

Sometimes you are given a set of data with which to work, so the number of data points is fixed and you have to work with what you have.

---

**CASE STUDY 10**

The BPD Human Resources Vice President has received complaints from some employees in the Denver office that there seems to be a bias against employees who have no college degree. They told her that the degreed employees get higher percentage raises for doing the same job. They reported that the average raise given to nondegreed employees was 2.9% while the average raise given to degreed employees was 3.7%. She verified those figures, and asked you to investigate the basis for recent salary raises given to the 40 employees in the Denver office.

---

## 17.3 BUILDING THE MODEL

We will be covering just the highlights of multiple linear regression in a descriptive statistics context, so the compensation practitioner can comfortably conduct simple analyses and properly interpret the results. For multiple linear regression in a sampling context with statistical inference, see Appendix A. For more in-depth analyses, a practicing statistician familiar with compensation and human resources issues should be consulted.

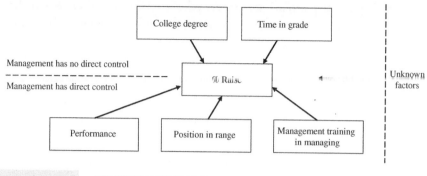

**FIGURE 17.1    PERCENT RAISE MODEL**

We will use the five steps in the model building process shown in Figure 6.3. As we start, it is important to keep in mind the following suggestion.

> You should use both tabular tools and graphical tools in your analysis, as one method may show something the other doesn't.

1. *Specify the Problem.*

   Your task is to investigate the reasons for the difference in average salary increases. This is important because these complaints may indicate that there may be some factors that are not job related that were the reasons for some of the raises. If this is true, then she wants the situation addressed and corrected.

2. *Generate Critical Factors.*

   You meet with the Denver HR management team and after a spirited brainstorming session, come up with the following model shown in Figure 17.1.

   You have data on all the factors except whether or not the managers had training in managing. Hence, the model variables you will be investigating are shown in Table 17.1.

**TABLE 17.1    MODEL VARIABLES**

| Variable | Symbol | Units |
|---|---|---|
| % Raise | $y$ | Percent |
| Performance | $x_1$ | 5 = outstanding; 1 = unsatisfactory |
| Position in range | $x_2$ | Salary as % of midpoint |
| College degree | $x_3$ | 0 = no; 1 = yes |
| Time in grade | $x_4$ | Years |

The dependent variable, $y$, is the problem variable, namely, % raise. Its variability is what we are trying to explain. The $x$'s are the independent variables, and are what we are using to try to explain the variability of the dependent variable.

3. *Identify the Relationships, If Any, Between the Factors and the Problem.*

We gather the following data in Table 17.2 on the 40 employees under study. The data are also given in Data Set 38. The reader might want to enter these data in a computer that has statistical software and follow along.

Note that the college degree variable, $x_3$, is an indicator variable (also called a dummy variable or a categorical variable) consisting of 0s and 1s (a binary variable) to designate whether or not the employee has a college degree. The value of 1 is usually assigned to the category of interest as compared to a base, or reference category. In this case, the base category is having no degree and the category of interest is having a college degree. This coding scheme makes the corresponding coefficient of that variable easy to interpret.

When there are more than two categories, special coding rules apply, discussed in Appendix A.

We scan the data by calculating the average, minimum, and maximum for each variable and nothing seems out of line. No raises go over 6%. The salaries are all within 80–120% of midpoint, the limits of our salary structure. If there were extreme values, or outliers, we would need to ensure they were valid, as one outlier can skew the results in a particular direction, thereby leading to inflated regression coefficients. We could also create histograms of each variable, but decide not to.

We separately plot % raise, the $y$-variable, against each of the $x$-variables to identify if any relationship exists, and if so, what kind. We do this to gain a better understanding of the starting points for our models. See Figures 17.2–5.

There is a relationship between % raise and performance level. It is positive, linear, and not perfect. It is what we would expect (or hope for), as our guidelines give higher percent raises to higher levels of performance, all other things being equal.

There is a relationship between % raise and position in range. It is negative, linear, and not perfect. There is a lot of scatter. It is what we would expect as our guidelines give lower percent raises for higher positions in range, all other things being equal.

It is hard to tell if there is a relationship between % raise and degree, due to the overlapping data points, even though the tabular data indicated those without a degree averaged 2.9% raise and those with a degree averaged 3.7%.

**TABLE 17.2    DATA FOR RAISE INVESTIGATION**

| Emp. No. | $y$ (% Raise) | $x_1$ (Performance) | $x_2$ (Position in Range) | $x_3$ (Degree) | $x_4$ (Time in Grade) |
|---|---|---|---|---|---|
| 1 | 5.5 | 5 | 89 | 1 | 6 |
| 2 | 1.6 | 3 | 119 | 0 | 4 |
| 3 | 5.8 | 5 | 81 | 1 | 1 |
| 4 | 2.5 | 3 | 100 | 1 | 1 |
| 5 | 4.6 | 4 | 90 | 1 | 3 |
| 6 | 4.0 | 4 | 83 | 1 | 3 |
| 7 | 2.0 | 3 | 118 | 0 | 5 |
| 8 | 4.0 | 5 | 114 | 0 | 10 |
| 9 | 1.5 | 2 | 94 | 1 | 9 |
| 10 | 3.6 | 3 | 93 | 1 | 9 |
| 11 | 4.0 | 4 | 106 | 0 | 11 |
| 12 | 4.3 | 4 | 89 | 1 | 3 |
| 13 | 3.6 | 4 | 99 | 1 | 10 |
| 14 | 1.7 | 3 | 111 | 0 | 7 |
| 15 | 3.3 | 3 | 87 | 1 | 5 |
| 16 | 2.8 | 4 | 108 | 0 | 10 |
| 17 | 3.0 | 4 | 116 | 0 | 12 |
| 18 | 4.5 | 4 | 86 | 1 | 4 |
| 19 | 1.5 | 3 | 109 | 0 | 6 |
| 20 | 3.4 | 4 | 98 | 1 | 8 |
| 21 | 2.5 | 4 | 117 | 0 | 9 |
| 22 | 3.0 | 4 | 116 | 0 | 10 |
| 23 | 0.0 | 1 | 113 | 0 | 3 |
| 24 | 5.0 | 5 | 104 | 0 | 12 |
| 25 | 4.2 | 5 | 96 | 1 | 7 |
| 26 | 2.8 | 3 | 100 | 0 | 7 |
| 27 | 2.5 | 4 | 108 | 0 | 10 |
| 28 | 3.2 | 4 | 97 | 1 | 10 |
| 29 | 5.9 | 5 | 85 | 0 | 15 |
| 30 | 0.0 | 2 | 113 | 0 | 4 |
| 31 | 5.0 | 4 | 86 | 1 | 3 |
| 32 | 4.9 | 4 | 92 | 0 | 13 |
| 33 | 2.5 | 3 | 95 | 0 | 8 |
| 34 | 3.7 | 3 | 87 | 1 | 2 |
| 35 | 4.3 | 4 | 86 | 0 | 14 |
| 36 | 3.3 | 3 | 92 | 1 | 7 |
| 37 | 4.9 | 4 | 91 | 1 | 5 |
| 38 | 4.7 | 4 | 88 | 0 | 10 |
| 39 | 0.0 | 1 | 94 | 1 | 7 |
| 40 | 2.8 | 3 | 100 | 1 | 10 |
| Average | 3.31 | 3.6 | 98.8 | 0.5 | 7.3 |
| Minimum | 0.0 | 1 | 81 | 0 | 1 |
| Maximum | 5.9 | 5 | 119 | 1 | 15 |

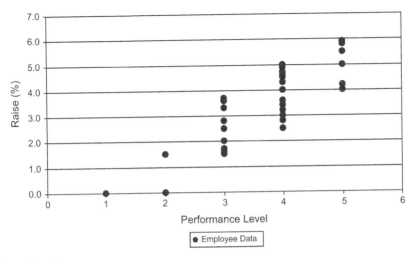

**FIGURE 17.2    PERCENT RAISE AND PERFORMANCE**

There does not appear to be any relationship between % raise and time in grade.

This preliminary graphical examination uncovered two fruitful areas to start the investigation to explain % raise, namely, performance and position in range.

4. *Quantify the Relationship and Analyze.*

This is where we conduct the regressions. We will use the variable designations already described in Table 17.1.

As we stated previously, model building is a trial-and-error process. We *build* a model with different combinations of x-variables to gain

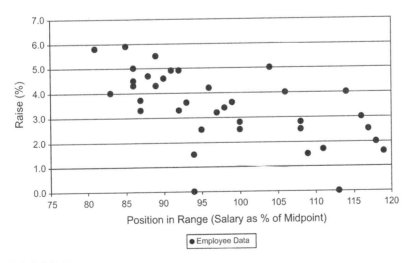

**FIGURE 17.3    PERCENT RAISE AND POSITION IN RANGE**

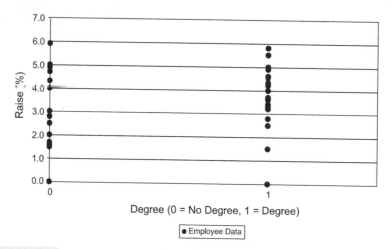

**FIGURE 17.4    PERCENT RAISE AND DEGREE**

understanding of the various relationships, rather than just throwing in all the x-variables at once and being done with it.

In building a model we must decide in what order to enter each x-variable in the model. There are several ways to do it. Here are the four most common.

- Specify the order of entry, the order usually based on the statement of the problem and your subject matter knowledge. Here you regress on the first x-variable of choice, then add the second, then the third, and so on.

- Forward stepwise entry. Here the criterion for entry is purely data driven, using statistical criteria such as the best improvement in the coefficient of determination, $r^2$. The first x-variable entered is the one

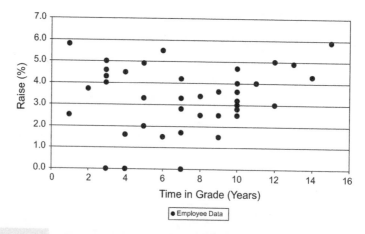

**FIGURE 17.5    PERCENT RAISE AND TIME IN GRADE**

**TABLE 17.3     VARIABLES FOR ALL-SUBSETS REGRESSIONS**

| | $x_1$ (Performance) | $x_2$ (Position in Range) | $x_3$ (Degree) | $x_4$ (Time in Grade) |
|---|---|---|---|---|
| One $x$-variable | | | | |
| Model 1 | X | | | |
| Model 2 | | X | | |
| Model 3 | | | X | |
| Model 4 | | | | X |
| Two $x$-variables | | | | |
| Model 5 | X | X | | |
| Model 6 | X | | X | |
| Model 7 | X | | | X |
| Model 8 | | X | X | |
| Model 9 | | X | | X |
| Model 10 | | | X | X |
| Three $x$-variables | | | | |
| Model 11 | X | X | X | |
| Model 12 | X | X | | X |
| Model 13 | X | | X | X |
| Model 14 | | X | X | X |
| Four $x$-variables | | | | |
| Model 15 | X | X | X | X |

with the highest correlation with $y$. Subsequent $x$-variables are selected using analogous criteria. At each step, the overall regression is checked for improvement in $r^2$ and for certain statistical "cutoff" criteria that tell when to stop adding variables

- Enter all $x$-variables into the model at the same time.
- Conduct "all-subsets" regressions. Here the $y$-variable is regressed against all possible combinations of the $x$-variables. This option includes all "specify the order of entry" models as well as entering all $x$-variables in the model at the same time. However, it is a lot of work unless the software has this as a programmed option.

If we were to conduct all-subsets regressions, there would be, with four $x$-variables, 15 models ($2^4 - 1$), with the indicated $x$-variables shown in Table 17.3.

Standard statistical software will usually give you all these options, so that you can conduct the regressions with ease. We will not go into how that is done here, but will simply report the results.

For this example we will specify the order of entry. The first $x$-variable will be performance.

## First $x$-Variable

Even though we will start off with performance, it is always useful, to gain insights, to do all the simple linear regressions. The results are shown in Table 17.4.

We will interpret these terms, using problem interpretations.

- *Model 1: % Raise Versus Performance.*

  Here, the intercept of $-1.39$ does not have a problem interpretation as we do not have a zero performance level. For the slope, the predicted average % raise increases by 1.30% for each performance level. The model has explained 72.5% of the variability of % raise by performance. The "average" difference between the actual % raise and the predicted average % raise is approximately 0.80%.

  If we were to use the equation to predict, for example, the average % raise for a 4-performer, we would get

$$y' = -1.39 + (1.30)(4) = -1.39 + 5.20 = 3.81\%$$

- *Model 2: % Raise Versus Position in Range.*

  Here also, the intercept does not have a problem interpretation as we do not have employees with salaries at 0% of midpoint. For the slope, the predicted average % raise decreases by 0.08% for every % of midpoint increase. The model has explained 37.9% of the variability of % raise by position in range. The "average" difference between the actual % raise and the predicted average % raise is approximately 1.20%.

  If we were to use the equation to predict, for example, the average % raise for a person with a salary at 80% of midpoint (at the bottom of the salary range), we would get

$$y' = 11.38 - (0.08)(80) = 11.38 - 6.40 = 4.98\%$$

**TABLE 17.4**     **REGRESSION COEFFICIENTS FOR FOUR SIMPLE LINEAR REGRESSIONS**

| % Raise | | $b_0$ | Performance $b_1$ | Performance $x_1$ | Position in Range $b_2$ | Position in Range $x_2$ | Degree $b_3$ | Degree $x_3$ | Time in Grade $b_4$ | Time in Grade $x_4$ | $r^2$ | SEE |
|---|---|---|---|---|---|---|---|---|---|---|---|---|
| Model 1 | $y' =$ | $-1.39$ | $+ 1.30$ | $x_1$ | | | | | | | 0.725 | 0.80 |
| Model 2 | $y' =$ | $11.38$ | | | $-0.08$ | $x_2$ | | | | | 0.379 | 1.20 |
| Model 3 | $y' =$ | $2.94$ | | | | | $+ 0.75$ | $x_3$ | | | 0.064 | 1.47 |
| Model 4 | $y' =$ | $2.81$ | | | | | | | $+ 0.07$ | $x_4$ | 0.027 | 1.50 |

If we were to use the equation to predict, for example, the average % raise for a person with a salary at 120% of midpoint (at the top of the salary range), we would get

$$y' = 11.38 - (0.08)(120) = 11.38 - 9.60 = 1.78\%$$

The difference of 3.20% between the bottom and the top of the salary range makes sense with respect to the merit matrix guidelines that were given to the managers.

- *Model 3: % Raise Versus Degree.*

   The predicted average % raise is 2.94% for those with no degree and an increase of 0.75% raise for those with a degree. The model has explained 6.4% of the variability of % raise by degree. The "average" difference between the actual % raise and the predicted average % raise is approximately 1.47%.

   Here is how to use and interpret an equation with an indicator variable. In Model 3, if we substitute $x_3 = 0$ in the equation, we will get the predicted average % raise for an employee with no degree, since the indicator variable, $x_3$, has a value of 0 for no degree. The result is 2.94%.

$$y' = 2.94 + (0.75)(0) = 2.94 + 0 = 2.94\%$$

If we substitute $x_3 = 1$ in the equation, we will get the predicted average % raise for an employee with a degree, since the indicator variable, $x_3$, has a value of 1 for an employee with a degree. The result is 3.69%.

$$y' = 2.94 + (0.75)(1) = 2.94 + 0.75 = 3.69\%$$

- *Model 4: % Raise Versus Time in Grade.*

   The predicted average % raise is 2.81% for no time in grade and increases by 0.07% for each additional year in grade. The model has explained 2.7% of the variability of % raise. The "average" difference between the actual % raise and the predicted average % raise is approximately 1.50%.

   If we were to use the equation to predict, for example, the average % raise for a person with 10 years in grade, we would get

$$y' = 2.81 + (0.07)(10) = 2.81 + 0.70 = 3.51\%$$

If we had to choose just one of these four models, the first one would be the likely candidate because it has the highest coefficient of determination, $r^2$, among the four, with a value of 72.5%.

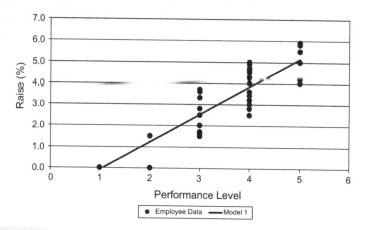

PREDICTED AVERAGE % RAISE AND PERFORMANCE

Note in Table 17.4 that $r^2$ and SEE have an inverse relationship. For the same number of data points and the same number of $x$-variables, the higher the amount of variability the model explains as measured by $r^2$, the closer the data points are to the model and hence the lower the SEE.

Figures 17.6–9 are the plots of the four models. We can see that the line describes the trend very nicely in each one, even though there is a lot of scatter.

As a side note for Figure 17.8, the line goes through the mean of the % raises of all those with no degree and the mean of the % raises of all those with a degree. This is a characteristic of a plot where the $x$-variable is an indicator variable.

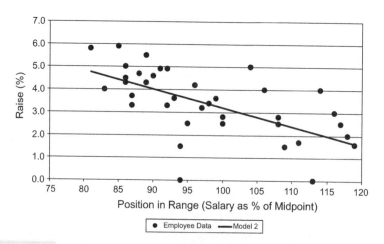

PREDICTED AVERAGE % RAISE AND POSITION IN RANGE

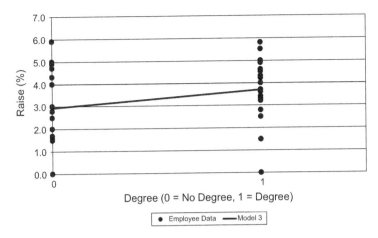

**FIGURE 17.8**    **PREDICTED AVERAGE % RAISE AND DEGREE**

## Second $x$-Variable

Now we will start to build the model with an additional $x$-variable. Since performance, $x_1$, is the main driver of % raise, and since position in range, $x_2$, is a secondary driver, we will start with Model 1 and add $x_2$ to Model 1. We end up with Model 5. If our primary focus instead had been on position in range, $x_2$, and we added performance, $x_1$, to Model 2, we would still end up with Model 5.

Table 17.5 shows the results of the regression with two variables, Model 5, along with the original regressions of each of the $x$-variables alone, Models 1 and 2.

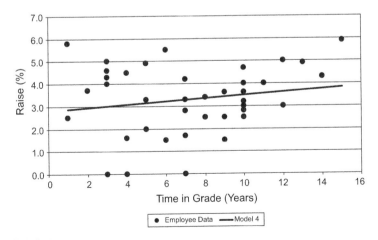

**FIGURE 17.9**    **PREDICTED AVERAGE % RAISE AND TIME IN GRADE**

**TABLE 17.5** REGRESSION RESULTS FOR PERFORMANCE AND POSITION IN RANGE

| % Raise | $b_0$ | Performance $b_1$ | $x_1$ | Position in Range $b_2$ | $x_2$ | Degree $b_3$ | $x_3$ | Time in Grade $b_4$ | $x_4$ | $r^2$ | SEE |
|---|---|---|---|---|---|---|---|---|---|---|---|
| Model 1 | $y' = -1.39 +$ | 1.30 | $x_1$ | | | | | | | 0.725 | 0.80 |
| Model 2 | $y' = 11.38$ | | | $- 0.08$ | $x_2$ | | | | | 0.379 | 1.20 |
| Model 5 | $y' = 5.03 +$ | 1.15 | $x_1$ | $- 0.06$ | $x_2$ | | | | | 0.916 | 0.45 |

If we were to use the equation for Model 5 to predict, for example, the average % raise for a person with a performance level of 4 and with a salary at 90% of midpoint, we would get

$$y' = 5.03 + (1.15)(4) - (0.06)(90) = 5.03 + 4.60 - 5.40 = 4.23\%$$

In multiple regression, which is when there is more than one $x$-variable, the coefficients have a special meaning, as mentioned previously. Let's look at Model 5.

$b_0$ is the intercept and it is the value of $y'$ when *all* of the $x_i$'s are zero. In this case, the predicted average % raise is 5.03% when the performance level is zero *and* when the position in range is 0% of midpoint. Since we don't have a zero level of performance or a salary at 0% of midpoint, the intercept does not have a problem interpretation. It is simply a locater.

In general for any variable $x_i$, $b_i$ is the slope associated with $x_i$. It is also called a regression coefficient. It is the change in $y'$ when $x_i$ increases by 1 *holding the other $x_i$'s constant*, or equivalently stated, *taking the other $x_i$'s into account*.

So for Model 5, $b_1$ is the slope associated with $x_1$. It is the change in $y'$ when $x_1$ increases by 1 *holding the other $x_i$'s constant*, or equivalently stated, *taking the other $x_i$'s into account*. For $b_1$, the predicted average % raise increases by 1.15% with each additional performance level *holding position in range constant* or *taking into account position in range*.

Comparing Model 5 to Model 1, why is the impact of performance less when you take position in range into account? The coefficient for performance goes from 1.30 without taking into account position in range to 1.15 taking into account position in range. The answer is that performance and position in range are statistically related to some degree (i.e., they are correlated to some degree), and as such, since position in range also impacts % raise, performance doesn't have to carry the whole "influencing" burden.

$b_2$ is the slope associated with $x_2$. It is the change in $y'$ when $x_2$ increases by 1 *holding the other $x_i$'s constant*, or equivalently stated, *taking the other $x_i$'s into* *account.* For $b_2$, the predicted average % raise decreases by 0.06% with each % of midpoint increase *holding the performance level constant* or *taking into account the performance level.*

Comparing Model 5 to Model 2, why is the impact of position in range less (in absolute value) when you take performance into account? The coefficient for position in range goes from $-0.08$ without taking into account performance to $-0.06$ taking into account performance. The answer is that position in range and performance are statistically related to some degree (i.e., they are correlated to some degree), and as such, since performance also impacts % raise, position in range doesn't have to carry the whole "influencing" burden.

With the new model, the coefficient of determination, $r^2$, tells us we have explained 91.6% of the variability of % raises by the linear combination of performance and position in range. The SEE tells us that the "average" difference between the actual % raise and the predicted average % raise is 0.45%.

Notice that $r^2$ has increased from 72.5% for Model 1 (performance only), and from 37.9% for Model 2 (position in range only), to 91.6% for Model 5 (performance and position in range). Adding new x-variables to a model will never cause $r^2$ to go down. It may not go up much, but it will never decrease. So are these changes enough to consider keeping both performance and position in range in the same model? The answer is "it depends."

When examining an increase in $r^2$, you may consider the following rule of thumb the author has found useful (or create your own, based on your experience and the context of the problem). Different practitioners have different rules of thumb.

| Increase in $r^2$ | Keep additional variable? |
|---|---|
| $\geq 5\%$ | Yes, keep it |
| $\geq 1\% - <5\%$ | Maybe, a judgment call |
| $<1\%$ | No, do not keep it |

You are trying to build a model that describes your problem, and adding an additional variable that increases the explanatory power by less than 1% only adds unnecessary complexity to the model.

If the context of the problem has performance as the primary focus for explaining the variability of % raises and it is important that performance be in the model, then the question becomes, should we add position in range to the performance model?

If, on the other hand, the context of the problem has position in range as the primary focus for explaining the variability of % raises and it is important that position in range be in the model, then the question becomes, should we add performance to the position in range model?

**TABLE 17.6    DECISION TABLE**

| | | | |
|---|---|---|---|
| Model 1 $r^2$ (performance only) | 72.5% | Model 2 $r^2$ (position in range only) | 37.9% |
| Model 5 $r^2$ (performance + position in range) | 91.6% | Model 5 $r^2$ (position in range + performance) | 91.6% |
| Change in $r^2$ | +19.1% | Change in $r^2$ | +53.7% |
| Decision | Keep position in range | Decision | Keep performance |

Table 17.6 is what we get using the above rule of thumb.

Based on our rule of thumb, we will keep both variables in the model.

When adding a new variable to a model, the $r^2$ of the new variable typically does not "add" totally to the $r^2$ of the model. We see that, in this instance, 72.5% + 37.9% $\neq$ 91.6%.

Notice that SEE decreased to 0.45, increasing the precision of the model prediction.

## Standardized Coefficient

There is another kind of coefficient in multiple linear regression called the standardized coefficient. It is also known as a $\beta$-weight, a beta weight, or a standardized weight. ($\beta$, lower case Greek letter beta, is the symbol statisticians often use for the standardized coefficient.) Standardized coefficients allow us to assess the *relative* impact of each $x$-variable in the model. They are only appropriate to use when there is more than one $x$-variable.

A standardized coefficient is much the same as a regression coefficient. Both describe the unique impact of each independent variable on the dependent variable, taking into account the effect of all other independent variables in the model. A standardized coefficient, however, is calculated only after all the variables ($y$ and $x$'s) have been transformed to $z$-scores.

The equation becomes a standardized equation with $z$-scores. For two $x$-variables, it is

$$z'_y = \beta_1 z_1 + \beta_2 z_2$$

where $z'_y$ is the predicted average $z$-score of the $y$-variable, $z_1$ is the $z$-score of the $x_1$ variable, $z_2$ is the $z$-score of the $x_2$ variable, $\beta_1$ is the standardized coefficient of the first variable, and $\beta_2$ is the standardized coefficient of the second variable.

The intercept of a standardized equation is always zero.

The formula of the standardized coefficient for the $k$th $x$-variable is

$$\beta_k = (b_k)(\text{Std dev of } x_k)/(\text{Std dev of } y)$$

Most statistical software programs automatically give you the standardized coefficients. Some spreadsheet programs do not. From the regression we get the
following results for Model 5:

$$z'_y = 0.752z_1 - 0.447z_2$$

To identify the *approximate* relative impact of the two $x$-variables on the predicted
average of the $y$-variable in a given model, we compare the absolute values
of the standardized weights. In this instance, to identify the relative impact of
performance and position in range on predicted average % raise, we compare the
absolute values of the two corresponding coefficients.

The absolute value of the standardized coefficient of 0.752 for performance
is 1.68 times the absolute value of the standardized coefficient of 0.447 for position in range, $0.752/0.447 = 1.68$. This means that performance has about 1.68
times more influence on % raise than position in range, even though the regular
regression coefficient of 1.15 for performance is 19 times the (absolute) value of
the regular regression coefficient of 0.06 for position in range.

An alternative comparison, the inverse of what we just did, is that the absolute
value of the standardized coefficient of 0.447 for position in range is 59% that
of the absolute value of the standardized coefficient of 0.752 for performance,
$(100)(0.447)/(0.752) = 59\%$. This means that position in range is about 59% as
influential as performance on % raise, even though the (absolute) value of the
regular coefficient of 0.06 for position in range is only 5.2% the value of the
regular regression coefficient of 1.15 for performance.

The standardized coefficients allow us to make such a comparison because
the variables have all been converted to $z$-scores, each with a mean of 0 and a
standard deviation of 1, and each without the original units (% raise, performance
level, % of midpoint). The variables are now, in a sense, all on an equal basis.

When there is only one $x$-variable, the standardized coefficient is equal to
the correlation between the $y$-variable and the $x$-variable. It does not have the
same interpretation when there is more than one $x$-variable, as shown here.

**Standardized Coefficient Perspective**    There is a very crude parallel
that may help shed some perspective regarding the interpretation of standardized coefficients that indicates the relative influences of the $x$-variables on the
variability of the $y$-variable.

It is similar to being on a teeter-totter (seesaw). The farther you are from the
balance point, the less weight you need to move the board. The closer you are,
the more weight you need. The distance to the balance point is equivalent to the
range of values for an $x$-variable, and the weight is equivalent to its regression
coefficient. Their product is equivalent to the relative influence on the dependent
variable. This is shown in Table 17.7.

**TABLE 17.7** STANDARDIZED COEFFICIENTS PERSPECTIVE

| Variable | Range of Values | Regression Coefficient (Absolute Value) | Product | Teeter-Totter Relative Scale of Product (Absolute Value) | Standardized Coefficient (Absolute Value) | Relative Scale of Standardized Coefficient (Absolute Value) |
|---|---|---|---|---|---|---|
| Performance | $5-1=4$ | 1.15 | 4.60 | 100 | 0.752 | 100 |
| Position in range | $119-81=38$ | 0.06 | 2.28 | 50 | 0.447 | 59 |

To facilitate comparisons, we set a relative scale of the product for performance at 100 and adjust the product for position in range accordingly, $(100)(2.28)/(4.60)=50$ (rounded). We create a similar relative scale for the standardized coefficients. Performance is 100 and position in range is $(100)(0.447)/(0.752)=59$ (rounded).

We see that our teeter-totter analogy indicates the same ballpark magnitude of the estimates of the relative influences of the $x$-variables using the regular coefficients, 100 and 50, as do the standardized coefficients, 100 and 59.

### Third $x$-Variable

We will now add a third variable, degree, which when combined with the first two results in Model 11, as noted in Table 17.3. The regression results are given in Table 17.8, which also shows the antecedent models for perspective.

If we were to use the equation for Model 11 to predict, for example, the average % raise for a person with a performance level of 4, a salary at 90% of midpoint, and a degree, we would get

$$y' = 5.68 + (1.14)(4) - (0.06)(90) - (0.17)(1) = 5.68 + 4.56 - 5.40 - 0.17 = 4.67\%$$

**TABLE 17.8** REGRESSION RESULTS FOR PERFORMANCE, POSITION IN RANGE, AND DEGREE

| % Raise | $b_0$ | $b_1$ | $x_1$ | $b_2$ | $x_2$ | $b_3$ | $x_3$ | $b_4$ | $x_4$ | $r^2$ | SEE |
|---|---|---|---|---|---|---|---|---|---|---|---|
| Model 1 $y'=$ | −1.39 | + 1.30 | $x_1$ | | | | | | | 0.725 | 0.80 |
| Model 2 $y'=$ | 11.38 | | | − 0.08 | $x_2$ | | | | | 0.379 | 1.20 |
| Model 3 $y'=$ | 2.94 | | | | | + 0.75 | $x_3$ | | | 0.064 | 1.47 |
| Model 5 $y'=$ | 5.03 | + 1.15 | $x_1$ | − 0.06 | $x_2$ | | | | | 0.916 | 0.45 |
| Model 11 $y'=$ | 5.68 | + 1.14 | $x_1$ | − 0.06 | $x_2$ | − 0.17 | $x_3$ | | | 0.917 | 0.45 |

**TABLE 17.9    DECISION TABLE**

| | |
|---|---|
| Model 5 $r^2$ (performance + position in range) | 91.6% |
| Model 11 $r^2$ (performance + position in range + degree) | 91.7% |
| Change in $r^2$ | +0.1% |
| Decision | Do not keep degree |

The interpretations of the coefficients are similar as before. The intercept has no problem interpretation because we do not have a zero level of performance or a 0% of midpoint.

For $b_1$, the predicted average % raise increases by 1.14% with each additional performance level *holding both position in range and degree constant or taking into account both position in range and degree.*

For $b_2$, the predicted average % raise decreases by 0.06% with each % of midpoint increase *holding both the performance level and degree constant or taking into account both the performance level and degree.*

For $b_3$, the predicted average % raise decreases by 0.17% with a degree *holding both the performance level and position in range constant or taking into account both the performance level and position in range.*

We see that compared to Model 5, adding degree resulted in virtually no improvement in $r^2$ or SEE. So based on our rule of thumb on keeping an additional variable, we would not keep this variable, as shown in Table 17.9.

The analysis so far tells us that once we take both performance and position in range into account, having a degree makes no difference.

However, even if we decided based on an increase in $r^2$ to keep degree in the model, we note that the coefficient of degree changed sign from a positive (+0.75) in Model 3, where we regressed on degree only, to a negative (−0.17) in Model 11, where we regressed on performance, position in range, and degree. This means that if we take into account performance and position in range, having a degree results in an average of 0.17% less raise. This is contrary to the original complaint that degreed employees were getting higher raises.

Even though in this case there was no improvement in the model with the addition of degree, the change in sign of the coefficient is very important for our investigation.

A change in sign of a coefficient for a variable from one model to another is a signal that multicollinearity may be present.

## Multicollinearity

Multicollinearity is a high correlation between one $x$-variable and another $x$-variable or between one $x$-variable and a linear combination of other $x$-variables. Another way to view this is that multicollinearity exists when one

**TABLE 17.10** CORRELATION MATRIX

| | Correlation, $n = 40$ | | | | |
| | % Raise | Performance | Position in Range | Degree | Time in Grade |
|---|---|---|---|---|---|
| % Raise | 1 | | | | |
| Performance | 0.852 | 1 | | | |
| Position in range | −0.615 | −0.223 | 1 | | |
| Degree | 0.252 | 0.000 | −0.639 | 1 | |
| Time in grade | 0.164 | 0.296 | 0.160 | −0.467 | 1 |

$x$-variable starts to become a surrogate for another $x$-variable or for a linear combination of other $x$-variables. Mathematically, when there is high multi-collinearity, the denominators of terms used to calculate the coefficients get close to zero, resulting in an unstable model. In that case, small changes in the data set will cause wild variations in the results, rendering the model unsuitable for predicting.

Ideally, all the $x$-variables in a multiple regression model should be statistically independent of each other. That is, the correlations among the $x$-variables should be zero. This is a critical assumption when interpreting the results of a multiple regression. However, total independence is rarely the case.

Let us look at a correlation matrix, which displays the correlations among all the variables. The correlation matrix for these variables is shown in Table 17.10. Statistical software will produce this easily.

The correlation of a variable with itself is always 1.

We would like high correlations (in absolute value) between the $x$-variables and the $y$-variable, and low to no correlations (in absolute value) among the $x$-variables. Notice the correlation between degree and position in range. It has a value of −0.639. The negative sign means that degreed employees ($x_3 = 1$) tend to have low positions in range. But it is the large absolute value that gives us concern.

Let us look at a histogram of the position in range. Figure 17.10 shows that all the degreed employees are in the lower half of the range, and most of the nondegreed employees are in the upper half of the salary range.

Having or not having a degree is almost a surrogate for being in the lower or upper half of the salary range, respectively.

The issue here is that multicollinearity violates the assumption of independence of $x$-variables so much that the interpretations of the coefficients become invalid.

For example, when we say for $b_2$ in Model 11, "holding performance and degree constant, each additional percent of midpoint decreases the % raise by

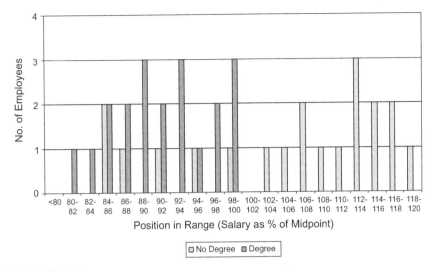

**FIGURE 17.10    DISTRIBUTION OF POSITION IN RANGE**

0.06," we have a contradiction, because we can't hold degree constant and at the same time increase percent of midpoint. They both move together. One is a surrogate for the other.

Similarly, when we say for $b_3$ in Model 11, "holding performance and position in range constant, having a degree decreases the % raise by 0.17," we have a contradiction, because we can't hold position in range constant and at the same increase the degree variable. They both move together. One is a surrogate for the other.

The solution to multicollinearity is that you should not enter two highly correlated x-variables, or an x-variable that is highly correlated with a linear combination of other x-variables, in the same model. In this case, choose position in range or degree, but not both. At the least, be very, very cautious in choosing both.

How high a correlation, in absolute value, is too high? It depends on the context of the problem and your experience. Some practitioners say that 0.90 or more correlation is too high. Others say that 0.60 or more correlation is too high. Some software will warn you at various thresholds.

The admonition here is that you should always look at the correlation matrix showing the two-way correlations among the x-variables to identify if there is a potential multicollinearity problem. This may influence how you enter the x-variables into the model.

If you enter one variable at a time in building the model, a change of sign or a severe shift in the value of the coefficient of a variable may indicate possible multicollinearity, caused by high correlations or close linear combinations.

**TABLE 17.11**     MERIT RAISE GUIDELINES

| Performance Level | Position in Range, % of Midpoint | | |
|---|---|---|---|
| | 80–93 | 94–106 | 107–120 |
| 5 | 5–6% | 4–5% | 3–4% |
| 4 | 4–5% | 3–4% | 2–3% |
| 3 | 3–4% | 2–3% | 1–2% |
| 2 | 2–3% | 1–2% | 0 |
| 1 | 0 | 0 | 0 |

For our example, we will use a threshold of 0.60. We could have used 0.80 or 0.70 as a threshold. Hence, we decide not to have both position in range and degree in the same model.

To summarize the issue of multicollinearity, the higher the absolute value of correlation between $x$-variables, the less confidence there is in the interpretation of the model. This notion bears repeating, stated another way.

The more independent the $x$-variables are among each other, the more valid are the interpretations of the model, including the regular coefficients and the standardized coefficients.

So where are we in addressing the issue that degreed employees get higher raises than nondegreed employees? The regression analysis tells us that once we take both performance and position in range into account, having a degree provides no advantage with respect to % raise. The histogram in Figure 17.10 gives us our clue as to why nondegreed employees received lower raises. They got lower raises because, all other things being equal (e.g., performance), they were higher in the range than the degreed employees. The managers simply followed the BPD merit matrix guidelines, shown in Table 17.11.

These are typical guidelines that accelerate the growth of an employee's pay in the lower half of the salary range and decelerate the growth of an employee's pay in the upper half of the salary range.

So now we have a tentative answer to the original issue.

But we still have one more variable to investigate, time in grade.

We will add time in grade to performance and position in range to get Model 12, and obtain the results shown in Table 17.12, along with the results from the antecedent models.

If we were to use the equation for Model 12 to predict, for example, the average % raise for a person with a performance level of 4, a salary at 90% of midpoint, and 10 years in grade, we would get

$$y' = 5.06 + (1.14)(4) - (0.06)(90) + (0.01)(10) = 5.06 + 4.56 - 5.40 + 0.10 = 4.32\%$$

**TABLE 17.12  REGRESSION RESULTS FOR PERFORMANCE, POSITION IN RANGE, AND TIME IN GRADE**

| % Raise | $b_0$ | Performance | | Position in Range | | Degree | | Time in Grade | | $r^2$ | SEE |
|---|---|---|---|---|---|---|---|---|---|---|---|
| | | $b_1$ | $x_1$ | $b_2$ | $x_2$ | $b_3$ | $x_3$ | $b_4$ | $x_4$ | | |
| Model 1  $y' =$ | $-1.39$ | $+ 1.30$ | $x_1$ | | | | | | | 0.725 | 0.80 |
| Model 2  $y' =$ | 11.38 | | | $- 0.08$ | $x_2$ | | | | | 0.379 | 1.20 |
| Model 4  $y' =$ | 2.81 | | | | | | | $+ 0.07$ | $x_4$ | 0.027 | 1.50 |
| Model 5  $y' =$ | 5.03 | $+ 1.15$ | $x_1$ | $- 0.06$ | $x_2$ | | | | | 0.916 | 0.45 |
| Model 12  $y' =$ | 5.06 | $+ 1.14$ | $x_1$ | $- 0.06$ | $x_2$ | | | $+ 0.01$ | $x_4$ | 0.916 | 0.45 |

**TABLE 17.13  DECISION TABLE**

| | |
|---|---|
| Model 5 $r^2$ (performance + position in range) | 91.6% |
| Model 12 $r^2$ (performance + position in range + time in grade) | 91.6% |
| Change in $r^2$ | None |
| Decision | Do not keep time in grade |

We note that the coefficient of determination of 91.6% and the SEE of 0.45 are the same as in Model 5, without the time in grade variable.

It appears that taking performance and position in range into account, time in grade has virtually no impact.

Using our rule on keeping a variable, our decision is not to keep it, as shown in Table 17.13.

We will stop here for now, but you should try other combinations not shown here, as sometimes you will find a surprising result hidden in the data.

5. *Evaluate the Model.* We have already done some evaluation along the way, but will present a complete evaluation now for the final model, Model 5 with the independent variables of performance and position in range. Following are the criteria generally used in evaluating a multiple regression model. In multiple regression, the multiple correlation (between $y$ and $y'$) is not a useful criterion.

## 17.4 MODEL EVALUATION

### Regression Coefficients

Do they make sense in the context of the problem? Are there any surprises? In this case, does it make sense that the predicted average % raise is increased by 1.15% for every increase in performance level holding position in range constant? Does it make sense that the predicted average % raise is decreased by 0.06% for every % of midpoint increase holding performance constant?

For this problem we will assume that the impacts of performance and position in range make sense. We may have been a little surprised that having a degree did not make a difference, given the complaint, but were not surprised that the combination of performance and position in range explained almost 92% of the variability of % raise. However, we were surprised at the high correlation between degree and position in range. That may bear further investigating. We didn't know what to expect about the effect of time in grade.

## Standardized Coefficients

Usually standardized coefficients create more of a revelation than anything else regarding the relative impacts of the $x$-variables on the variation of the $y$-variable. We learned that performance has about 1.68 times more influence on predicted average % raise than position in range.

## Coefficient of Determination

The higher the $r^2$, the better the model can explain the variability of the dependent variable, and the better the predictive ability of the model. In this problem, $r^2$ has a value of 91.6%. There are other factors that are not in our model that may explain the other 8.4%. Whether 91.6% is "good" or "good enough" has to be determined by the context of the problem and your experience. In the author's experience, if you can explain over 90% of the variability of any dependent variable with two independent variables, that is very good.

## Standard Error of Estimate

The lower the SEE, the more precise the predictions. In this problem, SEE has a value of 0.45. Whether this is "good" or "good enough" also has to be determined by the context of the problem and your experience. We will assume that this is a nice tight SEE.

## Multicollinearity

The lower the correlations are among the $x$-variables, the more valid are the interpretations of the model. We have eliminated degree as a factor in this problem, and the maximum correlation, in absolute value, among the remaining $x$-variables is 0.296. Again, whether this is too high or not has to be determined by the context of the problem and your experience. This particular value, in the author's experience, is not too high.

## Simplicity

The fewer the number of $x$-variables, the better, because the model will be easy to understand, explain, and accept. How many $x$-variables are too much? When you get much over three or four independent variables, it starts to get complex. But just like the other criteria, what is "too much" depends on the context of the problem and your experience. Some situations may warrant six or seven independent variables. In this problem, we have two $x$-variables in our final model. This is a very understandable model.

## Common Sense

Here is where you use your subject matter expertise along with the context of the problem and your experience. Does the model make sense? Does it make sense, for example, that performance and position in range affect % raise? Apply the commonsense criterion to all parts of the model.

To help the credibility of your model, the $x$-variables should have some logical tie to the $y$-variable. A high correlation does *not* imply cause and effect, but it does help if there is in fact some reasonable tie germane to the problem at hand.

## Acceptability

Will your model be accepted? Are the factors relevant to the situation? Did you get input during the brainstorming session from key decision makers? Did you use their input? Did you keep them informed of your progress and findings along the way?

## Reality

Finally, does the model work when it is applied? You never know whether it is any good until you apply it and make decisions from it. Your numbers may look fine, but you may have used incorrect factors, or had erroneous data, or made faulty assumptions. Reality is always the final arbiter.

## Decision

We have developed a model with which we are comfortable. The conclusion is that after performance and position in range are taken into account, having a degree did not give an advantage for % raise. The decision is that the

BPD Human Resources Vice President will communicate this conclusion to the affected employees and close the case.

## 17.5 MIXED MESSAGES IN EVALUATING A MODEL

Sometimes one criterion may indicate you have a good model while another might indicate that you do not. Here are some examples.

### $r^2$ Versus Common Sense

There used to be a high predictability of human births in English villages by the number of nesting storks. The $r^2$ was high but the common sense for a causal relationship between the variables is lacking.

### $r^2$ Versus Simplicity

We can keep on adding variables to drive up the $r^2$ of a model, but after a while it gets to be too complex to understand and describe. We might be tempted to consider age, number of dependents, location, work hazards, inflation, company size, federal regulations, medical liability costs, jury awards, incidence of illness, and so on in order to explain rising health care costs, but our model would then be too complex.

### Simplicity Versus Acceptability

Suppose you are examining a point factor job evaluation system, where one or two factors explain over 90% of the variation in predicted grade. However, for the system to be acceptable to management and to the employees who went through a lot of trouble to furnish a lot of job-related information, you had best include at least a few of the asked-for items.

In all cases, it is a judgment call on balancing the criteria of the "goodness" of a model, as all model building involves a series of trade-offs.

## 17.6 SUMMARY OF REGRESSIONS

For completeness and for further study on the reader's part, Table 17.14 with regular regression resulting in regular coefficients and Table 17.15 with regression on $z$-scores resulting in standardized coefficients contain the results of the "all-subsets" regressions. These include the models listed in Table 17.3 that we did not discuss.

**TABLE 17.14    REGRESSION RESULTS FOR ALL-SUBSETS REGRESSIONS**

| | | | Performance | | | Position in Range | | | Degree | | | Time in Grade | | | | |
|---|---|---|---|---|---|---|---|---|---|---|---|---|---|---|---|---|
| % Raise | | $b_0$ | | $b_1$ | $x_1$ | | $b_2$ | $x_2$ | | $b_3$ | $x_3$ | | $b_4$ | $x_4$ | $r^2$ | SEE |
| Model 1 | $y' =$ | −1.39 | + | 1.30 | $x_1$ | | | | | | | | | | 0.725 | 0.80 |
| Model 2 | $y' =$ | 11.38 | | | | − | 0.08 | $x_2$ | | | | | | | 0.379 | 1.20 |
| Model 3 | $y' =$ | 2.94 | | | | | | | + | 0.75 | $x_3$ | | | | 0.064 | 1.47 |
| Model 4 | $y' =$ | 2.81 | | | | | | | | | | + | 0.07 | $x_4$ | 0.027 | 1.50 |
| Model 5 | $y' =$ | 5.03 | + | 1.15 | $x_1$ | − | 0.06 | $x_2$ | | | | | | | 0.916 | 0.45 |
| Model 6 | $y' =$ | −1.76 | + | 1.30 | $x_1$ | | | | + | 0.75 | $x_3$ | | | | 0.789 | 0.71 |
| Model 7 | $y' =$ | −1.25 | + | 1.35 | $x_1$ | | | | | | | − | 0.04 | $x_4$ | 0.734 | 0.80 |
| Model 8 | $y' =$ | 13.72 | | | | − | 0.10 | $x_2$ | + | 0.71 | $x_3$ | | | | 0.412 | 1.18 |
| Model 9 | $y' =$ | 11.12 | | | | − | 0.09 | $x_2$ | | | | + | 0.11 | $x_4$ | 0.449 | 1.15 |
| Model 10 | $y' =$ | 1.59 | | | | | | | + | 1.25 | $x_3$ | + | 0.15 | $x_4$ | 0.165 | 1.41 |
| Model 11 | $y' =$ | 5.68 | + | 1.14 | $x_1$ | − | 0.06 | $x_2$ | + | 0.17 | $x_3$ | | | | 0.917 | 0.45 |
| Model 12 | $y' =$ | 5.06 | + | 1.14 | $x_1$ | − | 0.06 | $x_2$ | | | | + | 0.01 | $x_4$ | 0.916 | 0.45 |
| Model 13 | $y' =$ | −1.85 | + | 1.29 | $x_1$ | | | | + | 0.81 | $x_3$ | − | 0.02 | $x_4$ | 0.790 | 0.72 |
| Model 14 | $y' =$ | 12.07 | | | | − | 0.09 | $x_2$ | + | 0.28 | $x_3$ | + | 0.10 | $x_4$ | 0.453 | 1.16 |
| Model 15 | $y' =$ | 5.71 | + | 1.14 | $x_1$ | − | 0.06 | $x_2$ | + | 0.19 | $x_3$ | + | 0.00 | $x_4$ | 0.918 | 0.46 |

**TABLE 17.15 REGRESSION RESULTS WITH STANDARDIZED COEFFICIENTS FOR ALL-SUBSETS REGRESSIONS**

Regressions with $z$-Scores. Coefficients Are Standardized Coefficients

| % Raise | | Performance | | Position in Range | | Degree | | Time in Grade | |
|---|---|---|---|---|---|---|---|---|---|
| | | $\beta_1$ | $z_1$ | $\beta_2$ | $z_2$ | $\beta_3$ | $z_3$ | $\beta_4$ | $z_4$ |
| Model 1 | $z'_y = +$ | 0.852 | $z_1$ | | | | | | |
| Model 2 | $z'_y = $ | | | $-$ 0.615 | $z_2$ | | | | |
| Model 3 | $z'_y = $ | | | | | $+$ 0.252 | $z_3$ | | |
| Model 4 | $z'_y = $ | | | | | | | $+$ 0.164 | $z_4$ |
| Model 5 | $z'_y = +$ | 0.752 | $z_1$ | $-$ 0.447 | $z_2$ | | | | |
| Model 6 | $z'_y = +$ | 0.852 | $z_1$ | | | $+$ 0.252 | $z_3$ | | |
| Model 7 | $z'_y = +$ | 0.880 | $z_1$ | | | | | $-$ 0.097 | $z_4$ |
| Model 8 | $z'_y = $ | | | $-$ 0.767 | $z_2$ | $-$ 0.238 | $z_3$ | | |
| Model 9 | $z'_y = $ | | | $-$ 0.658 | $z_2$ | | | $+$ 0.269 | $z_4$ |
| Model 10 | $z'_y = $ | | | | | $+$ 0.421 | $z_3$ | $+$ 0.361 | $z_4$ |
| Model 11 | $z'_y = +$ | 0.743 | $z_1$ | $-$ 0.487 | $z_2$ | $-$ 0.059 | $z_3$ | | |
| Model 12 | $z'_y = +$ | 0.746 | $z_1$ | $-$ 0.451 | $z_2$ | | | $+$ 0.015 | $z_4$ |
| Model 13 | $z'_y = +$ | 0.839 | $z_1$ | | | $+$ 0.272 | $z_3$ | $+$ 0.043 | $z_4$ |
| Model 14 | $z'_y = $ | | | $-$ 0.712 | $z_2$ | $-$ 0.093 | $z_3$ | $+$ 0.234 | $z_4$ |
| Model 15 | $z'_y = +$ | 0.745 | $z_1$ | $-$ 0.488 | $z_2$ | $-$ 0.063 | $z_3$ | $-$ 0.008 | $z_4$ |

## 17.7 DIGGING DEEPER

It is important to explore all avenues when conducting a multiple regression analysis, as sometimes there are hidden gems in the data that are not obvious at first glance.

When examining the distribution of position in range, you discovered that all the degreed employees were below the midpoint and most of the nondegreed employees were above the midpoint, shown in Figure 17.10.

You decide to segregate the data set into two parts—one for nondegreed employees and one for degreed employees—and conduct the analyses for each part.

The possible models are shown in Table 17.16. To help relate these models to the ones we have analyzed already, we are using the same numbering scheme for the $x$-variables as before, shown in Table 17.1, and not use the variable $x_3$, as it was used for degree. In the same vein for ease of comparison to the original models listed in Table 17.3, we are using a similar numbering scheme for the various models, using the prefix N for nondegreed and D for degreed employees.

We will look at a series of models, comparing the nondegreed data with the degreed data for each one as we go along.

**TABLE 17.16    MODELS SEPARATED BY DEGREE STATUS**

### Models for Nondegreed Employees

| | $x_1$ (Performance) | $x_2$ (Position in Range) | $x_4$ (Time in Grade) |
|---|---|---|---|
| One $x$-variable | | | |
| Model N1 | X | | |
| Model N2 | | X | |
| Model N4 | | | X |
| Two $x$-variables | | | |
| Model N5 | X | X | |
| Model N7 | X | | X |
| Model N9 | | X | X |
| Three $x$-variables | | | |
| Model N12 | X | X | X |

### Models for Degreed Employees

| | $x_1$ (Performance) | $x_2$ (Position in Range) | $x_4$ (Time in Grade) |
|---|---|---|---|
| One $x$-variable | | | |
| Model D1 | X | | |
| Model D2 | | X | |
| Model D4 | | | X |
| Two $x$-variables | | | |
| Model D5 | X | X | |
| Model D7 | X | | X |
| Model D9 | | X | X |
| Three $x$-variables | | | |
| Model D12 | X | X | X |

As before, we separately plot % raise, the $y$-variable, against each of the $x$-variables to identify if any relationship exists, and if so, what kind. We do this to gain a better understanding of the starting points for our models. We use one symbol for the nondegreed data and another for the degreed data. On each plot we also show the regression line and equation and the coefficient of determination for each of the two groups of data.

The first plot, Figure 17.11, shows % raise versus performance.

We can see that for each group there is a positive, linear, and nonperfect relationship. Since the solid line for the degreed group is higher than the dashed line for the nondegreed group, this visually shows that if you just look at performance only, on average, the degreed employees get higher % raises than the nondegreed employees.

Next we look at Figure 17.12, which shows % raise versus position in range.

This plot visually shows the clustering of degreed employees in the lower half of the range with concomitant higher % raises and the bulk of the nondegreed employees in the upper half of the range with concomitant lower % raises.

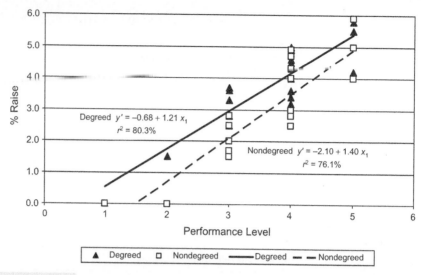

**FIGURE 17.11    PERCENT RAISE AND PERFORMANCE SEPARATED BY DEGREE STATUS**

We now look at Figure 17.13, which shows % raise versus time in grade.

For the degreed employees, there is a loose negative linear relationship, with a coefficient of determination of only 18.2%, and for the nondegreed employees there is a tight linear positive relationship with a coefficient of determination of 80.9%.

This is a surprise that there is such a high relationship between % raise and time in grade for the nondegreed employees. It appears that degree status is

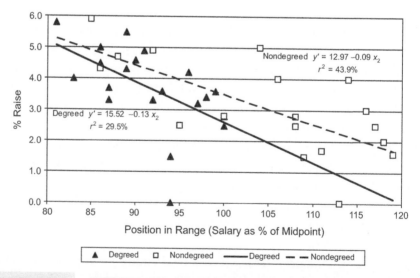

**FIGURE 17.12    PERCENT RAISE AND POSITION IN RANGE SEPARATED BY DEGREE STATUS**

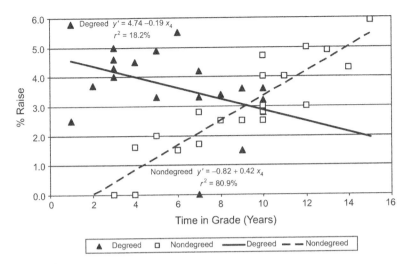

**FIGURE 17.13    PERCENT RAISE AND TIME IN GRADE SEPARATED BY DEGREE STATUS**

determining whether or not there is a relationship between % raise and time in grade. This explanation is complex and suggests that degree status acts as a contingent condition for the relationship between % raise and time in grade, as shown in Figure 17.14.

Sometimes there are relationships within the data set that are not apparent when you look at the data set as a whole, and this is a prime example of it. Figure 17.9 showed almost no relationship between % raise and time in grade when looking at the entire data set. But here in Figure 17.13 we see there is a relationship for the nondegreed employees.

Let us compare the one-variable models we have so far shown in Table 17.17.

For the nondegreed employees, there are two models with a high coefficient of determination: Model N1 of % raise versus performance with an $r^2$ of 76.1%, and Model N4 of % raise versus time in grade with an $r^2$ of 80.9%. If we had to choose just one of these to use, which would it be?

We use a rule of thumb similar to the one used to decide whether or not to keep an additional variable. As before, different practitioners have different rules of thumb.

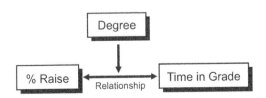

**FIGURE 17.14    MODEL OF DEGREE STATUS AFFECTING RELATIONSHIP BETWEEN RAISE AND TIME IN GRADE**

**TABLE 17.17**    SIMPLE LINEAR REGRESSION RESULTS SEPARATED BY DEGREE STATUS

### Regression Results for Nondegreed Employees

| % Raise | $b_0$ | Performance $b_1$ | $x_1$ | Position in Range $b_2$ | $x_2$ | Time in Grade $b_4$ | $x_4$ | $r^2$ | SEE |
|---|---|---|---|---|---|---|---|---|---|
| Model N1 $y' = -2.10$ | | $+ 1.40$ | $x_1$ | | | | | 0.761 | 0.80 |
| Model N2 $y' = 12.97$ | | | | $- 0.09$ | $x_2$ | | | 0.439 | 1.23 |
| Model N4 $y' = -0.82$ | | | | | | $+ 0.42$ | $x_4$ | 0.809 | 0.72 |

### Regression Results for Degreed Employees

| % Raise | $b_0$ | Performance $b_1$ | $x_1$ | Position in Range $b_2$ | $x_2$ | Time in Grade $b_4$ | $x_4$ | $r^2$ | SEE |
|---|---|---|---|---|---|---|---|---|---|
| Model D1 $y' = -0.68$ | | $+ 1.21$ | $x_1$ | | | | | 0.803 | 0.61 |
| Model D2 $y' = 15.52$ | | | | $- 0.13$ | $x_2$ | | | 0.295 | 1.16 |
| Model D4 $y' = 4.74$ | | | | | | $- 0.19$ | $x_4$ | 0.182 | 1.25 |

| Difference in $r^2$ | Use model with higher $r^2$? |
|---|---|
| $\geq 5\%$ | Yes. Use the model with a higher $r^2$ |
| $\geq 1\%$–$<5\%$ | Maybe, a judgment call |
| $<1\%$ | Flip a coin or use other criterion |

Using this rule, the decision is a judgment call, and we could choose either one. Hence, we can use criteria other than mathematical ones. If we just wanted to use a model for mathematical prediction purposes, we would probably use the one based on time in grade.

However, since basing % raises on time in grade goes against the BPD policy of rewarding performance, it appears that we need to do some further investigating.

For the degreed employees, there are no issues here. The models are what we would expect, with the highest coefficient of determination being with the model based on performance.

We now look at the correlation matrices for both groups, shown in Table 17.18, to see what we might discover.

One item that pops out is the very high correlation of 0.848 between performance and time in grade for the nondegreed employees. There is multicollinearity between those two $x$-variables, and we should not have them in the same model for predicting % raise.

The second item is that since there is a low correlation of 0.283 in absolute value between performance and time in grade for the degreed employees, perhaps degree status acts as a contingent condition for the relationship between performance and time in grade, as shown in Figure 17.15.

**TABLE 17.18    CORRELATION MATRICES SEPARATED BY DEGREE STATUS**

| Nondegreed Employees, Correlation, $n = 20$ | | | | |
| --- | --- | --- | --- | --- |
| | **% Raise** | **Performance** | **Position in Range** | **Time in Grade** |
| % Raise | 1 | | | |
| Performance | 0.873 | 1 | | |
| Position in range | −0.663 | −0.317 | 1 | |
| Time in grade | 0.899 | 0.848 | −0.605 | 1 |

| Degreed Employees, Correlation, $n = 20$ | | | | |
| --- | --- | --- | --- | --- |
| | **% Raise** | **Performance** | **Position in Range** | **Time in Grade** |
| % Raise | 1 | | | |
| Performance | 0.896 | 1 | | |
| Position in range | −0.543 | −0.283 | 1 | |
| Time in grade | −0.426 | −0.238 | 0.660 | 1 |

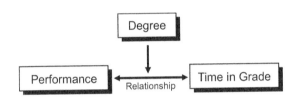

**FIGURE 17.15    MODEL OF DEGREE STATUS AFFECTING RELATIONSHIP BETWEEN PERFORMANCE AND TIME IN GRADE**

So now we have a different issue to investigate, namely, why, for the nondegreed employees, do the ones with more time in grade get predominantly higher performance levels, which, in turn, result in higher % raises? We have moved from the model shown in Figure 17.14 to the one in Figure 17.15 for our investigation.

We will close out our analysis by creating models with two variables, namely, % raise versus performance and position in range, shown in Table 17.19.

These results are very similar to the model based on the combined data, shown in Table 17.5.

There may be more things to discover in these data, but we will stop here.

## Summary

We started with a complaint by some of the nondegreed employees that the degreed employees received higher % raises on average than the nondegreed employees.

We resolved that issue by discovering that the nondegreed employees were generally higher in the salary ranges than the degreed employees, and, all other

**TABLE 17.19    REGRESSION RESULTS FOR PERFORMANCE AND POSITION IN RANGE SEPARATED BY DEGREE STATUS**

| | | Regression Results for Nondegreed Employees | | | | | | | | |
|---|---|---|---|---|---|---|---|---|---|---|
| | | Performance | | Position in Range | | Time in Grade | | | | |
| % Raise | $b_0$ | $b_1$ | $x_1$ | $b_2$ | $x_2$ | $b_4$ | $x_4$ | $r^2$ | SEE |
| Model N5   $y' =$ | 5.19 | + 1.18 | $x_1$ | $-$ 0.06 | $x_2$ | | | 0.927 | 0.45 |

| | | Regression Results for Nondegreed Employees | | | | | | | | |
|---|---|---|---|---|---|---|---|---|---|---|
| | | Performance | | Position in Range | | Time in Grade | | | | |
| % Raise | $b_0$ | $b_1$ | $x_1$ | $b_2$ | $x_2$ | $b_4$ | $x_4$ | $r^2$ | SEE |
| Model D5   $y' =$ | 6.62 | + 1.09 | $x_1$ | $-$ 0.07 | $x_2$ | | | 0.894 | 0.46 |

things being equal, received lower % raises based on the merit matrix guidelines that our managers were following.

But we uncovered perhaps a more serious issue, which is, for the nondegreed employees, the longer the time in grade, the higher the performance level, and this was not the case for the degreed employees. We would not expect a difference.

This is typical of multiple linear regression analyses. You start out looking for one thing and find some surprises.

What this analysis illustrates is that you should keep on digging in the data to discover and understand what is going on. It also illustrates that statistical analysis often identifies underlying issues that need to be addressed.

There are many ways to conduct an analysis. We have shown just one approach. There is no prescribed rule for how you should go about such an analysis other than to keep on digging until you can explain what is happening and until all your assumptions are validated.

Remember, modeling is a trial-and-error process. You are on a search mission to understand the data and discover and identify reality.

What is important in all of the modeling we have done is not to have as your only goal to develop a fancy model that explains everything with a high $r^2$ and low SEE, although that is nice and should always be a goal, but to identify issues that need addressing. The statistics and mathematics don't always solve problems. They often raise important issues.

## RELATED TOPICS IN APPENDIX A

A.14   Statistical Inference in Regression
        $t$-Statistic and Its Probability

F-Statistic and Its Probability
Mixed Messages in Evaluating a Model
A.15   Additional Multiple Linear Regression Topics
Adjusted $r^2$
Coding of Indicator Variables
Interaction Terms

# PRACTICE PROBLEMS

**17.1**  BPD has a sales call center in Atlanta, Georgia, with 36 sales associates who are paid a base salary plus a commission on the sales that they make. A year ago, a 2-week intensive training program was instituted for the employees that focused on understanding company policies, giving good customer service, understanding the products and services they were selling, and selling techniques. Only 22 employees took the training, as it was voluntary. You have been asked to determine if the training had an impact on the pay of the employees.

You gather what you think are the relevant data and are satisfied that they are valid. The variables you have are total pay, the dependent variable, and the independent variables of grade, time in grade, company service, and whether or not they attended the training program. These are listed in Table 17.20. The data are shown in Table 17.21, and also in Data Set 39.

Create the correlation matrix. In building the model, what independent variable would you choose first and why?

**TABLE 17.20   VARIABLES FOR CALL CENTER ANALYSIS**

| Variable | Symbol | Units |
|---|---|---|
| Pay | $y$ | Annual dollars |
| Grade | $x_1$ | 31, 32, 33 |
| Time in grade | $x_2$ | Years |
| Company service | $x_3$ | Years |
| Attended training | $x_4$ | 0 = no, 1 = yes |

**17.2**  What cautions should you observe when building the model and why?

**17.3**  Plot the dependent variable against each of the four independent variables. What kinds of relationships are there?

**17.4**  For each of the four independent variables, conduct a simple linear regression with pay as the dependent variable, and calculate the regression equation, the coefficient of determination, and the standard error of estimate. Interpret the terms.

**17.5**  Regress pay versus grade and attended training. Interpret the coefficients, the coefficient of determination, and the standard error of estimate.

**17.6**  Do you keep both variables in the model? Why or why not?

**17.7**  Regress pay versus grade, time in grade, and attended training. Interpret the coefficients, the coefficient of determination, and the standard error of estimate.

**TABLE 17.21**    DATA FOR CALL CENTER ANALYSIS

| Emp. No. | Pay | Grade | Time in Grade | Company Service | Attended Training |
|---|---|---|---|---|---|
| 1 | 38,326 | 32 | 4 | 5 | 1 |
| 2 | 34,691 | 31 | 1 | 1 | 0 |
| 3 | 38,716 | 31 | 1 | 1 | 1 |
| 4 | 35,672 | 31 | 2 | 3 | 0 |
| 5 | 39,877 | 31 | 2 | 2 | 1 |
| 6 | 36,263 | 31 | 1 | 1 | 0 |
| 7 | 40,281 | 31 | 3 | 4 | 1 |
| 8 | 36,683 | 32 | 3 | 5 | 0 |
| 9 | 42,121 | 32 | 5 | 7 | 1 |
| 10 | 39,125 | 31 | 3 | 3 | 0 |
| 11 | 44,139 | 31 | 2 | 2 | 1 |
| 12 | 40,676 | 32 | 4 | 5 | 0 |
| 13 | 45,189 | 32 | 5 | 8 | 1 |
| 14 | 37,562 | 32 | 2 | 5 | 0 |
| 15 | 39,818 | 32 | 1 | 4 | 1 |
| 16 | 41,835 | 32 | 1 | 4 | 1 |
| 17 | 38,321 | 32 | 1 | 5 | 0 |
| 18 | 43,334 | 32 | 4 | 7 | 1 |
| 19 | 39,927 | 32 | 4 | 6 | 0 |
| 20 | 44,864 | 32 | 2 | 5 | 1 |
| 21 | 42,121 | 32 | 5 | 7 | 0 |
| 22 | 46,529 | 32 | 6 | 8 | 1 |
| 23 | 42,606 | 33 | 6 | 9 | 0 |
| 24 | 47,631 | 32 | 2 | 4 | 1 |
| 25 | 48,441 | 32 | 2 | 5 | 1 |
| 26 | 45,070 | 32 | 5 | 8 | 0 |
| 27 | 49,436 | 32 | 3 | 6 | 1 |
| 28 | 50,364 | 32 | 3 | 7 | 1 |
| 29 | 48,976 | 32 | 1 | 4 | 0 |
| 30 | 54,747 | 33 | 6 | 9 | 1 |
| 31 | 44,557 | 32 | 1 | 5 | 1 |
| 32 | 44,116 | 33 | 3 | 8 | 0 |
| 33 | 51,164 | 33 | 3 | 7 | 1 |
| 34 | 48,905 | 33 | 2 | 6 | 1 |
| 35 | 54,495 | 33 | 3 | 7 | 1 |
| 36 | 53,186 | 33 | 4 | 8 | 1 |

**17.8** You have added time in grade to the previous model. Do you keep time in grade in the new model? Why or why not?

**17.9** For this last model with three independent variables, calculate the standardized beta weights.

**17.10** What are the relative influences of the three independent variables on the variability of pay? How does this perspective help with the decision on whether or not to keep time in grade in the model?

# Appendix

## A.1 Value Exchange Theory

As many of the statistical and quantitative analyses focus on specific aspects of compensation and human resources, and as their results often influence organization policies and strategies, it is important to step back and look at the bigger picture and the notion of an organization achieving its goals. One perspective is the value exchange theory [3].

An overarching theoretical perspective in the employer–employee relationship is the concept of value exchange.

### Achieving Organization Goals

An organization has many resources to achieve its goals. Even though these resources include material, capital, and people, it is only people who make decisions about and do things with the material, the capital, and the people. In other words, an organization's goals are accomplished only through people.

Hence, the major challenge of an organization is to attract, retain, motivate, and align the kinds and numbers of people it needs to achieve its goals. It does this through a value exchange.

### Value Exchange

A value exchange is where the company and the employee each give value to the other in exchange for value received in order to achieve their respective self-interests. This notion can be summarized by the phrase:

*Value given for value received*

*Statistics for Compensation: A Practical Guide to Compensation Analysis,* By John H. Davis
Copyright © 2011 John Wiley & Sons Inc.

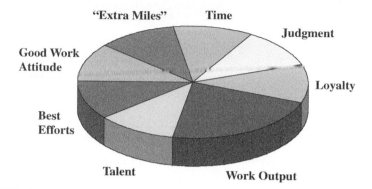

**FIGURE A.1**   THE EMPLOYEE GIVES AND THE EMPLOYER RECEIVES

The pie charts in Figures A.1 and A.2 highlight *some* of the items involved in the exchange.

Many of the items in both figures are not quantitatively measurable, but we know they are present and they are very important to the company.

The items in Figure A.2 may differ from one employee to the next as far as what is of value. Indeed, even the relative sizes of the pieces differ between employees and for individuals over time. For example, a relatively new employee may value growth opportunities more than one who is near retirement. Likewise, an individual might feel pay is very important today, but tomorrow when a new baby joins the family, benefits become more important.

## A Fair Value Exchange Is a Good Deal

A fair exchange is a good deal for all parties involved. Conceptually the employer and the employee are traders in a peer relationship, each giving value to receive value to further their respective self-interests.

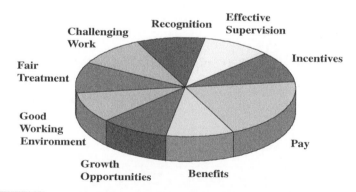

**FIGURE A.2**   THE EMPLOYER GIVES AND THE EMPLOYEE RECEIVES

Often there are trade-offs for both parties, for example, balancing short-term interests and goals with long-term ones. And most of the time the values involved in the exchange are not explicit—they are "just there," such as an open management climate, or a culture that encourages entrepreneurship, or the enthusiasm a person brings to the job. They are part of the subconscious valuations an employer and an employee make when considering all aspects of the relationship.

If the exchange is fair, the employer achieves its business goals and strategies, and at the same time, the employee achieves his or her own business-related personal goals.

A fair value exchange engenders a high degree of employee engagement and is a win–win situation for all concerned.

## A.2  Factors Determining a Person's Pay

As discussed above, pay is just one factor in the exchange between an employer and an employee. We want to focus here just on pay.

There are many factors that go into deciding how much to pay a person. Some of these are system factors and some are individual factors. They all influence an organization's ability to attract, retain, motivate, and align the kinds and numbers of employees needed to achieve the organization's goals. These are shown in Figure A.3.

**FIGURE A.3    FACTORS DETERMINING A PERSON'S PAY**

## System Factors

There are typically five major system factors that apply in general to all jobs in the organization.

1. *Internal Value of the Job* (e.g., via grades, points, whole job ranking). In every organization, there is a formal or informal value order of the jobs. The president's job is of higher value than the engineer's job, which in turn is of higher value than the clerk's job. No matter who are in the jobs, there is a value hierarchy. This is because different jobs require different levels of knowledge, skills, and abilities, and have different levels of responsibility.

   Included in this factor is how critical the job is to achieving company goals. The more critical the job, the more value that is placed on it.

2. *Market Pay for Similar Jobs/Skills.* What other organizations pay for similar jobs and skills is an important external perspective. This information usually is obtained from salary surveys. You want your pay to be in the same range as the "market" for those jobs. You don't have to pay exactly at the market but your pay should be in the ball park so that it is not a dissatisfier.

3. *Pay Philosophy* (what the company *wants* to pay for). There are two aspects to the pay philosophy. The first is *where* the company wants to pay with respect to the market. Does it want the pay for its jobs to be at the average pay or median of the market? Does it want to target its pay to a certain percentile of the market? Does it want to target its pay a certain percent above or below the market?

   The second aspect is *what* the company wants to pay for. What does it want to reward and reinforce? This may be performance, education, experience, teaming behavior, customer service, innovation, etc., or some combination of these.

4. *What the Company can Afford.* A key factor is the affordability of labor costs. A company has to be profitable to be able to afford high salaries for its employees. But a company cannot afford to pay excessively for a long time, for it will soon go out of business. Indeed, some companies have had to have pay cuts when the only alternative was bankruptcy and subsequent loss of all jobs. And in the extreme, if it cannot afford even the minimum at which people will work, then the business may not be viable at that point and must close its doors.

5. *Balance with Other Factors.* Pay is not the only factor that a company offers its employees. Some of the factors are outlined in Section A.1, where various values that are part of the exchange between the employee and the employer are described. Pay was just one factor.

Some of the other factors include incentives, benefits, growth opportunities, working environment, challenging work, up-to-date equipment, discounts on the company products, and so on. A start-up organization may offer low pay with high growth opportunities. Another organization may offer higher pay with low benefits. One airline has lower pay but nice flight benefits along with an exciting work environment.

These five factors constitute system factors that influence the pay of all jobs, regardless of the people in the jobs. The system factors typically will result in the salary range for a job. The company will pay at least the minimum of the range and no more than the maximum of the range for a job. Within the range, that part of the pay philosophy that defines what the company wants to reward and reinforce identifies the individual factors that influence the pay for an individual. The remaining three factors that influence pay are based on the individual performing the job.

## Individual Factors

The three factors described here are all interrelated. For example, higher education and skills along with more experience may result in higher performance levels, all other things being equal.

1. *Performance.* Performance is the most predominant factor that companies say they want to reward and reinforce. It is a topic for another discussion as to what performance is, how it is observed and measured, and how it is rewarded. But in the usual case, performance is defined as the degree to which job expectations are met. Sometimes both individual and team performance are factors.

2. *Education and Skills.* Education includes both formal educations, such as at a college or a training institute, and informal training programs that many companies offer. Along with this are skills that are often recognized by a certification, such as certain computer programming languages. Sometimes a company will give a pay raise upon achieving an advanced college degree.

3. *Experience.* Usually, the more experience a person has, the better the decisions that he or she will make. There is a maturity of judgment that often comes with experience. In the beginning of a job, a person moves up rapidly on the learning curve, where the impact of experience is most obvious in performance improvement. After several years, the learning curve tapers off, and the additional experience does not translate into improved performance.

   It is obvious that there is a great deal of balancing of these factors in determining a person's pay and no one factor is more important than another.

There is no single "right" combination of factors as each company is different, just as individual human beings are different.

What is important is that these factors be recognized and that decisions made about them are done explicitly and communicated to all employees.

## A.3 TYPES OF NUMBERS

In the mathematics that we use in compensation, there are four kinds of numbers. Since each kind has its own properties, it is important to know what kind of numbers you are working with in order to perform only the proper mathematical operations on them. For example, it does not make sense to take the average of employee numbers. You can do it, but it doesn't have any meaning.

As we go down the following list, each type includes the properties of the previous kind.

1. Nominal number

2. Ordinal number

3. Interval number

4. Ratio, or cardinal number

### Definitions and Properties

1. *Nominal Number.* A nominal number is simply an identifier, a label, or a name. The word nominal comes from the Latin word *nomen*, which means name. Examples include the following:

   - Employee number
   - Job number in a salary survey
   - Company number in a benefits survey
   - Department number
   - Number on a soccer jersey
   - Passport number

   These are just labels, or identifiers. Any arithmetic operations with them are meaningless. For example, it does not make sense to calculate the 75th percentile of the jersey numbers of a basketball team.

2. *Ordinal Number.* An ordinal number indicates the rank, or position, or order of an item within a group. Examples include the following:

   - Rank of company revenue: Largest, next largest, and so on
   - Rank of salaries: Highest paid, next highest paid, and so on

- A performance rating system, where 5 is outstanding, 4 is exceeds expectations, 3 is meets expectations, 2 is below expectations, and 1 is unsatisfactory
- Rank of height: Tallest, next tallest, and so on
- Winners in a horse race: 1st place, 2nd place, 3rd place, and so on

In a horse race, we know that the first place horse was faster than the second place horse, but not by how much. We do not know if the first place horse won by a nose, a head, or a length. All we know is that it was the fastest, but not how fast.

With ordinal numbers, differences do *not* have a quantitative meaning. For example, in the above performance rating system, we know that a 5 performer is better than a 4, and that a 4 performer is better than a 3. Even though the respective numerical difference in each case is 1, we cannot say that the performance difference between a 5 performer and a 4 performer is the same as the performance difference between a 4 performer and a 3 performer.

In addition, to say that the difference in performance between a 5 and a 3 is twice as much as the difference in performance between a 5 and a 4 does not make sense, even though the numerical difference of 2 is twice as much as the numerical difference of 1.

You can perform the arithmetic, but the answers do not make sense.

Ordinal numbers indicate only relative positions, and not absolute or quantitative amounts.

Note that ordinal numbers also are identifiers. For example, the horse that placed 1st place is also the 1st place horse.

3. *Interval Number.* An interval number is one where differences do have a quantitative meaning, but there is no absolute zero that means "nothing." The designation of a zero is arbitrary. An example is

- Calendar year

With interval numbers, differences do have a quantitative meaning. For example, the difference between the years 2002 and 2005 is 3 years, and that difference has a quantitative meaning. It is an absolute amount. It is also the same as the difference between 2009 and 2012.

However, the designation of the year "0" is arbitrary.

There are many different calendars throughout history and even today. The Gregorian calendar is the *de facto* international standard, and its zero is set to correspond with the birth of Christ. In the Islamic calendar, zero is set to correspond to Muhammad's emigration from Mecca to Medina.

Other calendar systems include the Hindu calendar, the Buddhist calendar, the Mayan calendar, the Persian calendar, the Chinese calendar, the

Ethiopian calendar, and many others. Each has its own zero year, and the zero years have changed over time for some of the calendars.

Since there is no real zero, it doesn't make sense to divide one number by another and have a meaning. For example, it doesn't make sense to say that the year 2000 is twice as much as the year 1000. And this is true no matter what calendar you are using.

Note that interval numbers also indicate rank, or an order relationship. For example, the year 2013 is later than the year 2012. In addition, they also are identifiers. We can refer to a particular year as, for example, 2013.

4. *Ratio Number, or Cardinal Number.* A ratio number, or cardinal number, indicates how much or how many of what is being measured, and where zero means nothing of the item. It is the kind of number we use most of the time. Examples include the following:

- Salary
- Raise (in dollars or percent)
- Revenue
- Number of . . .
  - employees
  - departments
  - companies
  - products
  - surveys
  - countries
- Height (inches, centimeters) and weight (pounds, kilograms)
- Volume of water (cubic feet, cubic meters)

With a ratio number, not only do differences between numbers make sense, but there is also meaning when you divide one number by another one to get a ratio. For example, a salary of 100,000 is twice as much as a salary of 50,000 ($100,000/50,000 = 2$). Or a raise of 5,000 divided by a salary of 50,000 is 10% $(5,000/50,000)(100) = (0.10)(100) = 10\%$.

In addition, zero means nothing of the item being measured. A raise of zero means you got nothing. Zero revenue means there was no revenue.

A ratio number includes the properties of the other three types of numbers. Differences have a quantitative meaning (one department has 12 more employees than another), the numbers indicate an order relationship (the department with 40 employees is bigger than the department with 28 employees), and they are labels (the 40 employee department).

Table A.1 is a summary of these four types of numbers.

**TABLE A.1     FOUR TYPES OF NUMBERS**

| Properties | Nominal | Ordinal | Interval | Ratio, Cardinal |
|---|---|---|---|---|
| Identifier, label | X | X | X | X |
| Indicates rank or order | | X | X | X |
| Differences have quantitative meaning | | | X | X |
| Ratios have quantitative meaning and there is a true zero | | | | X |

As you can see, it is important to know what kind of numbers you are working with, so that you use only those mathematical operations that are appropriate for those particular numbers.

## Histograms with All Four Types of Measurements

The categories for a frequency distribution and histogram can be any of the four types of numbers, or measurements. Here are examples of histograms with all four types. The corresponding frequency distributions are not shown here.

1. *Nominal Measurement.* Since there is no order relation among nominal measurements you have to decide in what order the categories are presented. There are three usual ways. The first way is arbitrary, as shown in Figure A.4. Here the histogram was used to decide where to put efforts to decrease employee turnover.

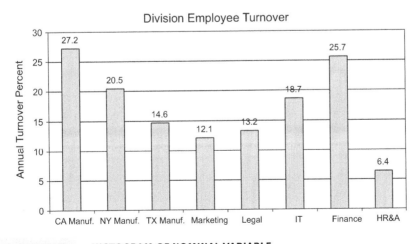

**FIGURE A.4     HISTOGRAM OF NOMINAL VARIABLE**

The second way is to show the category with the highest frequency first, the category with the next highest frequency next, and so on. This is called a Pareto chart, as in Figure A.5. This histogram was used to help decide budget allocation for compensation communications.

The third way is to show the most important (to the problem being addressed) category first, the next most important category next, and so on (example not shown).

Technically with nominal measurements, these charts would properly be called bar charts rather than histograms. For the purpose of statistics in compensation practices, we usually do not make that subtle differentiation. What is important is the display with the information it confers, not what it is called.

2. *Ordinal Measurement.* Here, there is an order relationship among the categories, and so the lowest rating is to the left and the highest is to the right. The histogram in Figure A.6 was used to confirm the expected distribution of performance ratings.

3. *Interval Measurement.* The calendar is an interval measurement. The histogram in Figure A.7 was used to show the cyclical nature of BPD's day care usage at headquarters, to help justify temporary help during the summer months.

4. *Ratio Measurement.* The histogram in Figure A.8 was used to identify the magnitude of upcoming retirements in one of the BPD manufacturing plants.

The histogram in Figure A.9 was used as part of a project to identify concerns with BPD's internal training program.

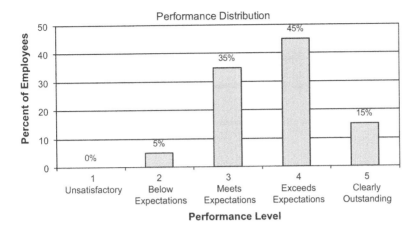

**FIGURE A.6    HISTOGRAM OF ORDINAL VARIABLE**

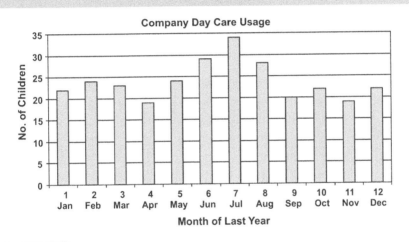

**FIGURE A.7    HISTOGRAM OF INTERVAL VARIABLE**

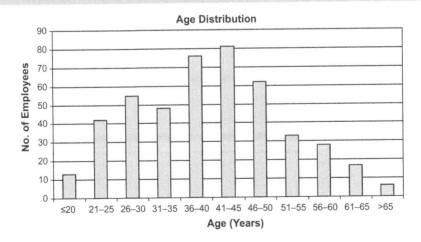

**FIGURE A.8    HISTOGRAM OF RATIO VARIABLE**

**FIGURE A.9**    **HISTOGRAM OF RATIO VARIABLE**

## A.4 SIGNIFICANT FIGURES

It is important to be honest when reporting a statistic so that it does not appear to be more precise than the data from which it was calculated. The precision of a number is represented by the number of significant figures in it. The number of significant figures in a measurement is the number of digits that are known with some degree of confidence.

Trailing zeros are not significant unless so designated. One example is the estimated revenue of a company. The company may report $235,000,000. Unless otherwise noted, there are only three significant figures in this number.

In arithmetic operations, the answer contains no more significant figures than the least precisely known number. Hence, you should know the precision of your starting numbers.

Keep in mind, though, that when taking the average of numbers, the divisor is exact as it represents the number of data points, and is theoretically precise to an infinite number of decimal places. So the number among those whose average is being calculated that has the least number of significant figures is the one dictating the number of significant figures in the answer.

Sometimes it gets a little tricky. Consider the following estimated revenues of six departments in Table A.2.

The department whose estimated revenue has the least number of significant figures that is involved in calculating the average is Department F, with two figures. Hence, the average should be reported with only two significant figures, and rounded to 25,000,000.

It should be noted that if Department F had reported that its estimate was good to three significant figures, then the average would be reported with three significant figures, namely 24,900,000.

**TABLE A.2    ESTIMATES OF DEPARTMENT REVENUE**

| Department | Estimated Department Revenue | No. of Significant Figures |
|---|---|---|
| A | 2,435,000 | 4 |
| B | 13,500,000 | 3 |
| C | 21,400,000 | 3 |
| D | 22,761,444 | 8 |
| E | 33,300,000 | 3 |
| F | 56,000,000 | 2 |
| Calculated average | 24,899,407 | |
| Reported average | 25,000,000 | 2 |
| Calculated median | 22,080,722 | |
| Reported median | 22,000,000 | 2 |

The department whose estimated revenue has the least number of significant figures that is involved in calculating the median is Department C, with three figures. Hence, on first glance the median should be reported with three significant figures, and rounded to 22,100,000. However, as the group as a whole is only good to two significant figures, the median should be reported as 22,000,000.

Again, if Department F had reported that its estimate was good to three significant figures, then the median would be reported with three significant figures, as that would be the group precision. In this case, the reported median would be 22,100,000.

As a side note, one should question the "exactness" of Department D's estimate.

## A.5  SCIENTIFIC NOTATION

Sometimes on a worksheet or a calculator you see a number such as $3.71E - 6$ or $1.23E + 7$. This is a shorthand notation that is used when there is not enough space in the worksheet cell or calculator display to display the entire number. When you see $E - n$, where $n$ is a number, it means to move the decimal point $n$ places to the left. $3.71E - 6$ means $0.00000371$. When you see $E + n$, where $n$ is a number, it means to move the decimal point $n$ places to the right. $1.23E + 7$ means 12,300,000.

Often, but not always, the number of digits shown before the "E" is the number of significant figures. For example, $3.4521E + 8$ represents 345,210,000 and there are five significant figures. The number $3.452E + 8$ represents 345,200,000 and there are four significant figures.

An example where the number of significant figures is truncated is with the number 6,437,100, which has five significant figures. However, it may be represented as $6.437E + 6$, which shows that it has been rounded to show only the first four significant figures because that is all the room there is in the display.

## A.6 ACCURACY AND PRECISION

Accuracy and precision are often used as synonyms. However, they are different when applied to groups of numbers and subsequent calculations. We will illustrate concerning using a survey average to estimate the true market pay for a job.

*Accuracy* is how close on average your results are to the intended target. For example, you want a salary survey average to be equal to the true average market pay for a job.

*Precision* is how close the results are to each other. For example, you would like different salary survey results for a given job to be close to each other.

Figure A.10 illustrates the four possible combinations of accuracy and precision. The center of the target represents the true market average pay.

*Not Accurate, Not Precise* The results are scattered, and their average is far away from the center of the target.

*Accurate, Not Precise* Even though the results are scattered, the average is at the center of the target.

*Not Accurate, Precise* The results are very close to each other, but they and their average are far away from the center of the target.

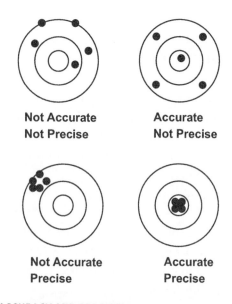

**Not Accurate**
**Not Precise**

**Accurate**
**Not Precise**

**Not Accurate**
**Precise**

**Accurate**
**Precise**

**FIGURE A.10**     **ACCURACY AND PRECISION**

*Accurate, Precise* The results are very close to each other, and they and their average are at the center of the target.

## Which Is More Important?

Obviously we would like our results to be both accurate and precise. But, if you have to choose between the two, *accuracy is more important than precision*. We would like our survey results, on average, to reflect the true market values of the jobs.

When different survey results are close to each other, this precision gives us more confidence in the results. But lack of precision does not indicate bad data. It just means that the data are spread out and that you should investigate why.

## A.7 COMPOUND INTEREST–ADDITIONAL

The basic formula relating the four values listed below can be rearranged to solve for any one of them given the other three.

Let $PV$ = present value

$FV$ = future value

$i$ = interest rate per period, expressed as a decimal

$n$ = number of periods

Then the four formulas relating these four terms are as follows.

$FV = PV(1 + i)^n$

$PV = FV/(1 + i)^n$

$i = (FV/PV)^{1/n} - 1$

$n = (\log(FV/PV))/\log(1 + i)$

## Other Formulas

Other formula based on the notion of compound interest have been derived that relate regular payments on a loan to present value, future value, interest rate, and number of periods.

One area this is used in is to calculate the monthly mortgage payments on your home.

Let $PMT$ = constant payment

$PV$ = present value

$i$ = interest rate per period, expressed as a decimal

$n$ = number of periods

CF = compounding factor = $(1 + i)^n$

Then PMT = $(PV)(i)(CF)/(CF - 1)$

### Example A.1

Suppose you borrowed \$140,000 to buy a house on a 30-year loan at a fixed rate of 6% annual interest. What are your monthly payments?

PV = 140,000

$i$ = 6%/12 = 0.5% = 0.005 per month

$n$ = (30)(12) = 360 months

Then CF = $(1 + 0.005)^{360}$ = 6.022575

And PMT = $(140,000)(0.005)(6.022575)/(6.022575 - 1)$ = \$839.37 (rounded) per month.

## A.8 RULE OF 72

The Rule of 72 is an approximation used to relate the number of years it will take a given amount of money to double to the required annual interest rate. It goes like this.

Let

$N$ = Number of years to double

$I$ = Percent interest rate per year

Then

$(N)(I) \approx 72$

where $\approx$ means equals approximately

Note that $I$ is a percent, and not the decimal equivalent.

### Example A.2

If your salary increases at a rate of 6% per year, about how many years will it take for your salary to double? The answer is 12 because (6)(12) = 72. Or,

$$N \approx 72/I = 72/6 = 12$$

*Example A.3*

About what annual interest rate would it take to double your investment in 8 years? The answer is 9% because $(8)(9) = 72$. Or,

$$I \approx 72/N = 72/8 = 9$$

More rigorously, the Rule of 72 states that the number of periods it will take a given amount to double times the periodic growth rate is approximately 72.

## Derivation of the Rule of 72

Here we will derive the Rule of 72, using some notions of calculus and some approximations. The Rule of 72 is based on the following formula:

$$FV = PV(1 + i)^n$$

and on an approximation to natural logarithms.

First let us rearrange the formula as follows.

$$FV/PV = (1 + i)^n$$

If we are interested in doubling the present value to a future value, then $FV/PV = 2$ and we get

$$2 = (1 + i)^n$$

Now we ask the question, "What relation exists between the values of $i$ and $n$ to make the equation a true statement?" To get an answer, we first take the natural logarithm of both sides.

$$\ln(2) = \ln((1 + i)^n) = (n)\ln(1 + i)$$

Substituting the approximate value of $\ln(2)$ to 4 decimal places in the equation, we get

$$0.6931 \approx (n)\ln(1 + i)$$

At this point, we will use some notions of calculus for approximating a function. Using the Maclaurin's series expansion of a function about the origin, for the function $\ln(1 + x)$, we get

$$\ln(1 + x) = x - x^2/2 + x^3/3 - x^4/4 + x^5/5 - x^6/6 + \ldots$$

If $x$ is our interest rate $i$, we get

$$\ln(1 + i) = i - i^2/2 + i^3/3 - i^4/4 + i^5/5 - i^6/6 + \ldots$$

For small values of $i$, all the terms past the first term in the right-hand side become exceedingly small, and we can use the first term for a close approximation.

$$\ln(1 + i) \approx i \text{ for small values of } i.$$

We now have

$$0.6931 \approx (n)(i)$$

Multiply both sides by 100 to get $i$ as a percent

$$69.31 \approx (n)(i\%)$$

The last approximation is to change 69.31 to 72 because a lot more numbers divide evenly into 72 than divide into 69.31.

$$72 \approx (n)(i\%)$$

This completes the derivation of the Rule of 72.

The two major approximations both go in the same direction, as the number 72 is a little more than 69.31, and $i$ is a little more than $\ln(1 + i)$ for values of $i$ less than 1.00 (100%), and this helps make the rule a reasonably accurate one.

## A.9 NORMAL DISTRIBUTION

The normal distribution, also known as the Gaussian distribution, is one of the most widely used theoretical distributions in statistical analysis. It is used in all fields, such as physical sciences, engineering, medicine, social sciences, business, and so on.

Properties include the following:

- Completely specified by the mean and the standard deviation.
- Continuous
- Symmetric
- Bell-shaped
- 68.3% of the distribution falls within 1 standard deviation of the mean.
- 95.4% of the distribution falls within 2 standard deviations of the mean.
- 99.7% of the distribution falls within 3 standard deviations of the mean.

The probability density function for the normal distribution is given by

$$f(x) = \frac{1}{\sigma\sqrt{2\pi}} e^{-\frac{1}{2}\left(\frac{x-\mu}{\sigma}\right)^2}$$

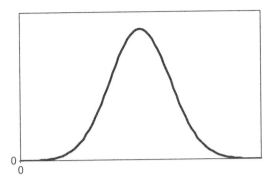

**FIGURE A.11     NORMAL DISTRIBUTION**

where $\mu$ is the mean and $\sigma$ is the standard deviation.

Figure A.11 shows a plot of this function.

There are two types of data that follow a normal distribution.

1. The distribution of many phenomena in nature that can be thought of as the result of many independent influences, which can be closely approximated with the normal distribution. Examples include the following:

   - heights of men of similar ethnicity
   - body temperature
   - blood pressure of adults
   - yield of corn per acre
   - number of apples on an apple tree
   - observation errors in an experiment

   To determine if a set of data is normally distributed, one must conduct one or more of a variety of standard statistical tests, and even then, the best answer one can get is that the data are normally distributed with a certain probability.

2. The averages of repeated sampling tend to a normal distribution. See Central Limit Theorem below. This allows normal distribution theory to be used in testing of hypotheses and in constructing confidence intervals.

   The mathematical properties of the normal distribution and its theoretical basis were first investigated by Pierre Laplace (1749–1827), Abraham de Moivre (1667–1745), and Carl Gauss (1777–1855).

## Central Limit Theorem

This theorem states that given a population of whatever type (uniform, bell-shaped, squiggle-shaped, continuous, discrete, weird, etc.) with finite values of the mean $\mu$ and variance $\sigma^2$, the distribution of the means of random samples of size $n$ from that population approaches a normal distribution with mean $\mu$ and variance $\sigma^2/n$ as the sample size gets large.

This theorem is why the normal distribution is so important in all fields where statistical inference is used.

### Distribution of Salary Survey Data

In the author's experience, salary survey data do not have a normal distribution. Indeed, most of the time it is not even bell-shaped. Figure A.12 shows some examples from a private study done on hundreds of jobs. These are representative of the distributions encountered. The histograms have been smoothed and data converted to $z$-scores for easy comparison.

There are positively skewed distributions, negatively skewed distributions, uniform distributions, bimodal distributions, trimodal distributions, bell-shaped distributions, and distributions that defy description. But as you can see from these examples, there are no nice smooth normal distributions.

## A.10 LINEAR REGRESSION TECHNICAL NOTE

Linear regression in the context of statistical modeling is a regression model that is linear in the coefficients, namely that the dependent variable is a sum of terms that are each an independent variable multiplied by a coefficient.

In modeling a nonlinear relationship, we make logarithmic transformations of the variables or add higher powers of the $x$-variable. In both cases, we can then apply linear regression mathematics. So long as the final model is *linear in the coefficients*, we can apply simple or multiple linear regression to these nonlinear relationships. For simplicity of describing the relationships, we refer to these instances as nonlinear regression.

It is easy to see how linear regression mathematics are used in the logarithmic transformation models when the equations are listed in a table and the coefficients are lined up.

| Model | Equation |
| --- | --- |
| Linear | $y = a + bx$ |
| Exponential | $\log y = a + bx$ |
| Logarithmic | $y = a + b \log x$ |
| Power | $\log y = a + b \log x$ |

Similarly for polynomial models, multiple linear regression mathematics are used. Using the symbols $a$, $b$, $c$, $d$, and $e$ instead of $b_0$, $b_1$, $b_2$, $b_3$, and $b_4$ for the coefficients, we get the following.

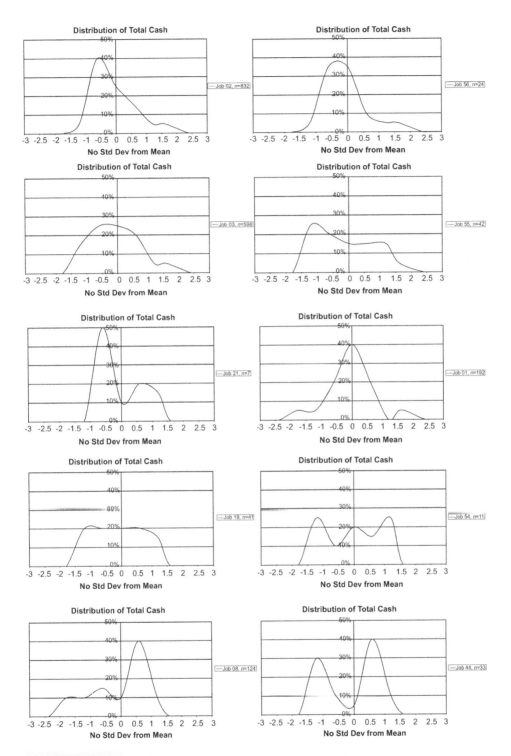

**FIGURE A.12    DISTRIBUTIONS OF SALARY SURVEY DATA**

| Model | Equation |
|---|---|
| Multiple linear | $y = a + bx + cy + dz + ew$ |
| Quadratic | $y = a + bx + cx^2$ |
| Cubic | $y = a + bx + cx^2 + dx^3$ |
| Spline | $y = a + bx + cx^2 + dx^3 + ep^3$ |

## A.11 FORMULAS FOR REGRESSION TERMS

Table A.3 contains formulas for terms that appear on typical regression outputs from statistical software. Sometimes there are alternate formulas you may encounter that are mathematically equivalent.

## A.12 LOGARITHMIC CONVERSION

To convert a common logarithm (base 10), denoted by log, to and from a natural logarithm (base e), denoted by ln, use the following formulas.

$$\ln x = (\log x)(\ln 10)$$

or

$$\ln x = (\log x)/(\log e)$$

$$\log x = (\ln x)/(\ln 10)$$

or

$$\log x = (\ln x)(\log e)$$

## A.13 RANGE SPREAD RELATIONSHIPS

Most of the time, we describe the range of pay for a job in terms of the range spread, which is the percent that the maximum is more than the minimum. For example, when we say that the range spread of a job is 60%, we mean that the maximum is 60% more than the minimum.

But sometimes we describe the range of pay for a job in terms of a ±percent off the midpoint. For example, we say that the maximum and minimum are each 23% from the midpoint, or that they are ±23% from the midpoint.

In the first case (60%), the reference point for the percent is the *minimum*. In the second case (23%), the reference point for the percent is the *midpoint*.

## TABLE A.3    FORMULAS FOR REGRESSION TERMS

| | |
|---|---|
| Slope for simple linear regression | $b = \dfrac{\sum(x - \bar{x})(y - \bar{y})}{\sum(x - \bar{x})^2}$ or $b = \dfrac{n\left(\sum xy\right) - \left(\sum x\right)\left(\sum y\right)}{n\left(\sum x^2\right) - \left(\sum x\right)^2}$ |
| Intercept for simple linear regression | $a = \bar{y} - b\bar{x}$ |
| ANOVA | Acronym for ANalysis Of VAriance. A table where the total variance in the $y$-variable (SS total) is analyzed by partitioning it into two components—the regression, which is the variance "explained" by the model (SS regression) and the residual, which is the "unexplained" variance (SS residual). <br><br> For linear regressions, SS regression + SS residual = SS total |
| SS | Sum of squares. This is a measure of the variability of various components of the model. |
| SS total | Also know as SS about the mean. It represents the variability of the data points from the mean. <br><br> SS total $= \sum(y - \bar{y})^2$ |
| SS regression | It represents the variability of the model from the mean. <br><br> SS regression $= \sum(y' - \bar{y})^2$ |
| SS residual | Also known as SS error. It represents the variability of the data points from the predicted values of the model. <br><br> SS residual $= \sum(y - y')^2$ <br><br> This is the value of the minimum sum of squares in the least squares criterion. |
| Degrees of freedom | Sometimes denoted by df. Any sum of squares has associated with it a number called its degrees of freedom. This number indicates how many independent pieces of information involving the $n$ independent data points are available and used to compile the sum of squares. <br>    *df regression* is equal to the number of $x$-variables, and also equal to $p - 1$, where $p$ is the number of parameters in the model being estimated. For simple linear regression (just one $x$-variable) df regression is 1. <br>    *df residual* is equal to $n - p$, where $n$ is the number of data points and $p$ is the number of parameters in the model being estimated. For simple linear regression (just one $x$-variable) $p = 2$, for the intercept $a$ and the slope $b$. <br>    *df total* is equal to $n - 1$. <br><br> Note that df regression + df residual = df total. |
| MS | Mean square. This is a measure of "average" variability, and is calculated by dividing each sum of squares by its corresponding degrees of freedom. |
| MS total | MS total = SS total/df total <br><br> As a side note, if the mean square were calculated for total, it would be <br><br> $$\dfrac{\sum(y - \bar{y})^2}{n - 1}$$ <br> If you then took at its square root, it would be the sample standard deviation of the $y$-variable, a measure of the variability of the $y$-variable about its mean. |

*(continued)*

| TABLE A.3 | *(CONTINUED)* |
|-----------|---------------|

| | |
|---|---|
| MS regression | MS regression $=$ SS regression/df regression $= \dfrac{\sum (y' - \bar{y})^2}{p - 1}$ |
| MS residual | Also known as MS error |
| | MS residual $=$ SS residual/df residual $= \dfrac{\sum (y - y')^2}{n - p}$. |
| Standard error of estimate | Also known as the standard error. It is calculated by taking the square root of the mean square residual. |
| | $SEE = \sqrt{\dfrac{\sum (y - y')^2}{n - p}}$ |
| Coefficient of determination, $r^2$ | $r^2 = 1 - \dfrac{SS\ residual}{SS\ total} = 1 - \dfrac{\sum (y - y')^2}{\sum (y - \bar{y})^2}$ |
| | For simple linear regressions, this is mathematically equivalent to |
| | $r^2 = \dfrac{\left(\sum (x - \bar{x})(y - \bar{y})\right)^2}{\left(\sum (x - \bar{x})^2\right)\left(\sum (y - \bar{y})^2\right)}$ |
| Correlation, $r$ | Also known as the correlation coefficient. It is the square root of the coefficient of determination. For a regression, it is sometimes called multiple $r$, and is the correlation between $y$ and $y'$. |
| | $r = \sqrt{r^2}$ |
| Adjusted $r^2$ | Adjusted $r^2 = 1 - \left(\dfrac{n-1}{n-p}\right)\left(\dfrac{SS\ residual}{SS\ total}\right)$ |
| | where $n$ is the number of data points (observations) and $p$ is equal to the number of parameters (or coefficients in the least squares equation) estimated. For simple linear regression, $y = a + bx$, there are two parameters we are estimating: $a$ and $b$, and so $p = 2$. |
| Standardized coefficient | Also known as the beta weight. For the $i$th $x$-variable, $x_i$, it is |
| | $\beta_i = \dfrac{(b_i)(\text{Std. dev. of } x_i)}{\text{Std. dev. of } y}$ |
| | where $b_i$ is the "regular" regression coefficient for $x_i$. |
| Standard error of coefficient | The standard deviation of the regression coefficient |
| | For the $i$th $x$-variable, $x_i$, the standard deviation for $b_i$ is |
| | $SE_i = SEE\left(\sqrt{c_{ii}}\right)$ |
| | where SEE is the standard error of estimate and $c_{ii}$ is the diagonal element in the matrix $(x'x)^{-1}$, the inverse of the matrix $x'x$ in the normal equation (in matrix form) $x'xb = x'y$. |
| $t$-statistic | The number of standard errors a coefficient is from zero. |
| | For the $i$th $x$-variable, $x_i$, the $t$-statistic for $b_i$ is |
| | $t_i = \dfrac{b_i}{SE_i}$ |
| | where $b_i$ is the regression coefficient and $SE_i$ is its standard error. |
| $F$-statistic | The ratio of MS regression to MS residual |
| | $F = \dfrac{MS\ regression}{MS\ residual}$ |

**TABLE A.4**     RELATIONSHIPS BETWEEN RANGE SPREAD AND± FROM MIDPOINT

| Range Spread (%) | ± from Midpoint (%) | ± from Midpoint (%) | Range Spread (%) |
|---|---|---|---|
| 30 | 13 | 10 | 22 |
| 35 | 15 | 12 | 27 |
| 40 | 17 | 14 | 33 |
| 45 | 18 | 16 | 38 |
| 50 | 20 | 18 | 44 |
| 55 | 22 | 20 | 50 |
| 60 | 23 | 22 | 56 |
| 65 | 25 | 24 | 63 |
| 70 | 26 | 26 | 70 |
| 75 | 27 | 28 | 78 |
| 80 | 29 | 30 | 86 |

There is a mathematical relationship between the two ways of stating the range of pay.

Let   $R =$ Range Spread, the decimal equivalent of the percent

$D = \pm$ Difference from the midpoint, the decimal equivalent of the percent

Then

$$D = \frac{R}{R + 2} \quad \text{and} \quad R = \frac{2D}{1 - D}$$

Table A.4 shows the relationships between typical values of $R$ and $D$. The values are rounded to the nearest percent.

## Overlap

The overlap is the percent of a range that is overlapped by the next lower range, shown in Figure A.13.

$$\text{Overlap} = \frac{\text{Max } A - \text{Min } B}{\text{Max } B - \text{Min } B} \times 100$$

Suppose the ranges have the values shown.

$$\text{Overlap} = \frac{104 - 80}{120 - 80} \times 100 = \frac{24}{40} \times 100 = 0.60 \times 100 = 60\%$$

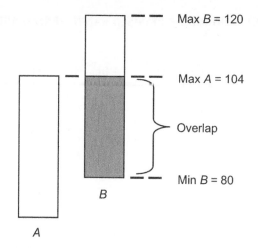

There is a formula that relates overlap to range spread and midpoint progression.

$$\text{Let} \quad L = \text{Overlap as a decimal}$$
$$R = \text{Range Spread as a decimal}$$
$$P = \text{Midpoint Progression as a decimal}$$

$$\text{Then} \quad L = \frac{R - P}{R(1 + P)}$$

For example,

$$\text{Let} \quad \text{Range Spread} = 50\%, \ R = 0.50$$
$$\text{Midpoint Progression} = 12\%, \ P = 0.12$$

$$\text{Then} \quad L = \frac{0.50 - 0.12}{(0.50)(1 + 0.12)} = \frac{0.38}{0.56} = 0.68$$

There is a 68% overlap.

## A.14 Statistical Inference in Regression

Statistical inference, or inferential statistics, is making inferences or generalizations about a population based on the results from a properly constructed sample from that population. In non–human resources settings, for example, we infer the outcome of an election based on a sample of voters, or we infer the success of a new product on the market based on a sample of test marketing to potential buyers.

Sometimes in human resources there is sampling involved, such as in an attitude survey among employees, where, for example, you use serial sampling to obtain a sample. An example of serial sampling is when you arrange employee names in some order—alphabetically by last name or numerically by employee number—and take every *n*th one for your sample.

In these cases, there is a sampling process that allows the use of sophisticated statistics to help make sound generalizations and subsequent decisions. There are a variety of sampling processes, such as drawing names from a hat or using a random number generator, serial sampling, cluster sampling, and stratified sampling, among others.

Unless such valid sampling is done in a sampling situation where you are trying to estimate population statistics from sample statistics, it is not appropriate to use inferential statistics. Unfortunately, inferential statistics are sometimes used erroneously in a non-sampling, that is, strictly descriptive, situation to evaluate a model. This is sometimes done in pay discrimination analyses.

Recall the definition of statistics: a branch of mathematics concerned with the measurement of uncertainty. This is what inferential statistics is all about—making inferences about a population based on the results from a properly constructed sample. However, if you have a population, such as data on all of your employees, you have certainty, and hence the use of inferential statistics is not appropriate.

If you do encounter a sampling situation, it is wise to seek help from a statistician as to the proper way to conduct such an analysis. Seek help *before* you take your sample, to ensure the design is valid. In such a case, using statistical inference is appropriate.

The following discussion applies, then, to a situation in which a valid sample is taken from a population and inferences are made concerning the goodness of the resulting model.

We have already discussed evaluation criteria for descriptive models. From the chapters on linear models and multiple linear regression, we have used the following criteria.

- Appearance
- Coefficient of determination
- Correlation
- Standard error of estimate
- Common sense
- Regression coefficients
- Standardized coefficients
- Multicollinearity

- Simplicity
- Acceptability
- Reality

These apply to both descriptive and inferential situations. To inferential situations, we will add two more.

- *t*-statistic and its probability
- *F*-statistic and its probability

The assumptions underlying these two criteria can best be illustrated with a simple linear regression model shown in Figure A.14. The population you are making inferences about is represented by the "True line" $y = a + bx$. The sample is represented by the observed values, $(x_i, y_i)$. The assumptions are as follows:

1. For each value of $x_i$, there is a subpopulation of $y$-values.
2. The residuals from the true line $(y_i - a - bx_i)$ are normally distributed about the true line.
3. The distribution of the residuals has a mean, $\mu$, of zero and a standard deviation, $\sigma$, which are the same for all subpopulations of the $y$-values.

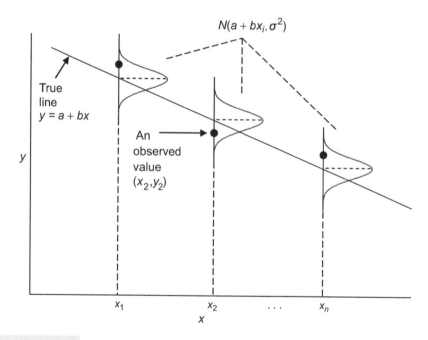

**FIGURE A.14** **SIMPLE LINEAR REGRESSION MODEL**

4. The residuals for different values of $x_i$ are independent (i.e., not correlated).

The more that these assumptions are met, the more valid the inferences are about the $t$- and $F$-statistics.

## $t$-Statistic and Its Probability

The $t$-statistic and its probability are used when we are using a sample model as an estimate for the true model of the population.

If you have all the data from a population and are describing them with a model, the resultant model *is* the population model. You do not have to estimate it; you have it.

In a sampling situation, due to the nature of the sampling process itself, there is variability in the raw data selected, and as such there is variability in the resulting models derived from the sample raw data.

This means that the slopes and intercepts will vary from one sample to another. Hence, the model coefficients derived from a sample are *estimates* of the coefficients of the true population model.

When we are examining an independent variable in a model, we want to know if it impacts the dependent variable. If it does not, we would expect the associated regression coefficient, the slope, to be zero. That is, if an $x$-variable does not explain any of the variation of the $y$-variable (after taking into account the other $x$-variables for a multiple regression model), its regression coefficient will be zero. The two possibilities are shown in Table A.5.

If we have a sample, it could happen in the situation where the true coefficient is zero that the sample model could have a coefficient different from zero by chance. In this case, the $t$-statistic is used to calculate the odds of this happening.

In the output from the statistical software, there are usually two terms: the $t$-statistic and its associated probability. The $t$-statistic for a variable is the number of standard deviations the estimated coefficient is from zero. The larger the absolute value of $t$, the more evidence there is that the variable does indeed impact the $y$-variable.

The associated probability is the probability that the value of the estimated coefficient is not zero due to chance when the true value is zero. Another way of

**TABLE A.5     TRUE RELATION FOR THE POPULATION**

| True Relation for the Population | |
|---|---|
| **$x$-Variable** | **Regression Coefficient** |
| Does not impact $y$-variable | Zero |
| Does impact $y$-variable | Different from zero |

expressing this is that it is the probability of a "false positive" for the coefficient. It is the odds that the model says that the particular $x$-variable matters when it actually does not.

What one usually focuses on is the probability of the $t$-statistic rather than the value of the $t$-statistic itself. The probability is either calculated or looked up in a table behind the scenes. It is a function of the $t$-statistic and the degrees of freedom of that statistic.

When the probability of a false positive is very low, we say that the variable is statistically significant at a certain level, often termed as the $\alpha$-level (alpha-level) and keep it in the model. When it is too high, we say that the variable is not statistically significant and exclude it from the model.

What are appropriate cut-off levels? Although much of the world (academic, legal) seems to regard a 0.05 or 0.10 $\alpha$-level as one of the "sacred" markers of statistical significance, it really depends on the situation. In some cases, such as medical research being used to support release of a potentially dangerous new drug, we might want a more stringent level. On the other hand, when we are doing exploratory work in human behavior, we might be happy with more relaxed limits.

## *F*-Statistic and Its Probability

The $F$-statistic and its probability are analogous to the $t$-statistic and its probability. Whereas the $t$-statistic applies to an individual coefficient, the $F$-statistic applies to all the coefficients at once.

It is calculated as the mean square regression divided by the mean square residuals (both defined in Section A.11).

$$F = \text{MS regression/MS residuals}$$

The numerator is an indication of variability explained by the model. The denominator is an indication of the variability *not* explained by the model. Hence, $F$ is the ratio of an explanatory measure to a nonexplanatory measure. The larger the ratio, the more evidence there is that the model is explaining some of the variability observed in the dependent variable.

In a sampling situation, the associated probability is the probability that at least one of the regression coefficients, or a linear combination of them, is different from zero by chance when in reality they are all equal to zero. In other words, it is the probability of a "false positive" for the entire model. It is the odds that the model says it matters when it really does not.

If the probability is very small, there is something in the model that is explaining some of the variability of the dependent variable, but you do not know *which* terms are useful and which are not. It reveals only that your model is better than no model at all, which is to say it is better than merely using the mean of the $y$-variable to describe the situation.

As with the *t*-statistic, there is no single standard cut-off probability for concluding that the chance of making a "not zero when it really is zero" error is significant. Typical cut-off probabilities are 5% and 10%, but the cut-off level is your choice.

Once you decide on the level you are comfortable with, this becomes your "level of significance." The models that have *F*-statistic probabilities below this level are said to be "statistically significant at that given level."

## Mixed Messages in Evaluating a Model

For a descriptive statistics situation in modeling, we had the following examples of mixed messages.

- $r^2$ versus common sense
- $r^2$ versus simplicity
- Simplicity versus acceptability

In a sampling situation, we will add two more examples.

- Additional variable is statistically significant versus tiny improvement in $r^2$
- $r^2$ versus significance of the model

**Additional Variable Is Statistically Significant Versus Tiny Improvement in $r^2$**   Suppose we regress salary on grade and obtain an $r^2$ of 0.77 to two decimal places. Now suppose we then add performance level and obtain an $r^2$ of 0.78 to two decimal places and get a *t* probability of 0.04 for the new variable. We have added a significant variable that barely increased the power of the model.

**$r^2$ Versus Significance of the Model**   Sometimes we get a statistically significant model as indicated by the *F* probability, but it does not explain a whole lot. We might regress salary versus service and obtain an *F* probability of 0.04, considered statistically significant at usually acceptable levels, but an $r^2$ of only 0.03. We would have a statistically significant model that explains only 3% of the variation of the dependent variable. This sometimes happens when there are a lot of data points with a definite trend but lots of scatter.

# A.15 ADDITIONAL MULTIPLE LINEAR REGRESSION TOPICS

## Adjusted $r^2$

The adjusted $r^2$ is an adjustment of $r^2$ that takes into account the effect of small sample sizes in a regression model. It is often used in multiple linear regression when comparing different models. The formula is shown in Table A.3.

Because of the least squares criterion, the line and subsequent $r^2$ are both influenced by the square of the distance of all points from the regression line, including "outliers."

When the number of data points is small, a few outliers will exert a relatively large influence on both the line and $r^2$. The line will be "pulled" toward the outliers and $r^2$ will be slightly inflated, both as compared to what would result with a larger sample size. Theoretically, the results with a larger sample size are closer to the "truth."

Hence, an adjustment is made to $r^2$ to estimate what it would be if a larger sampling condition had occurred. There is no comparable adjustment to the line itself.

For most compensation problems, the differences are negligible, so using the "regular" $r^2$ is acceptable.

## Coding of Indicator Variables

An indicator variable is a binary variable with values of 0 or 1, where 1 indicates the presence of a designated characteristic within a category and 0 indicates its absence. When there are more than two characteristics within a category, special rules must be followed in creating indicator variables.

The value 0 is used to designate the reference or base against which the other characteristics will be compared.

**One Category, Three Characteristics**   Suppose we are modeling total pay of sales personnel and there is just one category, say type of plan, and suppose, and there were three plans in operation: salary only, salary plus commission, and commission only. We decide that the reference point in this instance is salary only.

With three characteristics (i.e., choices of plans), there will be two indicator variables, designated here as $x_1$ and $x_2$. With $n$ characteristics there will be $n - 1$ indicator variables for a given category. Table A.6 shows how the coding works for this example.

**TABLE A.6**   **CODING OF INDICATOR VARIABLES—ONE CATEGORY, THREE CHARACTERISTICS**

|  | $x_1$ | $x_2$ |
|---|---|---|
| Salary only | 0 | 0 |
| Salary plus commission | 1 | 0 |
| Commission only | 0 | 1 |
| Not allowed | 1 | 1 |

| | $x_1$ | $x_2$ |
|---|---|---|
| No degree | 0 | |
| Has degree | 1 | |
| Did not attend | | 0 |
| Did attend | | 1 |

**TABLE A.7** CODING OF INDICATOR VARIABLES—TWO CATEGORIES, TWO CHARACTERISTICS EACH

$x_1 = 0$ if not salary plus commission, 1 if salary plus commission

$x_2 = 0$ if not commission only, 1 if commission only

So if both variables have a value of zero, that means that salary only was the plan. If $x_1$ is 1 and $x_2$ is 0, that means salary plus commission was the plan. If $x_1$ is 0 and $x_2$ is 1, that means that commission only was the plan. Since an employee cannot be on more than one plan, having both variables with a value of 1 is not allowed.

**Two Categories, Two Characteristics Each**    If there are two categories, each one with only two characteristics, then there will be two indicator variables—one for each category. For example, one category, $x_1$, could be college degree (0 = no degree, 1 = degree) and another category, $x_2$, could be attendance at orientation (0 = did not attend, 1 = did attend). The coding schemes for each category are independent of each other, as shown in Table A.7.

## Interaction Terms

Interaction terms are often used in conjunction with indicator variables, although they can be used with regular variables as well.

Suppose we have found that the best simple linear model for productivity ($y$) is given by

$$y = b_0 + b_1 x_1$$

where $x_1$ is years of experience.

We want to know if productivity is also affected by an employee having a certification, so we introduce an indicator variable, $x_2$ into the model which will indicate whether or not an employee is certified.

$$y = b_0 + b_1 x_1 + b_2 x_2$$

where $x_1$ is years of experience; $x_2 = 0$ if not certified, 1 if certified.

We run our regression and fit this model to the data (not shown) and obtain the following.

$$y' = 20 + 5x_1 + 30x_2$$

What does the coefficient of the indicator variable mean?

For employees without a certification ($x_2 = 0$) the model is

$$y' = 20 + 5x_1 + 30(0) = 20 + 5x_1$$

This is the equation of a line with an intercept of 20 and a slope of 5.

For employees with a certification ($x_2 = 1$) the model is

$$y' = 20 + 5x_1 + 30(1)$$

Rearranging terms, we have

$$y' = (20 + 30) + 5x_1 = 50 + 5x_1$$

This is the equation of a line with an intercept of 50 and a slope of 5.

Both lines are shown in Figure A.15.

What does this say about the productivity differences? The presence of certification shifts the line upward by 30 units/day, but keeps the same slope.

Using an indicator variable in this manner allows one to see if the intercept term of the model should be different for those items having the characteristic of interest than it is for those without. It asks if the line should be shifted either up (positive $b_2$) or down (negative $b_2$) because of the presence of the specified characteristic.

If you suspect that a characteristic will affect the value of the dependent variable, you can check this out initially by looking at a plot of the data and

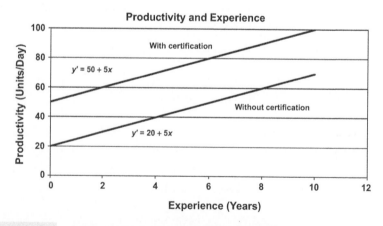

**FIGURE A.15    PRODUCTIVITY AND EXPERIENCE MODELS**

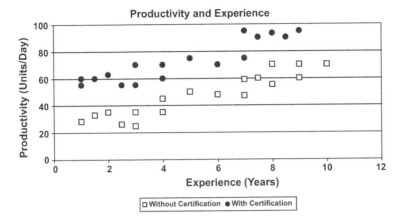

**FIGURE A.16    PRODUCTIVITY AND EXPERIENCE DATA**

noting which data points have the characteristic in question. If the characteristic makes a difference, the data points should segregate themselves accordingly, as shown in Figure A.16.

From this chart, it is obvious that the data points do segregate.

Now, suppose the plot of the data points looks like Figure A.17.

There is a segregation, but will using an indicator variable as before adequately model this situation? The answer is no, because the slopes of the two groups are also different. There is obviously some interaction between the experience and the presence of certification; the more experience, the greater certification seems to affect productivity. We can reflect this in the model statement by using an interaction term. An interaction term is a term that is formed

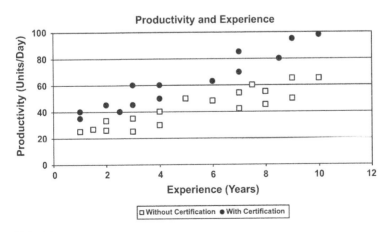

**FIGURE A.17    PRODUCTIVITY AND EXPERIENCE DATA**

by multiplying together two (or more) independent variables together. For this example, we will use the following model:

$$y = b_0 + b_1 x_1 + b_2 x_2 + b_3 x_1 x_2$$

$x_1 =$ years of experience

$x_2 = 0$ if not certified, 1 if certified

$x_1 x_2 =$ the interaction term

We run our regression and fit this model to the data (not shown) and obtain the following.

$$y' = 20 + 4x_1 + 10x_2 + 3x_1 x_2$$

For employees without a certification ($x_2 = 0$) the model is

$$y' = 20 + 4x_1 + 10(0) + 3x_1(0) = 20 + 4x_1$$

This is the equation of a line with an intercept of 20 and a slope of 4.
    For employees with a certification ($x_2 = 1$) the model is

$$y' = 20 + 4x_1 + 10(1) + 3x_1(1)$$

Rearranging, $y' = (20 + 10) + (4 + 3)x_1 = 30 + 7x_1$
    This is the equation of a line with an intercept of 30 and a slope of 7.
    Both lines are added to the chart and we see the separation of both the intercepts and the slopes in Figure A.18.

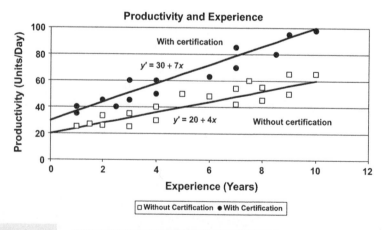

**FIGURE A.18      PRODUCTIVITY AND EXPERIENCE MODELS**

Thus, both the slope and intercept can be different for employees with certification than they are for employees without certification.

So when modeling with an indicator variable, if all you use is an indicator variable, then what you have identified is simply a difference in intercepts between the two subsets of data and are left with the untested assumption that the slopes are the same.

If, when using an indicator variable, you also create an interaction variable, you will be able to test for both the intercepts and the slopes to see if they are different and by how much for the two subsets.

Hence, when using an indicator variable, it is advisable to include an interaction term.

# Glossary

Various terms as they are used in this text are defined here, many of which are defined in the text itself but are compiled here for the reader's convenience. It should be noted that definitions evolve over time, and sometimes there is not agreement among authors or users on some of them.

Many words have more than one definition or connotation, often depending on the context in which they are used. We have limited these definitions to the context of statistics in compensation practices.

**Abscissa**    The horizontal or $x$-axis in a two-dimensional Cartesian coordinate system.

**Absolute value**    For a positive number, the positive number itself; for a negative number, the positive equivalent after changing the minus sign to a plus sign.

**Accuracy**    The degree of closeness to the truth of an average of measurements.

**Adjusted $r^2$**    An adjustment of $r^2$ that takes into account the effect of small sample sizes in a regression model. It is often used in multiple linear regression when comparing different models.

**Antilog**    The inverse of a logarithmic transformation, and consists of raising the base to the power of the logarithm. For example, in common logs (base 10), if the log of a number were 3, then the antilog would be $10^3 = 1,000$.

**Arithmetic mean**    See mean.

**Average**    See mean.

**Axis**    A straight line that is a reference line of a Cartesian coordinate system.

**Bell-shaped distribution**    A symmetric distribution in which the associated histogram has a peak in the middle and a tail at each end. It has the shape of a vertical cross section of a bell.

**Bimodal distribution**    A distribution in which the associated histogram has two modes or humps with high frequencies.

**Binary variable**    A variable with only two values.

**Cardinal number**    See ratio number.

*Statistics for Compensation: A Practical Guide to Compensation Analysis,* By John H. Davis
Copyright © 2011 John Wiley & Sons Inc.

**Cartesian coordinate system**    A system of mutually perpendicular coordinate axes [8].

**Categorical variable**    A variable whose values are categories or nominal numbers.

**Chebyshev's inequality**    A statistical theorem that states, among other things, that in any data set at least 75% of the data will fall within two standard deviations of the mean. Chebyshev's inequality is also known as Chebyshev's theorem, named after the Russian mathematician P.L. Chebyshev (1821–1894) (aka P.L. Tchebysheff) and states that at least $(1 - 1/k^2)(100)\%$ of any set of data falls within $k$ standard deviations of their mean. The two-sigma rule follows when $k = 2$.

**Coefficient of determination**    A quantitative measure of the association between two variables and is the proportion of variability of one variable that can be attributed to the variability of the other variable. The coefficient of determination is represented by the symbol $r^2$ and has values between 0 and 1. It is equal to 1 minus the quotient sum of squares residual divided by sum of squares total.

**Coefficient of variation**    For a set of data, it is a measure of relative variability and is equal to the standard deviation divided by the mean, and then converted to a percent.

**Common logarithm**    A logarithm to the base 10.

**Compensation**    The branch of human resources dealing with the elements of pay provided by an employer to its employees for services rendered. Elements of pay include base pay, variable pay, and stock.

**Compound interest**    The interest earned on the principal and on the accumulated interest earned previously that was not withdrawn.

**Compounding**    The process of earning compound interest.

**Compounding factor**    A factor that relates the present value to the future value. Mathematically, it is $(1 + i)^n$, where $i$ is the interest rate per period and $n$ is the number of periods.

**Convenience sample**    A sample that is not random.

**Coordinate**    A number in an ordered $n$-tuple that indicates the position of a data point with respect to a corresponding axis.

**Coordinate axis**    See axis.

**Correlation**    A qualitative measure of the association between two variables and is the square root of the coefficient of determination. The correlation is represented by the symbol $r$ and has values between $-1$ and $+1$. If the correlation is positive, it means the high values of one variable are associated with the high values of the other variable, and the low values of one variable are associated with the low values of the other variable. If the correlation is negative, it means the high values of each variable are associated with the low values of the other variable.

**Correlation coefficient**  See correlation.

**Critical factors**  Variables that may influence or impact the problem variable. These are the independent variables in a model.

**Cubic model**  A third-degree polynomial model, expressed by the equation $y = a + bx + cx^2 + dx^3$.

**Data**  A collection of data points.

**Data point**  The measured characteristic(s) of an element. For example, the sales of a company is a data point for a single variable analysis. The sales and profits of a company comprise a data point for a two-variable analysis, expressed as an ordered pair of numbers, $(x, y)$, where $x$ is the value of the $x$-variable and $y$ is the value of the $y$-variable. They are always displayed with $x$ first and $y$ second and enclosed in parentheses. The sales, profits, and number of employees of a company comprise a data point for a three-variable analysis, expressed as $(x, y, z)$. For $n$ variables, a data point is expressed as an ordered $n$-tuple.

**Data set**  Data with a commonality. For example, one data set might be the sales of companies in a particular location, and another data set might be the sales of companies in another location.

**Decile**  Any one of nine percentiles that divide an ordered data set into 10 equal parts. The 1st decile is the 10th percentile, the 2nd decile is the 20th percentile, ..., and the 9th decile is the 90th percentile.

**Decision model**  A representation of the process that starts with data and ends with a solution.

**Degree of polynomial**  The highest exponent in a polynomial equation.

**Degrees of freedom**  A number associated with a sum of squares that indicates how many independent pieces of information involving the $n$ independent data points are available and used to compile the sum of squares.

**Dependent variable**  A variable that is affected by an independent variable(s), usually denoted as the $y$-variable. In model building, it is the problem variable. Also called the response variable or the criterion variable.

**Descriptive statistics**  A branch of statistics that summarizes and describes variables and the strength and nature of relationships between them.

**Distribution**  See frequency distribution.

**Dummy variable**  See indicator variable.

**$e$**  Also known as Euler's constant, it is the base of natural logarithms. The value of $e$ is approximately 2.71828. It is used in many fields such as mathematics, science, and engineering. It is one of the five most important numbers in mathematics, namely, the additive identity 0, the multiplicative identity 1, the constant $\pi$, the constant $e$, and the imaginary unit $i$ (square root of $-1$).

**Element**  An entity on which a measurement of one or more of its characteristics is taken. For example, a company is an element and its sales and profits are measured.

**Equation**  A symbolic representation of a relationship between variables.

**Exponent**  The power to which a variable is raised. For example, the exponent of $x^3$ is 3.

**Exponential model**  A logarithmic transformation model in which the $y$-variable is transformed into logarithms, expressed by the equation $\log y = a + bx$.

**Extreme value**  See outlier.

**Frequency distribution**  A classification of a set of data into certain categories, with subsequent counting and sometimes percentage calculations that indicate the amount in each category.

**Functional form**  The type of model that describes a trend between two variables.

**Future value**  How much a given amount of money now is worth in the future.

**Grand mean**  The point of the means $(\bar{x}, \bar{y})$. Also called the centroid.

**Histogram**  A bar chart display of a frequency distribution, with each bar representing a category and the height of each bar indicating the amount in that category.

**Human resources**  The function of an organization dealing with the management of people employed within the organization.

**Independent variable**  A variable that affects the dependent variable, usually denoted as an $x$-variable. In model building it is a critical factor. Also called an explanatory variable or a predictor variable.

**Indicator variable**  A binary variable with values of 0 or 1, where 1 indicates the presence of a designated characteristic and 0 indicates its absence.

**Inferential statistics**  A branch of statistics in which conclusions or generalizations are made about a population based on the results from a properly constructed sample from that population.

**Intercept**  In linear regression, the value of the dependent variable when all the independent variables equal zero. For a simple linear regression, the intercept is the symbol $a$ in the equation $y = a + bx$ and it is the value of $y$ when $x$ equals zero.

**Interest**  The amount earned from the application of an interest rate to a base amount.

**Interest rate**  A rate applied to a principal for the use of money for a specified term. If you are a borrower, it is the rate the bank charges you for the use of its money. If you are a saver, it is the rate you get in return for the bank's use of your money.

**Interquartile range**  A measure of variability equal to the 75th percentile minus the 25th percentile. The range of the middle half of an ordered data set.

**Interval number**  A number on a scale where differences have quantitative meaning but where there is no zero that means "nothing." The designation of a zero is arbitrary.

**Inverse relationship**    See negative relationship.

**Join**    See knot.

**Knot**    In a spline model, the value where two cubic polynomials are joined.

**Least squares line**    The line that minimizes the sum of squared deviations of the data points from the line.

**Linear model**    A first-degree polynomial model or straight line model, expressed by the equation $y = a + bx$.

**Linear regression**    In the context of statistical modeling, a regression that fits a model that is linear in the coefficients, namely that the dependent variable is equal to a sum of terms that are each an independent variable multiplied by a coefficient, to the data.

**Linear scale**    A scale in which a constant difference between numbers on the scale is represented by a constant distance no matter where you are on the scale.

**log**    See logarithm.

**Logarithm**    The exponent or power to which a base is raised to produce a certain value. Often shortened to log for a common logarithm (base 10) or ln for a natural logarithm (base $e$).

**Logarithmic model**    A logarithmic transformation model in which the $x$-variable is transformed into logarithms, expressed by the equation $y = a + b \log x$.

**Logarithmic scale**    A scale in which a constant ratio between numbers on the scale is represented by the same distance no matter where you are on the scale.

**Logarithmic transformation model**    A two-variable model in which one or both of the variables is transformed into logarithms.

**Log–log plot**    A plot in which both axes are on a logarithmic scale.

**Mathematics**    The science of measurement [5].

**Maturity curve**    A description of the relationship between pay and experience, usually for people with similar educational backgrounds and doing similar types of work, such as nonsupervisory BS engineers conducting research or first-level supervisory PhD scientists [2].

**Mean**    For a set of data, a measure of central tendency and is a value calculated by dividing the sum of the data by the number of data points.

**Measure**    See measurement.

**Measure of central tendency**    For a set of data, a measure of location that indicates where the central part of the data is located.

**Measure of location**    For a set of data, a measure that indicates where a certain part of the data is located.

**Measure of variability**    For a set of data, a measure that indicates its variability.

**Measurement**    The identification of the quantity of specified units of a characteristic of an entity. More broadly, measurement is the process of assigning a number to an attribute (or phenomenon) according to a rule or set of rules. The term can also be used to refer to the result obtained after performing the process.

**Median**   For a set of data, a measure of central tendency and is a value that half of the data are less than or equal to. Also known as the 50th percentile.

**Method of least squares**   For a given set of data and a given model, fitting the model to the data by choosing the coefficients of the model equation that minimize the sum of squared deviations of the data points from the model. That minimum sum of squared deviations is the sum of squares residual.

**Midpoint**   The point halfway between the minimum and the maximum of the pay range for a job or a grade. Mathematically, it is the average of the minimum and maximum.

**Midpoint progression**   The percentage increase from one midpoint of a grade in a salary structure to the midpoint of the next higher grade.

**Mode**   For a set of data, a measure of central tendency and is the value with the highest frequency. For a frequency distribution, it is the category with the highest frequency.

**Model**   An abstraction of reality used to solve problems in an objective manner. It consists of the problem (the $y$-variable), critical factors that impact the problem (the $x$-variables), and the relationships among the variables (expressed mathematically as an equation).

**Model-building**   The process of building a model.

**MOTH**   An acronym standing for Minus One Times Hundred. It describes the steps used to convert a ratio to a percent difference.

**Multicollinearity**   In multiple linear regression, a high correlation between one $x$-variable and another $x$-variable or between one $x$-variable and a linear combination of other $x$-variables.

**Multiple linear regression**   A linear regression with more than one $x$-variable, expressed by the equation $y = b_0 + b_1x_1 + b_2x_2 + b_3x_3 + \cdots + b_kx_k$.

**Multiple $r$**   In a regression, the correlation between $y$ and $y'$.

**Natural logarithm**   A logarithm to the base $e$.

**Negative relationship**   For a set of data, a relationship between two variables in which the high values of each variable are associated with the low values of the other variable.

**Negatively skewed distribution**   A skewed distribution in which the associated histogram has a hump on the right side and a tail on the left side.

**Nominal number**   A number that is used as an identifier, or label, or name.

**Nonlinear model**   A model that is not linear.

**Normal distribution**   A theoretical continuous bell-shaped distribution. Also known as Gaussian distribution.

**Number**   A mental symbol that integrates units into a single larger unit (or subdivides a unit into fractions) with reference to the basic number of "1," which is the basic mental symbol of "unit." Thus, "5" stands for | | | | | [5].

**Number of significant figures**   The number of digits in a number that are known to be true with some degree of confidence.

**Order of magnitude**    In a set of numbers, a designation of the range of values when one of the numbers is 10 times the value of another.

**Ordinal number**    A number that indicates rank, position, or order of an item within a group.

**Ordinate**    The vertical or $y$-axis in a two-dimensional Cartesian coordinate system.

**Origin**    The point in a Cartesian coordinate system where all the axes have a value of zero. In a two-dimensional system, it is the point $(0, 0)$.

**Outlier**    A data point whose value appears to deviate markedly from other data points in the data set in which it occurs. It is numerically distant from the rest of the data. There is no rigid statistical criterion of what constitutes an outlier; determining whether a data point is an outlier is ultimately a judgment decision.

**Overlap**    The percent of a salary range of a grade that is overlapped by the salary range of the next lower grade.

**P90/P10**    For a set of data, a measure of variability and is equal to the 90th percentile divided by the 10th percentile.

**Percent**    A representation of the amount of a particular quantity relative to a reference quantity, expressed as a proportion or ratio that is multiplied by 100.

**Percentage**    See percent.

**Percentile**    For a set of data, a measure of location and is a value that a given percentage of the data is less than or equal to.

**Percentile bar**    A graphical display of percentiles in a bar with certain (usually the standard) percentiles indicated.

**Perfect relationship**    A relationship in which the data points, when plotted, fall perfectly on a straight or simple curved line.

**Period**    The event for one application of interest, usually an amount of time.

**Pie chart**    A pie-shaped graphical display where the pieces designate categories and the sizes of the pieces are proportional to the amount in the categories.

**Plot**    A visual display on a Cartesian coordinate system showing the relationship between variables.

**Point**    See data point.

**Polynomial equation**    An equation that is the summation of polynomial terms.

**Polynomial model**    A model that is expressed by a polynomial equation.

**Polynomial term**    A variable, usually denoted by a letter such as $x$ or $y$, that is raised to a nonnegative whole number power and that may be multiplied by a coefficient. Examples are $2x$, $5x^2$, $-4x^3$, and $x^4$. Theoretically, a constant is a polynomial term, because a variable raised to the 0th power is 1. For example, $6x^0 = (6)(1) = 6$.

**Population**    The part of reality that you are analyzing and describing or about which you are making inferences or decisions. It is a collection of elements you want to describe or about which you want to make an inference.

**Positive relationship**    For a set of data, a relationship between two variables in which the high values of one variable are associated with the high values of the other variable, and the low values of one variable are associated with the low values of the other variable.

**Positively skewed distribution**    A skewed distribution in which the associated histogram has a hump on the left side and a tail on the right side.

**Power model**    A logarithmic transformation model in which both variables are transformed into logarithms, expressed by the equation $\log y = a + b \log x$.

**Precision**    The degree of closeness to each other of measurements.

**Present value**    How much a given amount of money in the future is worth now (the present).

**Principal**    The original amount on which interest is earned.

**Problem variable**    The dependent variable in a model.

**Quadratic model**    A second-degree polynomial model expressed by the equation $y = a + bx + cx^2$.

**Quartile**    As a point, any one of three percentiles that divide an ordered data set into four equal parts. The 1st quartile is the 25th percentile, the 2nd quartile is the 50th percentile, and the 3rd quartile is the 75th percentile.

**Quartile**    As a range, any one of four equal parts of an ordered data set. The 1st quartile is the range of data from the minimum up to the 25th percentile, the 2nd quartile is the range of data from the 25th percentile up to the 50th percentile, the 3rd quartile is the range of data from the 50th percentile up to the 75th percentile, and the 4th quartile is the range of data from the 75th percentile up to and including the maximum.

*r*    See correlation.

$r^2$    See coefficient of determination.

**Random sample**    A sample of size $n$ in which every possible sample of that size has an equal chance of being selected.

**Range**    For a salary structure, see range spread.

**Range**    For a set of data, a measure of variability and is equal to the difference between the maximum value and the minimum value.

**Range spread**    The percent that the maximum of a salary range for a grade or job is more than the minimum.

**Ratio number**    A number that indicates the quantity (how much, how many) of the item(s) being measured, and where zero means nothing of the item. Zero is fixed rather than arbitrary.

**Rectangular coordinate system**    See Cartesian coordinate system.

**Regression**    Fitting a model to data using the method of least squares.

**Relationship**    The type of association between two variables.

**Reverse percentile**   A percentile in a data set corresponding to a given value.

**Rule of 72**   A rule that states the number of periods it will take a given amount to double times the periodic growth rate is approximately 72.

**Sample**   Part of a population and ideally representative of it.

**Scale**   A series of numbered marks along a line at regular or graduated intervals.

**Scatter chart**   See plot.

**Scatter graph**   See plot

**Scatter plot**   See plot.

**Scattergram**   See plot.

**Scientific method**   The process to construct an objective (i.e., accurate, reliable, consistent, and nonarbitrary) representation of the world.

**Scientific notation**   A shorthand notation that indicates where the decimal point of a number is located. For example, 3.71E-6 means that you should move the decimal place to the left six places and get 0.00000371. Similarly, 1.23E+7 means that you should move the decimal place to the right seven places and get 12,300,000.

**Semi-log plot**   A plot in which one of the axes is on a logarithmic scale and the other is on a linear scale.

**Significant figure**   A digit in a number that is known to be true with some degree of confidence.

**Simple linear regression**   A linear regression of a straight line, expressed by the equation $y = a + bx$.

**Simple mean**   See mean.

**Simpson's paradox**   A statistical paradox in which the trends observed in several groups are reversed when the groups are combined or when the trend observed in a combination of groups is reversed when examining the groups separately. Usually caused by unequal group sizes combined with a lurking variable.

**Skewed distribution**   A nonsymmetric distribution in which the associated histogram has a hump on one side and a tail on the other.

**Slope**   For a simple linear model, the change in the value of $y$ for a unit change in $x$, represented by the symbol $b$ in the equation $y = a + bx$. For a multiple linear model, the change in the value of $y$ for a unit change in the corresponding $x$ holding the other $x$'s constant.

**Spline model**   A polynomial regression model that has one cubic model for the lower range of $x$-variables joined smoothly at a value called the knot with another cubic model for the upper range of $x$-variables, expressed by the equation $y = a + bx + cx^2 + dx^3 + ep^3$, where $p = 0$ if $x < k$, and $k - x$ for $x \geq k$, where $k$ is the knot. Some spline models are more complex with more than one knot. By joining smoothly at the knot is meant that the first and second derivatives are continuous at that point.

**Standard deviation**    For a set of $n$ data points, a measure of variability of the data points from the mean and is the square root of the "average" squared deviation of the data points from the mean. For a population, the average is calculated by dividing by $n$. For a sample the average is calculated by dividing by $n - 1$.

**Standard error of estimate**    For a set of $n$ data points and a model with $p$ parameters, a measure of variability of the data points from the predicted values and is the square root of the "average" squared deviations of the data points from the predicted values. The average is calculated by dividing by $n - p$.

**Standard percentiles**    The 10th, 25th, 50th, 75th, and 90th percentiles.

**Statistical inference**    See inferential statistics.

**Statistics**    The branch of mathematics concerned with the measurement of uncertainty. It consists of a body of methods for obtaining, organizing, summarizing, analyzing, interpreting, presenting, and acting upon the mathematical attributes of items of interest. It consists of a collection of mathematical techniques that provide information from data for decisions.

**Sum of Squares Error**    Sometimes denoted by SS Error or SSE. See sum of squares residual.

**Sum of Squares Regression**    In a regression model, a measure of the variation of the $y$-variable that is explained by or attributed to the model. It is equal to the sum of squares total minus the sum of squares residual. Sometimes denoted by SS Regression.

**Sum of Squares Residual**    In a regression model, a measure of the variation of the $y$-variable that is not explained by or attributed to the model. It is equal to the sum of squared deviations of the data points from the predicted values. Sometimes denoted by SS Residual. This is also the minimum sum of squares obtained by the least squares method.

**Sum of Squares Total**    A measure of the total variation of a set of data and is equal to the sum of the squared deviations of the data points from their mean. Sometimes denoted by SS Total.

**Symmetric distribution**    A distribution in which the left side of the associated histogram is a mirror image of the right side.

**Trimmed mean**    For a set of data that has been ordered, a measure of central tendency that is the mean of the middle data points after first excluding a certain number of lowest values and the same number of highest values.

**Truth**    That which corresponds to reality.

**Two-sigma rule**    A rule that states that in any data set, at least 75% of the data will fall within two standard deviations of the mean. See Chebyshev's inequality.

**Uniform distribution**    A symmetric distribution in which the associated histogram has the same frequency in each category.

**Unweighted average**    See unweighted mean.

**Unweighted mean**    For a set of means, the simple mean of these means. For salary survey situations, sometimes called the company-weighted mean.

**Variable**    Any measured characteristic of an element that differs for at least two different elements.

**Variance**    For a set of $n$ data points, a measure of variability of the data points from the mean and is equal to the "average" squared deviation of the data points from the mean. For a population the average is calculated by dividing by $n$. For a sample the average is calculated by dividing by $n - 1$.

**Weighted average**    See weighted mean.

**Weighted mean**    For a set of means, the mean of these means weighted by the number of data points in each one. For salary survey situations, sometimes called the incumbent-weighted mean.

**$x$-variable**    See independent variable.

**$y$-variable**    See dependent variable.

**$z$-score**    The number of standard deviations a particular value is from the mean.

# References

1. Campbell, S. K. (1987) *Applied Business Statistics*. New York: Harper & Row.
2. Davis, J. H. (1991) Maturity Curves: How Companies Use Them Successfully. *Perspectives in Total Compensation*, Vol. 2, No. 10. Scottsdale, Arizona: American Compensation Association.
3. Davis, J. H. (1992) *Sound Compensation Practices: A Theoretical Foundation*. Richardson, Texas: Davis Consulting.
4. Davis, J. H. and Koechel, J. F. (2006) Fundamentals of Salary Surveys. *Survey Handbook and Directory: A Guide to Finding and Using Salary Surveys*. Scottsdale, Arizona: WorldatWork.
5. Rand, A. (1990) *Introduction to Objectivist Epistemology*, Expanded 2nd Edition, Edited by Harry Binswanger and Leonard Peikoff. Meridian Books.
6. Simpson, E. H. (1951) The Interpretation of Interaction in Contingency Tables. *Journal of the Royal Statistical Society Series B* **13**: 238–241.
7. Stigler, S. M. (1986) *The History of Statistics: The Measurement of Uncertainty Before 1900*. Cambridge: Belknap Harvard.
8. Thomas, G. B., Finney, R. L., Weir, M. D., and Giordana, F. R. (2003) *Thomas' Calculus*, 10th Edition. Addison-Wesley Publishing Company.
9. Yule, G. U. (1903) Notes on the Theory of Association of Attributes in Statistics. *Biometrika* **2**: 121–134.

## GENERAL STATISTICAL REFERENCES

10. Draper, N. R. and Smith, H. (1998) *Applied Regression Analysis*, 3rd Edition. New York: John Wiley & Sons, Inc.
11. Freund, J. E. and Perles, B. M. (2007) *Modern Elementary Statistics*, 12th Edition. Englewood Cliffs: Prentice Hall.
12. Huff, D. and Geis, I. (1993) *How To Lie with Statistics*. W. W. Norton.

# Answers to Practice Problems

## CHAPTER 2 BASIC NOTIONS

2.1  Total number of personnel is 1,400. Percent of office and clerical personnel is

$$(420/1{,}400) \times 100 = 0.30 \times 100 = 30\%.$$

2.2  Number living in Texas is

$$12\% \times 9{,}500 = (12/100) \times 9{,}500 = 0.12 \times 9{,}500 = 1{,}140.$$

2.3  Percent of goal is

$$(18{,}000{,}000/15{,}000{,}000) \times 100 = 1.20 \times 100 = 120\%.$$

2.4  Let $x$ = number of companies in the industry.

$$32 = (40\%)x = (40/100)x = 0.40x$$
$$x = 32/0.40$$
$$x = 80$$

2.5  No, because the 20% is applied against two different bases. When you had a reduction in force, you laid off 20% of 1,000, or 200, resulting in 800 remaining. When you subsequently increased your workforce, you added 20% of 800, or 160, resulting in a total of 960. If you wanted to return to the original workforce size, you would need to add 25% of 800, or 200.

2.6  % from survey mean $= \dfrac{72{,}000 - 78{,}000}{78{,}000} \times 100 = -7.7\%$ (rounded)

Your midpoint is 7.7% below the survey mean.

% from midpoint $= \dfrac{78{,}000 - 72{,}000}{72{,}000} \times 100 = 8.3\%$ (rounded)

The survey mean is 8.3% above your midpoint.

They are different because they have different reference points.

*Statistics for Compensation: A Practical Guide to Compensation Analysis,* By John H. Davis
Copyright © 2011 John Wiley & Sons Inc.

The second one is preferable—the survey mean is 8.3% above your midpoint—because you will be taking action on your midpoint by adjusting it upward by 8.3%.

2.7 The raise would probably be more meaningful in a positive way to the administrative assistant, because his raise was 10%, while the vice president's raise was only 1%. Sometimes we make absolute comparisons and sometimes we make relative comparisons. In the case of salary raises, the relative comparisons are often (but not always) perceived as more meaningful.

If they each received a 10% raise, they might think they would have been rewarded equivalently.

2.8 With annual raises of 6%, $FV = 70,000 \times 1.06^3 = \$83,371$ (rounded). With semiannual raises of 3%, $FV = 70,000 \times 1.03^6 = \$83,584$ (rounded).

2.9 $PV = 50,000/1.0075^{48} = \$34,931$ (rounded)

2.10 No. of offices $= 15 \times 1.10^5 = 24$ (rounded)

# CHAPTER 3 FREQUENCY DISTRIBUTIONS AND HISTOGRAMS

3.1 We selected categories with a width of €1,000. You may have selected another width. The frequency distribution is shown in Table S.1 and the histogram is shown in Figure S.1.

The distribution is negatively skewed, with a tail to the left.

**TABLE S.1** **FREQUENCY DISTRIBUTION OF TECHNICIAN SALARIES**

| Annual Salary (Euros) | No. of Technicians |
| --- | --- |
| ≤43K | 0 |
| 43K–44K | 1 |
| 44K–45K | 1 |
| 45K–46K | 2 |
| 46K–47K | 2 |
| 47K–48K | 3 |
| 48K–49K | 4 |
| 49K–50K | 4 |
| 50K–51K | 6 |
| 51K–52K | 3 |
| 52K–53K | 3 |
| 53K–54K | 1 |
| >54K | 0 |
| **Total** | **30** |

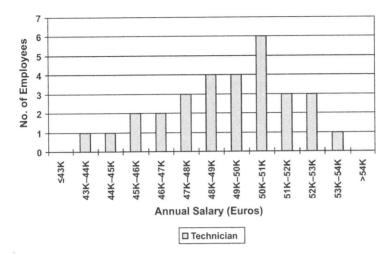

**FIGURE S.1**     **HISTOGRAM OF DISTRIBUTION OF TECHNICIAN SALARIES**

3.2  For the production assistants, the frequency distribution is shown in Table S.2 and the histogram is shown in Figure S.2. The distribution is positively skewed, with a tail to the right.

3.3  The comparison of the absolute distributions of the two employee populations is shown in Table S.3 and Figure S.3. We had to expand the category scale to accommodate both population, but still keeping the same interval width of €1,000. We note that the bars of the staff assistant

**TABLE S.2**     **FREQUENCY DISTRIBUTION OF PRODUCTION ASSISTANT SALARIES**

| Annual Salary (Euros) | No. of Production Assistants |
|---|---|
| ≤46K | 0 |
| 46K  47K | 3 |
| 47K–48K | 5 |
| 48K–49K | 11 |
| 49K–50K | 17 |
| 50K–51K | 15 |
| 51K  52K | 11 |
| 52K–53K | 6 |
| 53K–54K | 4 |
| 54K–55K | 3 |
| 55K–56K | 3 |
| 56K–57K | 2 |
| >57K | 0 |
| **Total** | **80** |

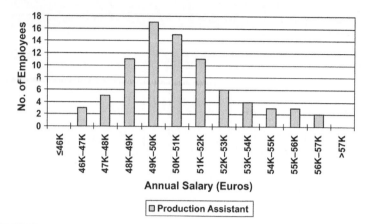

**FIGURE S.2**   HISTOGRAM OF DISTRIBUTION OF PRODUCTION ASSISTANT SALARIES

population are much higher than those of the technician population, due to the larger population size of the staff assistants.

3.4   The comparison of the relative distributions of the two employee populations is shown in Table S.4 and Figure S.4. We note that the bars of the two populations are of similar magnitudes in height, since they are on an "equal" basis. That is, the relative distributions each have to add up to 100%.

**TABLE S.3**   ABSOLUTE FREQUENCY DISTRIBUTIONS OF TECHNICIAN SALARIES AND PRODUCTION ASSISTANT SALARIES

| Annual Salary (Euros) | No. of Technicians | No. of Production Assistants |
| --- | --- | --- |
| ≤43K | 0 | 0 |
| 43K–44K | 1 | 0 |
| 44K–45K | 1 | 0 |
| 45K–46K | 2 | 0 |
| 46K–47K | 2 | 3 |
| 47K–48K | 3 | 5 |
| 48K–49K | 4 | 11 |
| 49K–50K | 4 | 17 |
| 50K–51K | 6 | 15 |
| 51K–52K | 3 | 11 |
| 52K–53K | 3 | 6 |
| 53K–54K | 1 | 4 |
| 54K–55K | 0 | 3 |
| 55K–56K | 0 | 3 |
| 56K–57K | 0 | 2 |
| >57K | 0 | 0 |
| **Total** | **30** | **80** |

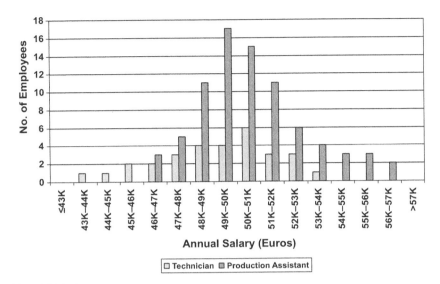

**FIGURE S.3    HISTOGRAM OF ABSOLUTE FREQUENCY DISTRIBUTIONS OF TECHNICIAN SALARIES AND PRODUCTION ASSISTANT SALARIES**

The fact that the distributions are of different shapes (in both the absolute distribution comparison and the relative distribution comparison) may warrant an investigation as to why they are different.

3.5   See Table S.5 and Figure S.5.

3.6   See Table S.6 and Figure S.6.

**TABLE S.4    RELATIVE FREQUENCY DISTRIBUTIONS OF TECHNICIAN SALARIES AND PRODUCTION ASSISTANT SALARIES**

| Annual Salary (Euros) | % of Technicians | % of Production Assistants |
|---|---|---|
| ≤43K | 0 | 0 |
| 43K–44K | 3 | 0 |
| 44K–45K | 3 | 0 |
| 45K–46K | 7 | 0 |
| 46K–47K | 7 | 4 |
| 47K–48K | 10 | 6 |
| 48K–49K | 13 | 14 |
| 49K–50K | 13 | 21 |
| 50K–51K | 20 | 19 |
| 51K–52K | 10 | 14 |
| 52K–53K | 10 | 8 |
| 53K–54K | 3 | 5 |
| 54K–55K | 0 | 4 |
| 55K–56K | 0 | 4 |
| 56K–57K | 0 | 3 |
| >57K | 0 | 0 |
| **Total** | **100** | **100** |

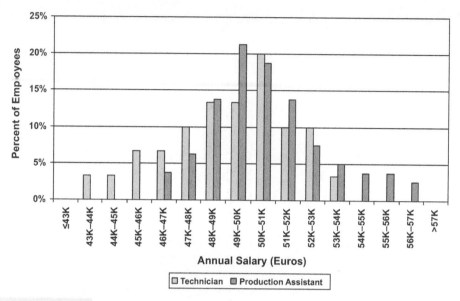

**FIGURE S.4**     HISTOGRAM OF RELATIVE FREQUENCY DISTRIBUTIONS OF TECHNICIAN SALARIES AND PRODUCTION ASSISTANT SALARIES

These results are different because there are a number of employees with ages on the boundaries of the categories, and a shift of 1 year in the category definition results in a shift of the number of employees in the respective categories. The point illustrated is that one should try different boundaries to see if this would be an issue and to understand the data.

3.7 See Table S.7 and Figure S.7.

These results are different because the grouping of the data into 10-year categories masks large differences between some of the adjacent 5-year categories. The point illustrated is that one should try different

| TABLE S.5 | FREQUENCY DISTRIBUTION OF AGES |
|---|---|
| **Age** | **No. of Employees** |
| 20–29 | 11 |
| 30–39 | 35 |
| 40–49 | 21 |
| 50–59 | 45 |
| 60–69 | 20 |
| 70–79 | 41 |
| **Total** | **173** |

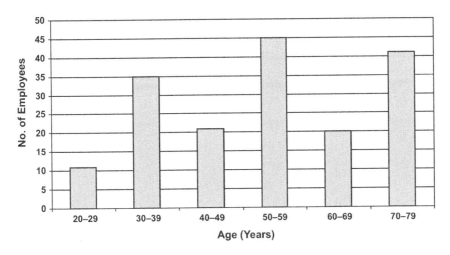

**FIGURE S.5    HISTOGRAM OF DISTRIBUTION OF AGES**

**TABLE S.6    FREQUENCY DISTRIBUTION OF AGES**

| Age | No. of Employees |
|-----|------------------|
| 21–30 | 20 |
| 31–40 | 27 |
| 41–50 | 37 |
| 51–60 | 29 |
| 61–70 | 32 |
| 71–80 | 28 |
| **Total** | **173** |

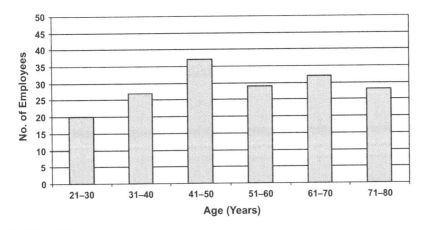

**FIGURE S.6    HISTOGRAM OF DISTRIBUTION OF AGES**

| TABLE S.7 | FREQUENCY DISTRIBUTION OF AGES |
|---|---|
| **Age** | **No. of Employees** |
| 21–25 | 8 |
| 26–30 | 12 |
| 31–35 | 24 |
| 36–40 | 3 |
| 41–45 | 2 |
| 46–50 | 35 |
| 51–55 | 27 |
| 56–60 | 2 |
| 61–65 | 2 |
| 65–70 | 30 |
| 71–75 | 16 |
| 76–80 | 12 |
| **Total** | **173** |

category definitions to see if this would be an issue and to understand the data.

3.8 (a) Skewed positively, with a tail to the right. (b) Uniform. (c) Bimodal. (d) Bell-shaped.

3.9 The histogram is shown in Figure S.8.

An issue might be that both the London office and the Brussels office each have about five times the total workforce that Geneva does. Hence, a difference of one employee in Geneva amounts to 10% of the workforce, whereas a difference of one employee in London or Brussels amounts

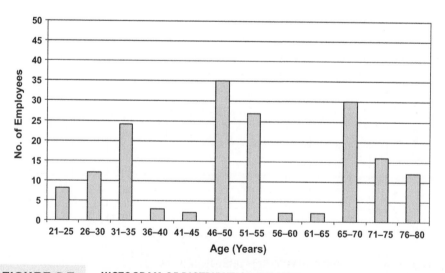

FIGURE S.7    HISTOGRAM OF DISTRIBUTION OF AGES

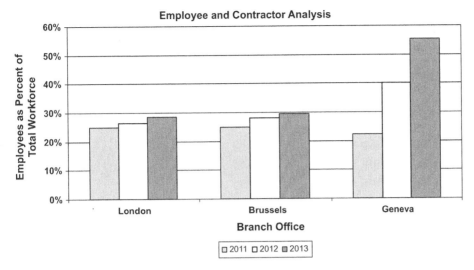

**FIGURE S.8    HISTOGRAM OF WORKFORCE POPULATIONS IN THREE LOCATIONS**

to only about 2% of the workforce. When comparing items of vastly different size, consider whether a comparison of absolute differences or relative differences makes the most sense.

3.10    The issue here is that the vertical axis ($y$-axis) has been truncated and starts at 20% and, in addition, has been stretched out, making the change in Geneva seem dramatic and more than it really is. This is an example of lying with statistics. If you presented this, your audience would question your honesty, thus undermining your reputation. This also exacerbates the issue regarding the comparison of a small office with two large offices.

# CHAPTER 4 MEASURES OF LOCATION

4.1

| | |
|---|---|
| Number | 99 |
| Mode | 50,000 |
| Median | 65,000 |
| Mean | 66,364 |
| Trimmed mean ($n = 89$) | 65,652 |

4.2    Skewed positively with a tail to the right, because the mean is more than the median. In addition, the mode is less than the median.

4.3

| | |
|---|---|
| Weighted mean | 96,200 (rounded to nearest 100) |
| Unweighted mean | 90,000 |

The weighted mean is greater because the company with the highest mean salary also has the largest number of incumbents.

**TABLE S.8**     **SALARY INCREASE GUIDELINES (MERIT MATRIX)**

| | Raise (%) | | |
| | Position in Range | | |
| Performance Level | Lower 1/3 | Middle 1/3 | Upper 1/3 |
|---|---|---|---|
| 5 | 6.5% | 5.5% | 4.5% |
| 4 | 5.5% | 4.5% | 3.5% |
| 3 | 4.0% | 3.0% | 2.0% |
| 2 | 2.0% | 1.5% | 1.0% |
| 1 | 0.0% | 0.0% | 0.0% |

4.4 One answer (of many possible ones) is given in Table S.8. If the managers follow these guidelines, they will spend 4.46% of salaries on salary increases, which is just below the budget of 4.5%.

4.5 One answer (of many possible ones) is given in Table S.9. If the managers follow these guidelines, they will spend 2.93% of salaries on salary increases, which is just below the budget of 3.0%.

4.6 The completed Table S.10 shows that by territory, Jane has a higher percentage of success than Virginia. But overall, the trend is reversed, and Virginia has a higher percentage of success than Jane. This is an example of Simpson's paradox. Here it occurred because of the combination of the lurking variable of "territory" and the very different percentage of attempts by territory by the two sales associates.

4.7 Mean = 30,563
$P10 = 28,770$
$P25 = 29,000$
$P50 = 30,050$
$P75 = 31,625$
$P90 = 33,230$

4.8 81st percentile (rounded).

**TABLE S.9**     **SALARY INCREASE GUIDELINES**

| Performance Level | Raise (%) |
|---|---|
| 5 | 4.5 |
| 4 | 3.5 |
| 3 | 2.0 |
| 2 | 0.0 |
| 1 | 0.0 |

**TABLE S.10     EXAMPLE OF SIMPSON'S PARADOX**

| | Jane | | | Virginia | | |
|---|---|---|---|---|---|---|
| | No. of Sales Attempts | No. of Sales Successes | % Success | No. of Sales Attempts | No. of Sales Successes | % Success |
| Old territory | 40 | 20 | 50.0 | 95 | 44 | 46.3 |
| New territory | 32 | 8 | 25.0 | 19 | 3 | 15.8 |
| Total | 72 | 28 | 38.9 | 114 | 47 | 41.2 |
| Percent of sales attempts in old territory | 55.6% | | | 83.3% | | |
| Percent of sales attempts in new territory | 44.4% | | | 16.7% | | |

4.9  P60 estimate is 361,916. An alternative target is the mean, which is 370,700. Due to the high positive skew, the 60th percentile is below the mean.

4.10  One issue is financial. From a financial standpoint, choosing the median over the mean will cost the company less over time, on the assumption that the distributions of salary survey data remain generally positively skewed.

Another issue is from an employee communication standpoint. You have to weigh their understanding of the two terms. Many employees understand the term "average" better than the term "mean." However, they do not understand that the mean is impacted by extreme values. On the other hand, many employees are familiar with the concept of median house prices. Some companies state that they are targeting the "middle of the competitive pay." In any case, you have to do some educating on whichever target you choose.

Another issue is combined with the first two, and that is from an employee perception standpoint. If BPD has been targeting the mean and switches to the median, will the employees know the implications financially, and if so, what will be the impact on employee engagement?

There may be other issues that these numbers raise, but these three are the main ones.

# CHAPTER 5 MEASURES OF VARIABILITY

5.1  You have a population.

5.2  You have a sample.

| TABLE S.11 | FREQUENCY DISTRIBUTION OF NETWORK ASSISTANT SALARIES |

| Annual Salary (000) | No. of Incumbents |
| --- | --- |
| ≤45 | 0 |
| 45–50 | 1 |
| 50–55 | 2 |
| 55–60 | 4 |
| 60–65 | 2 |
| 65–70 | 6 |
| 70–75 | 6 |
| 75–80 | 2 |
| 80–85 | 2 |
| 85–90 | 2 |
| 90–95 | 2 |
| 95–100 | 1 |
| >100 | 0 |

5.3 You have a random sample from which valid statistical inferences may be made about the attitude of all the employees.

5.4 You have a convenience sample. Any generalizations you make from the study have to be done in the context of which companies participated and how representative they are of all the designated companies.

5.5 The frequency distribution and histogram are shown in Table S.11 and Figure S.9. The various measures of variability are shown in Table S.12.

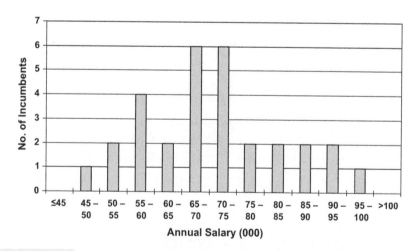

FIGURE S.9   HISTOGRAM OF FREQUENCY DISTRIBUTION OF NETWORK ASSISTANT SALARIES

**TABLE S.12** **MEASURE OF VARIABILITY OF NETWORK ASSISTANT SALARIES**

| | |
|---|---|
| No. of data points | 30 |
| Mean | 70,912 |
| Median | 69,850 |
| Sample standard deviation | 13,093 |
| Coefficient of variation | 18% |
| Maximum | 96,700 |
| Minimum | 47,100 |
| Range | 49,600 |
| P90 | 89,210 |
| P10 | 55,290 |
| P90/P10 | 1.61 |

5.6 The data are slightly skewed positively, with a tail to the right. This is indicated by the mean being slightly higher than the median and shown by the histogram.

The standard deviation means that the salaries vary on average from the mean of 70,912 by about 13,093. The coefficient of variation means that the salaries vary on average from the mean by about 18% of the mean. The range means that the span of all the salaries is 49,600. The P90/P10 means that the middle 80% of the data has a spread of 61% from the 10th percentile to the 90th percentile; that is, the 90th percentile is 61% more than the 10th percentile.

5.7 The two-sigma rule means, for this problem, that at least 75% of the data will fall between 44,726 and 97,098. Yes, it is true here. It is *always* true. In this example, 100% of the data falls between these limits, which is at least 75%.

5.8 Your bonus has a $z$-score of 2.5. This means that your bonus is 2.5 standard deviations above the mean.

5.9 At least 75% of the data are between 86,000 and 134,000.

5.10 The numerical results are in Table S.13.

They both obtained the same number of new customers, so without any other information, their results were equivalent. However, the $z$-scores give a different interpretation.

Ethan, with a $z$-score of 3.0, is among the top performers, if not *the* top performer, compared to the other sales representatives in the Middle East division. Jack, with a $z$-score of $-1.0$, is certainly below average among the other sales representatives in the Europe division.

**TABLE S.13** *z*-SCORES OF NEW CUSTOMERS

|  | Middle East | Europe |
|---|---|---|
| No. of Representatives | 24 | 30 |
| New customers |  |  |
| Mean | 100 | 150 |
| Standard deviation | 10 | 20 |

|  | Ethan | Jack |
|---|---|---|
| New customers | 130 | 130 |
| *z*-score | 3.0 | −1.0 |

Is Jack's performance equal to Ethan's because they had the same number of new customers or is his performance below that of Ethan because his *z*-score was lower?

What would it take to have a *z*-score of +3.0 in Europe? Applying the formula for *z*-score, we get 210.

$$3.0 = (x - 150)/20$$
$$60 = x - 150$$
$$210 = x$$

If you assume that the territory was more fertile in Europe in the first place (i.e., that it was "easier" in Europe) and that average performance in one region of the world is equivalent to average performance in another region of the world, you would say that obtaining 130 customers in the Middle East is equivalent to obtaining 210 new customers in Europe. And that obtaining 130 new customers in Europe was not as good as obtaining 130 new customers in the Middle East.

## CHAPTER 6 MODEL BUILDING

6.1 We want to *understand*, we want to *predict*, and we want to *control*.

6.2 A model is an abstraction of reality used to solve problems in an objective manner. The three components are as follows:

- The problem—the variable of interest
- Critical factors that impact the problem—other variables
- The relationships between the factors and the problem, and among the factors themselves

6.3 The problem is production bonuses. The critical factors are years of service, training, and team assignment. The relationships are the relationships between each of the three critical factors and bonuses, as well as the relationships between the three critical factors themselves.

6.4 The five steps in model building are as follows:

1. *Specify the Problem or Issue.* In this step you do two additional things.
   - Describe briefly why you are working on this problem.
   - Describe your terms mathematically.

2. *Generate Critical Factors That May Explain or Impact the Problem.*

3. *Identify the Relationship, If Any, Between Each Factor and the Problem.* This step has two parts.
   - Gather the data. You must ensure that the data are valid and "cleaned up." Otherwise your conclusions are meaningless.
   - Plot the points.

4. *Quantify the relationship and analyze.* This step includes calculating the coefficients of a mathematical equation that describes the relationship.

5. *Evaluate the model.*

6.5 The model is shown in Figure S.10.

Factors for which you have data or can get data on are training, hourly pay, team assignment, years of service, temperature, and precipitation.

6.6 (8, 37)

6.7 Age is $x$ and PTO is $y$. You are interested to determine if PTO is influenced by age.

6.8 It depends. If you are interested in predicting commissions as a function of sales, then sales is $x$ and commission is $y$. If you are interested in determining the motivational impact of commissions on sales, then commission is $x$ and sales is $y$.

6.9 The types of relationships are shown in Table S.14.

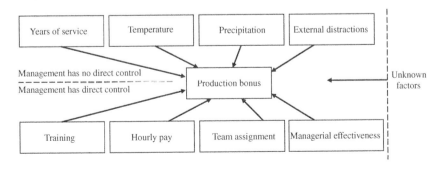

**FIGURE S.10    PRODUCTION BONUS MODEL**

**TABLE S.14**  **TYPES OF RELATIONSHIPS**

| Example | Is There a Relationship? | If So, What Kind? |
|---------|--------------------------|-------------------|
| A | Yes | Positive, linear, not perfect |
| B | Yes | Positive, exponential, not perfect |
| C | No | |
| D | Yes | Negative, linear, not perfect |
| E | Yes | Positive, maturity curve (or cubic or logarithmic), not perfect |
| F | Yes | Positive, power model, not perfect |

6.10  The method of least squares is a method of fitting a line (model) to data that produces the unique line that minimizes the sum of the squares of the vertical deviations between the $y$-values of the data points and the corresponding $y$-values predicted by the model. Its main advantages are that it is data driven and thus objective and credible, and it is widely accepted.

## CHAPTER 7 LINEAR MODEL

7.1  You plot the data after deciding that productivity is the $y$-variable and number of training courses taken and successfully completed is the $x$-variable, because you are interested in finding out if productivity is influenced by the training. After viewing the plot in Figure S.11, you

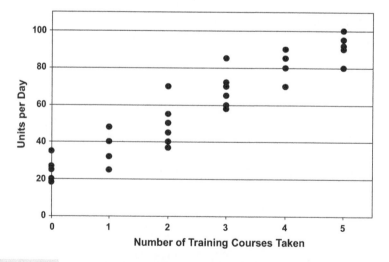

**FIGURE S.11**  **PLOT OF PRODUCTIVITY AND TRAINING**

**TABLE S.15    INTERPRETATION OF REGRESSION TERMS**

| Term | Symbol | Mathematical Interpretation | Problem Interpretation |
|------|--------|------------------------------|-------------------------|
| Intercept | $a$ | $y'$ has a value of 24.0 when $x$ is zero | The predicted average units per day is 24.0 for zero training courses taken |
| Slope | $b$ | $y'$ increases by 13.85 when $x$ increases by a value of 1 | The predicted average units per day increases by 13.85 for each additional training course taken |
| Coefficient of determination | $r^2$ | 88% of the variability in $y$ can be attributed to the relationship with $x$ | 88% of the variability in units per day can be attributed to the relationship with the number of training courses taken |
| Correlation | $r$ | +0.94. There is a strong association between $x$ and $y$ | +0.94. There is a strong association between training and units per day |
| Standard error of estimate | SEE | On average, $y$ varies from $y'$ by 8.9 | On average, the actual units per day varies from the predicted average units per day by about 8.9 |

conclude that there is a relationship that is positive, linear, and not perfect. You are now justified in fitting a straight line to the data.

7.2  $y' = 24.0 + 13.85x$

Predicted average units per day $= 24.0 + 13.85$ times the number of training courses taken

$r^2 = 0.88$
$r = 0.94$
$\text{SEE} = 8.9$

Average number of training courses taken is 2.5.
Average units per day is 58.6.

7.3  The interpretations are shown in Table S.15.

7.4  You draw the least squares line on your plot and change the title to reflect the conclusion of the relationship. This is shown in Figure S.12. Visually this is a good model.

7.5  Predicted average units per day $= 24.0 + (13.85 \times 3) = 65.55$

7.6  We have to use the prediction equation "in reverse," meaning we are given the $y'$ value and have to solve for $x$.

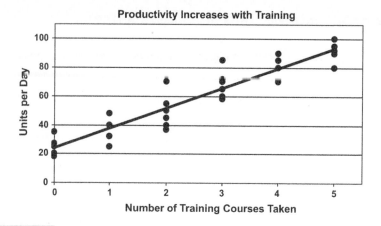

**FIGURE S.12    MODEL OF PRODUCTIVITY AND TRAINING**

$$80 = 24.0 + 13.85x$$
$$56.0 = 13.85x$$
$$56.0/13.85 = x$$
$$4.04 = x$$

The number of training courses needed to achieve the desired level of productivity is 4. This can also be seen graphically in Figure S.12.

7.7 The four major cautions of simple linear regression are as follows:

- A high association between variables does not imply cause and effect relationships.

- Coefficient of determination and correlation measure the strength of *linear* relationships only.

- Be cautious when using the model to make predictions too far away from the data points from which the model was derived.

- Have sufficient data points to be comfortable with your model and the conclusions.

7.8 The main purpose of plotting the data before conducting a regression is to identify what kind of relationship exists between the two variables. This verifies that you are using the appropriate model to describe the relationship.

7.9 You decide that the dependent variable is % employees using health facilities and the independent variable is age, because you want to find out if the % employees using health facilities is a function of age. The plot of the data is shown in Figure S.13.

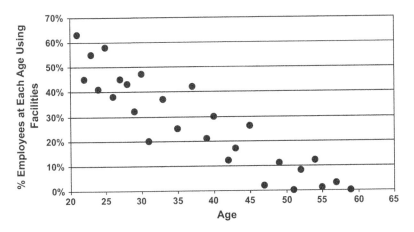

**FIGURE S.13**    PLOT OF % USING EXERCISE FACILITIES AND AGE

You conclude that the relationship is negative, linear, and not perfect. You decide to fit a straight line to the data. The equation of the resulting line is as follows:

$$y' = 82.6 - 1.46x$$

Predicted average % employees using exercise facilities = $82.6 - (1.46)(age)$

You plot the line on a chart shown in Figure S.14 and change the title to reflect the conclusion that younger employees tend to use the exercise facilities more than the older employees.

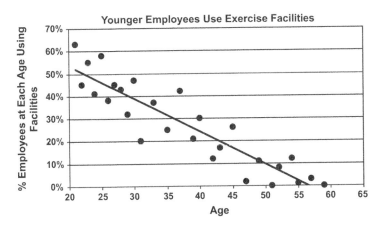

**FIGURE S.14**    MODEL OF % USING EXERCISE FACILITIES AND AGE

**TABLE S.16** INTERPRETATION OF REGRESSION TERMS

| Term | Symbol | Mathematical Interpretation | Problem Interpretation |
|---|---|---|---|
| Intercept | $a$ | $y'$ has a value of 82.6 when $x$ is zero | No interpretation |
| Slope | $b$ | $y'$ decreases by 1.46 when $x$ increases by a value of 1 | The predicted average % employees using exercise facilities decreases by 1.46 for each year older |
| Coefficient of determination | $r^2$ | 83% of the variability in $y$ can be attributed to the relationship with $x$ | 83% of the variability in % employees using exercise facilities can be attributed to the relationship with the age of the employees |
| Correlation | $r$ | +0.91. There is a strong association between $x$ and $y$ | +0.91. There is a strong association between % employees using exercise facilities and age |
| Standard error of estimate | SEE | On average, $y$ varies from $y'$ by 8.1 | On average, the actual % employees using exercise facilities varies from the predicted average % employees using exercise facilities by about 8.1 |

The full results of the regression and interpretations are shown in Table S.16.

7.10 There is no problem interpretation of the intercept because there are no employees with age zero. In this case, the intercept is simply a locator for the line.

## CHAPTER 8 EXPONENTIAL MODEL

8.1 You plot the data after deciding that recruiting cost is the $y$-variable and salary accepted is the $x$-variable, because you are interested in finding out if recruiting cost can be predicted by salary accepted. After viewing the plot in Figure S.15 you conclude that there is a relationship that is positive, exponential, and not perfect.

8.2 $$\log y' = a + bx$$
where
$$a = 3.468$$
$$b = 6.995\text{E-}06 \text{ or } 0.000006995$$

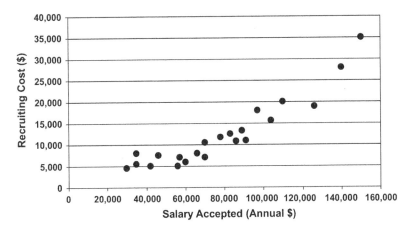

**FIGURE S.15**   PLOT OF RECRUITING COST AND SALARY ACCEPTED

and
$$r^2 = 0.893$$
$$r = 0.945$$
$$SEE = 0.0831$$

Substituting, we get

$$\log y' = 3.468 + 0.000006995x, \text{ or}$$

Predicted average log recruiting cost $= 3.468 + 0.000006995$ salary

Average starting salary accepted is 78,227
Average recruiting cost is 12,218.

8.3   See Figure S.16.

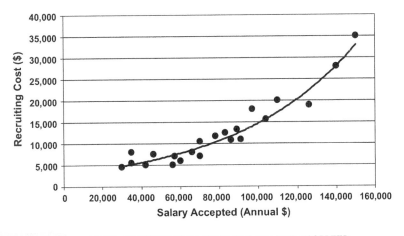

**FIGURE S.16**   MODEL OF RECRUITING COST AND SALARY ACCEPTED

**TABLE S.17   EVALUATION OF REGRESSION**

| Evaluation Criteria | Symbol | In Terms of Log Recruiting Cost | In Terms of Recruiting Cost |
|---|---|---|---|
| Appearance | | | Goes through middle of data and describes trend nicely |
| Coefficient of determination | $r^2$ | 89.3% of variability of log recruiting cost associated with salary accepted | 94.3% of variability of recruiting cost associated with salary accepted |
| Correlation | $r$ | +0.945—strong association | +0.971—strong association |
| Standard error of estimate | SEE | 0.0831 is "average" deviation of log recruiting cost from predicted average log recruiting cost | 21.1% is "average" deviation of recruiting cost from predicted average recruiting cost |
| Common sense | | | Makes sense based on experience |

8.4 The complete evaluation is in Table S.17.

Overall you are very much satisfied with this model and will present it with confidence as a basis for developing the recruiting budget.

8.5 See Table S.18.

8.6 You plot the data after deciding that assembly time is the $y$-variable and number of training courses is the $x$-variable, because you are interested in finding out if assembly time can be predicted by the number

**TABLE S.18   PREDICTED AVERAGE RECRUITING COST AND SALARY ACCEPTED**

| Salary Accepted | Predicted Average Recruiting Cost | Salary Accepted | Predicted Average Recruiting Cost |
|---|---|---|---|
| 30,000 | 4,763 | 100,000 | 14,706 |
| 40,000 | 5,595 | 110,000 | 17,276 |
| 50,000 | 6,573 | 120,000 | 20,296 |
| 60,000 | 7,721 | 130,000 | 23,842 |
| 70,000 | 9,071 | 140,000 | 28,009 |
| 80,000 | 10,656 | 150,000 | 32,904 |
| 90,000 | 12,518 | | |

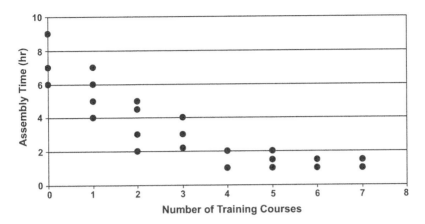

**FIGURE S.17    PLOT OF ASSEMBLY TIME AND TRAINING**

of training courses. After viewing the plot in Figure S.17 you conclude that there is a relationship that is negative, exponential, and not perfect.

8.7

$$\log y' = a + bx$$

where

$$a = 0.8176$$
$$b = -0.1232$$

and

$$r^2 = 0.799$$
$$r = 0.894$$
$$\text{SEE} = 0.1420$$

Substituting, we get

$\log y' = 0.8176 - 0.1232x$, or

Predicted average log assembly time

$= 0.8176 - (0.1232)(\text{number of training courses})$

Average assembly time is 3.5 hr.

Average number of training courses is 3.0.

8.8  You plot the line on a chart shown in Figure S.18 and change the title to reflect the conclusion of the chart, which is that assembly time decreases with increased training.

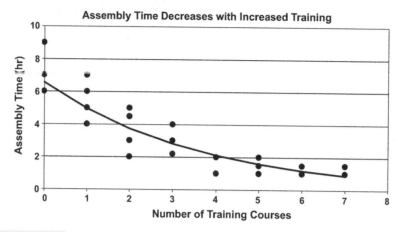

**FIGURE S.18** MODEL OF ASSEMBLY TIME AND TRAINING

8.9 The complete model evaluation is shown in Table S.19.

8.10 The predicted values are shown in Table S.20. The training manager will use this along with Figure S.18 to help decide what kind of policy she might want to recommend.

**TABLE S.19** **EVALUATION OF REGRESSION**

| Evaluation Criteria | Symbol | In Terms of Log Assembly Time | In Terms of Assembly Time |
|---|---|---|---|
| Appearance | | | Goes through middle of data and describes trend nicely |
| Coefficient of determination | $r^2$ | 79.9% of variability of log assembly time associated with number of training courses | 81.2% of variability of assembly time associated with number of training courses |
| Correlation | $r$ | +0.894—strong association | +0.901—strong association |
| Standard error of estimate | SEE | 0.1420 is "average" deviation of log assembly time from predicted average log assembly time | 38.7% is "average" deviation of assembly time from predicted average assembly time |
| Common sense | | | Makes sense based on experience |

TABLE S.20   PREDICTED AVERAGE ASSEMBLY
TIME AND TRAINING

| Number of Training Courses | Predicted Average Assembly Time (hr) |
| --- | --- |
| 0 | 6.6 |
| 1 | 4.9 |
| 2 | 3.7 |
| 3 | 2.8 |
| 4 | 2.1 |
| 5 | 1.6 |
| 6 | 1.2 |
| 7 | 0.9 |

# CHAPTER 9 MATURITY CURVE MODEL

9.1 You decide that the survey pay is the dependent variable and that YSBS is the independent variable, because you are interested in finding out if pay is a function of experience. You plot the data as shown in Figure S.19. After viewing the plot you conclude that there is a relationship that is positive, cubic or spline, and not perfect.

9.2 You first fit a cubic model to the data with statistical software and get the following equation:

$$y' = a + bx + cx^2 + dx^3$$

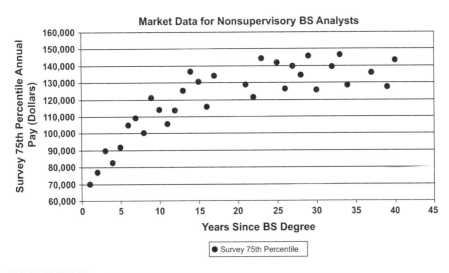

FIGURE S.19   PLOT OF SURVEY 75TH PERCENTILE PAY AND YSBS

where

$$a = 65{,}206$$
$$b = 6{,}825.9$$
$$c = -219.42$$
$$d = 2.3362$$

and

$$r^2 = 0.871$$
$$r = 0.933$$
$$SEE = 7{,}917$$

Substituting, we get

$$y' = 65{,}206 + 6{,}825.9x - 219.42x^2 + 2.3362x^3, \text{ or}$$

Predicted average survey 75th percentile of pay $= 65{,}206 + 6{,}825.9$ YSBS $- 219.42$ YSBS$^2$ $+ 2.3362$ YSBS$^3$

9.3 A plot of the cubic model is shown in Figure S.20.

9.4 The model evaluation is shown in Table S.21. Overall this is an excellent model, and it very nicely represents the predicted average survey 75th percentile of pay for the given years since BS degree.

9.5 Table S.22 shows the predicted average survey 75th percentile for the various values of YSBS, including those "in-between" years for which there was no 75th percentile reported.

9.6 You fit a spline model with a knot at YSBS $= 20$ to the data with statistical software and get the following equation:

$$y' = a + bx + cx^2 + dx^3 + ep^3$$

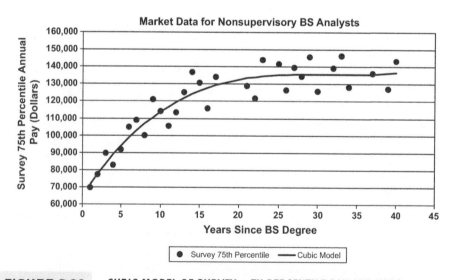

**FIGURE S.20　CUBIC MODEL OF SURVEY 75TH PERCENTILE PAY AND YSBS**

**TABLE S.21    EVALUATION OF CUBIC MODEL**

| Evaluation Criteria | Symbol | Cubic Model |
| --- | --- | --- |
| Appearance | | Goes through middle of data and describes trend nicely |
| Coefficient of determination | $r^2$ | 87.1% of the variability of survey 75th percentile of pay associated with YSBS |
| Correlation | $r$ | +0.933—strong association |
| Standard error of estimate | SEE | 7,917 "average" deviation of survey 75th percentile of pay from predicted average survey 75th percentile of pay—a reasonable variation |
| Common sense | | Makes sense with respect to pay levels and general shape of the relationship |

$$where \quad \begin{aligned} a &= 63,268 \\ b &= 7,632.5 \\ c &= -293.04 \\ d &= 4.1150 \\ e &= -3.4336 \end{aligned}$$

$$and \quad \begin{aligned} r^2 &= 0.872 \\ r &= 0.934 \\ SEE &= 8,033 \end{aligned}$$

**TABLE S.22    PREDICTED AVERAGE SURVEY 75TH PERCENTILES FOR CUBIC MODEL**

| YSBS | Predicted Average Survey Survey P75 Annual Pay Cubic Model | YSBS | Predicted Average Survey P75 Annual Pay Cubic Model | YSBS | Predicted Average Survey P75 Annual Pay Cubic Model |
| --- | --- | --- | --- | --- | --- |
| 1 | 71,815 | 14 | 124,173 | 27 | 135,532 |
| 2 | 77,999 | 15 | 126,110 | 28 | 135,590 |
| 3 | 83,772 | 16 | 127,818 | 29 | 135,602 |
| 4 | 89,148 | 17 | 129,312 | 30 | 135,582 |
| 5 | 94,142 | 18 | 130,605 | 31 | 135,544 |
| 6 | 98,767 | 19 | 131,711 | 32 | 135,501 |
| 7 | 103,037 | 20 | 132,646 | 33 | 135,468 |
| 8 | 106,966 | 21 | 133,421 | 34 | 135,459 |
| 9 | 110,569 | 22 | 134,052 | 35 | 135,488 |
| 10 | 113,859 | 23 | 134,553 | 36 | 135,568 |
| 11 | 116,851 | 24 | 134,937 | 37 | 135,714 |
| 12 | 119,557 | 25 | 135,219 | 38 | 135,940 |
| 13 | 121,993 | 26 | 135,413 | 39 | 136,259 |
| | | | | 40 | 136,687 |

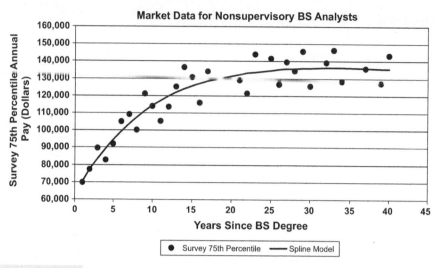

**FIGURE S.21** SPLINE MODEL OF SURVEY 75TH PERCENTILE PAY AND YSBS

Substituting, we get

$$y' = 63{,}268 + 7{,}632.5x - 293.04x^2 + 4.1150x^3 - 3.4336p^3, \text{ or}$$

Predicted average survey 75th percentile of pay = $63{,}268 + 7{,}632.5$ YSBS
$$-293.04\,\text{YSBS}^2 + 4.1150\,\text{YSBS}^3 - 3.4336(\text{YSBS} - 20)^3$$

9.7 A plot of the spline model is shown in Figure S.21.

9.8 The model evaluation is shown in Table S.23. Overall this is an excellent model, and it very nicely represents the predicted average survey 75th percentile of pay for the given years since BS degree.

**TABLE S.23** EVALUATION OF SPLINE MODEL

| Evaluation Criteria | Symbol | Spline Model |
| --- | --- | --- |
| Appearance | | Goes through middle of data and describes trend nicely |
| Coefficient of determination | $r^2$ | 87.2% of variability of survey 75th percentile of pay associated with YSBS |
| Correlation | $r$ | +0.934—strong association |
| Standard error of estimate | SEE | 8,033 "average" deviation of survey 75th percentile of pay from predicted average survey 75th percentile of pay—a reasonable variation |
| Common sense | | Makes sense with respect to pay levels and general shape of the relationship |

**TABLE S.24   PREDICTED AVERAGE SURVEY 75TH PERCENTILES FOR SPLINE MODEL**

| YSBS | Predicted Average Survey P75 Annual Pay Spline Model | YSBS | Predicted Average Survey P75 Annual Pay Spline Model | YSBS | Predicted Average Survey P75 Annual Pay Spline Model |
|---|---|---|---|---|---|
| 1 | 70,612 | 14 | 123,979 | 27 | 135,537 |
| 2 | 77,394 | 15 | 125,710 | 28 | 135,809 |
| 3 | 83,639 | 16 | 127,225 | 29 | 136,022 |
| 4 | 89,373 | 17 | 128,549 | 30 | 136,178 |
| 5 | 94,619 | 18 | 129,707 | 31 | 136,284 |
| 6 | 99,402 | 19 | 130,723 | 32 | 136,342 |
| 7 | 103,748 | 20 | 131,622 | 33 | 136,357 |
| 8 | 107,680 | 21 | 132,425 | 34 | 136,333 |
| 9 | 111,224 | 22 | 133,141 | 35 | 136,274 |
| 10 | 114,404 | 23 | 133,772 | 36 | 136,184 |
| 11 | 117,245 | 24 | 134,323 | 37 | 136,067 |
| 12 | 119,771 | 25 | 134,798 | 38 | 135,927 |
| 13 | 122,007 | 26 | 135,202 | 39 | 135,768 |
|  |  |  |  | 40 | 135,595 |

9.9   Table S.24 shows the predicted average survey median pay for the various values of YSBS, including those "in-between" years for which there was no 75th percentile.

9.10   Since the two models are virtually identical, the cubic model would be the appropriate one to use because it is simpler than the spline model.

# CHAPTER 10 POWER MODEL

10.1   You decide that CEO pay is the dependent variable and that company sales is the independent variable, because you are interested in finding out if pay is a function of sales. You plot the data in Figure S.22, observe a comet display, and conclude that a power model is appropriate.

You verify the use of a power model by converting both axes to logarithmic scales, as shown in Figure S.23.

Here the display indicates a linear trend on the log–log plot.

10.2

$$\log y' = a + b \log x$$

where

$$a = 1.656$$
$$b = 0.4380$$

and

$$r^2 = 0.830$$
$$r = 0.911$$
$$SEE = 0.165$$

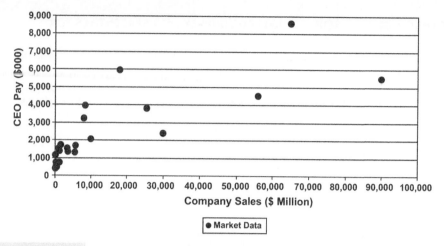

**FIGURE S.22**    CEO PAY AND COMPANY SALES ON LINEAR PLOT

Substituting, we get
$$\log y' = 1.656 + 0.4380 \log x$$

Predicted average log CEO pay $= 1.656 + 0.4380 \log \text{sales}$ ·

where CEO pay is in thousands of dollars and sales is in millions of dollars.

10.3   You calculate the predicted average pay using the following formula with the results shown in Table S.25 and plot the line on the log–log chart, shown in Figure S.24.

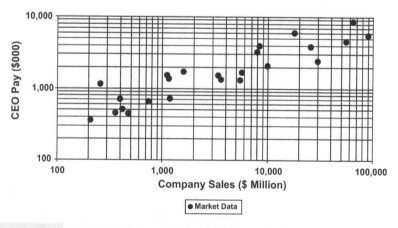

**FIGURE S.23**    CEO PAY AND COMPANY SALES ON LOG–LOG PLOT

**TABLE S.25    PREDICTED AVERAGE CEO PAY AND COMPANY SALES**

| Sales ($Million) | Predicted Average CEO Pay ($000) | Sales ($Million) | Predicted Average CEO Pay ($000) |
|---|---|---|---|
| 210 | 471 | 3,600 | 1,636 |
| 260 | 517 | 5,500 | 1,969 |
| 360 | 597 | 5,700 | 2,000 |
| 400 | 625 | 8,000 | 2,320 |
| 420 | 638 | 8,400 | 2,370 |
| 480 | 677 | 10,000 | 2,559 |
| 750 | 823 | 18,000 | 3,310 |
| 1,120 | 981 | 25,500 | 3,855 |
| 1,170 | 1,000 | 30,000 | 4,140 |
| 1,200 | 1,011 | 56,000 | 5,441 |
| 1,600 | 1,147 | 65,000 | 5,808 |
| 3,400 | 1,595 | 90,000 | 6,698 |

$$y' = 10^a x^b$$
$$y' = 10^{1.656} x^{0.4380}$$

10.4   The evaluation of the model is shown in Table S.26.

Overall you are very satisfied with this model and will use it with confidence as a basis for developing a market-based pay target for the base salary of your CEO.

10.5   Your company has sales of $15 billion or $15,000 million. To calculate the predicted average pay for a company of that size, you use the formula developed for the model.

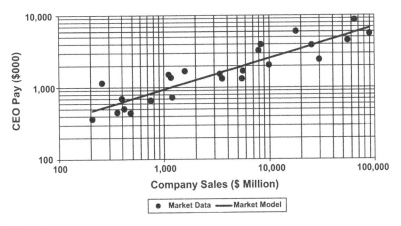

FIGURE S.24    MODEL OF CEO PAY AND COMPANY SALES

**TABLE S.26    EVALUATION OF POWER MODEL**

| Evaluation Criteria | Symbol | In Terms of Log CEO Pay | In Terms of CEO Pay |
|---|---|---|---|
| Appearance | | Goes through middle of data and describes trend nicely | |
| Coefficient of determination | $r^2$ | 83.0% of variability of log CEO pay is associated with log company sales | 74.9% of variability of CEO pay is associated with company sales |
| Correlation | $r$ | +0.911—strong association | +0.866—strong association |
| Standard error of estimate | SEE | 0.165 is "average" deviation of log CEO pay from predicted average log CEO pay | 46% is "average" deviation of CEO pay from predicted average CEO pay |
| Common sense | | | Makes sense based on experience |

$$y' = 10^{1.656}x^{0.4380}$$
$$y' = 10^{1.656} \times 13{,}000^{0.4380} = 3{,}056$$

The predicted average CEO base pay for a company of your size is $3,056,000.

10.6 When both variables vary by orders of magnitude, and when the plot on linear scales looks like a "comet plot."

10.7 A logarithmic transformation visually pulls in the high values and stretches out the low values.

10.8 Executive-level jobs, such as CEO, COO, CFO, and CIO.

10.9 The independent variable is usually company sales/revenue for a profit-making organization and budget for a nonprofit-making organization. The dependent variable is salary of the executive.

10.10 The least squares line on a log–log plot is a funny curved line on a chart with linear scales and is impossible to interpret and use, especially for low values of the variables that are scrunched up in the lower left corner of the plot (where the head of the "comet" is).

# CHAPTER 11 MARKET MODELS AND SALARY SURVEY ANALYSIS

11.1 See Table S.27.

11.2 The adjustment to the market-based salary structure is 1.0%.

11.3 See Table S.28.

**TABLE S.27    MARKET-BASED SALARY INCREASE BUDGET RECOMMENDATION**

| | |
|---|---|
| Market position, % from survey | −4.0% |
| Catch-up/fall back | 4.2% |
| Anticipated market movement | 3.0% |
| Pay policy | 0.0% |
| Market-based salary increase budget recommendation | 7.2% |

**TABLE S.28    MARKET-BASED SALARY INCREASE BUDGET RECOMMENDATION**

| | |
|---|---|
| Market position, % from survey | 3.0% |
| Catch-up/fall back | −2.9% |
| Anticipated market movement | 3.0% |
| Pay policy | 0.0% |
| Market-based salary increase budget recommendation | 0.1% |

11.4    The adjustment to the market–based salary structure is 5.0%.

11.5    See Table S.29.

11.6    The adjustment to the market–based salary structure is −1.0%.

11.7    See Table S.30.

11.8    The adjustment to the market–based salary structure is 3.0%.

**TABLE S.29    MARKET-BASED SALARY INCREASE BUDGET RECOMMENDATION**

| | |
|---|---|
| Market position, % from survey | −4.0% |
| Catch-up/fall back | 4.2% |
| Anticipated market movement | 1.5% |
| Pay policy | 0.0% |
| Market-based salary increase budget recommendation | 5.7% |

**TABLE S.30    MARKET-BASED SALARY INCREASE BUDGET RECOMMENDATION**

| | |
|---|---|
| Market position, % from survey | 0.0% |
| Catch-up/fall back | 0.0% |
| Anticipated market movement | 1.5% |
| Pay policy | 0.0% |
| Market-based salary increase budget recommendation | 1.5% |

11.9   Balance with benefits, business plans of the company, affordability, management style, maturity of the company, industry practices, acceptability and marketability to a board, union contracts, internal equity, government factors and regulations, past practices, turnover, inflation, state of the economy, comfort with whole numbers, what was read in *The Wall Street Journal*, and what was heard at the country club.

11.10   The salary market moves as anticipated, the employees get raises as anticipated, the labor supply and demand situation goes as anticipated, the business plans are implemented according to plan, the external competitive environment goes as anticipated, the political and regulatory environment is stable, and there are no disasters, natural or otherwise.

# CHAPTER 12 INTEGRATED MARKET MODEL: LINEAR

12.1   Survey A will be aged 2.5%. Survey B will be aged 3.0%.

12.2   See Table S.31.

12.3   The type of relationship is positive, linear, and not perfect. See Figure S.25.

12.4   Predicted average market pay $= -76.15 + (3.074)(\text{grade})$. $r^2 = 81\%$. SEE $= 1.41$.

12.5   See Figure S.26 and Table S.32.

12.6   See Table S.33 and Figure S.27.

**TABLE S.31**   AGED SURVEY DATA FOR CONSTRUCTION JOBS

| Job | Title | Grade | Survey | Hourly Base Salary, Market Unweighted Average | Aging Factor | Aged Hourly Base Salary, Market Unweighted Average |
|-----|-------|-------|--------|-----------------------------------------------|--------------|----------------------------------------------------|
| 1 | Carpenter 1 | 32 | A | 21.09 | 1.025 | 21.62 |
| 2 | Carpenter 2 | 33 | A | 22.75 | 1.025 | 23.32 |
| 3 | Carpenter 3 | 33 | A | 25.63 | 1.025 | 26.27 |
| 4 | Electrician 1 | 32 | B | 23.29 | 1.030 | 23.99 |
| 5 | Electrician 2 | 33 | B | 24.91 | 1.030 | 25.66 |
| 6 | Electrician 3 | 34 | B | 27.78 | 1.030 | 28.61 |
| 7 | HVAC Mechanic 1 | 32 | B | 19.90 | 1.030 | 20.50 |
| 8 | HVAC Mechanic 3 | 34 | B | 29.23 | 1.030 | 30.11 |
| 9 | Painter 1 | 31 | A | 19.28 | 1.025 | 19.76 |
| 10 | Painter 2 | 32 | A | 21.64 | 1.025 | 22.18 |
| 11 | Painter 3 | 33 | A | 26.68 | 1.025 | 27.35 |
| 12 | Plumber 1 | 32 | A | 21.39 | 1.025 | 21.92 |
| 13 | Plumber 2 | 33 | A | 23.34 | 1.025 | 23.92 |
| 14 | Plumber 3 | 34 | A | 26.05 | 1.025 | 26.70 |

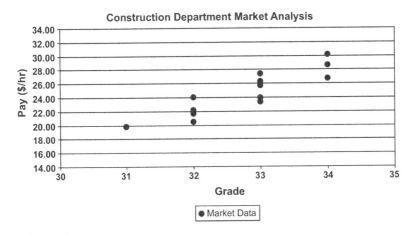

**FIGURE S.25    AGED HOURLY MARKET PAY AND GRADE**

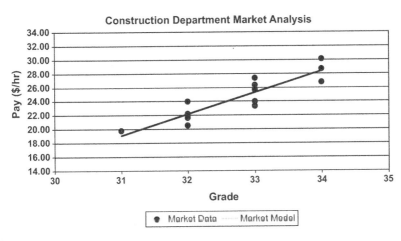

**FIGURE S.26    PREDICTED AVERAGE HOURLY MARKET PAY AND GRADE**

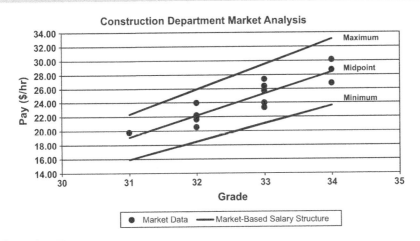

**FIGURE S.27    MARKET-BASED SALARY STRUCTURE**

**TABLE S.32**    **EVALUATION OF LINEAR MODEL**

| Term or Evaluation Criteria | Symbol | Mathematical Interpretation | Problem Interpretation or Evaluation |
|---|---|---|---|
| Appearance | | | Goes through middle of data and describes trend nicely |
| Intercept | $a$ | $y'$ has a value of $-76.15$ when $x$ is zero | There is no problem interpretation since there is no grade zero |
| Slope | $b$ | $y'$ increases by 3.074 when $x$ increases by a value of 1 | The predicted average market pay increases by \$3.074/hr for each higher grade |
| Coefficient of determination | $r^2$ | 81% of the variability in $y$ can be attributed to the relationship with $x$ | 81% of the variability in market pay can be attributed to the relationship with grade |
| Standard error of estimate | SEE | On average, $y$ varies from $y'$ by 1.41 | On average, the actual market pay varies from the predicted average market pay by about \$1.41/hr |
| Common sense | | | Makes sense with respect to pay levels and increases with grade |

**TABLE S.33**    **MARKET-BASED SALARY STRUCTURE**

| | Market-Based Salary Structure | | |
| Grade | Minimum | Midpoint | Maximum |
|---|---|---|---|
| 31 | 15.95 | 19.14 | 22.33 |
| 32 | 18.52 | 22.22 | 25.93 |
| 33 | 21.08 | 25.29 | 29.51 |
| 34 | 23.64 | 28.37 | 33.10 |

The midpoint progression from one grade to the next is a constant dollar amount and is equal to the slope of the midpoint line. The progression is \$3.07 (rounded).

12.7 See Table S.34 and Figure S.28.

12.8 See Table S.35. Some values may seem off by 1 due to rounding.

**TABLE S.34   EMPLOYEE PAY AND MARKET-BASED SALARY STRUCTURE**

| Emp. No. | Grade | Pay | Minimum | Midpoint | Maximum | Pay as % of Midpoint | Dollars Under Min | Dollars Over Max |
|---|---|---|---|---|---|---|---|---|
| 1 | 31 | 14.56 | 15.95 | 19.14 | 22.33 | 76 | 0.39 | – |
| 2 | 31 | 15.92 | 15.95 | 19.14 | 22.33 | 83 | 0.03 | – |
| 3 | 31 | 16.67 | 15.95 | 19.14 | 22.33 | 87 | – | – |
| 4 | 31 | 17.11 | 15.95 | 19.14 | 22.33 | 89 | – | – |
| 5 | 31 | 18.71 | 15.95 | 19.14 | 22.33 | 98 | – | – |
| 6 | 31 | 18.79 | 15.95 | 19.14 | 22.33 | 98 | – | – |
| 7 | 31 | 19.26 | 15.95 | 19.14 | 22.33 | 101 | – | – |
| 8 | 31 | 19.87 | 15.95 | 19.14 | 22.33 | 104 | – | – |
| 9 | 31 | 20.77 | 15.95 | 19.14 | 22.33 | 109 | – | – |
| 10 | 31 | 21.71 | 15.95 | 19.14 | 22.33 | 113 | – | – |
| 11 | 31 | 22.97 | 15.95 | 19.14 | 22.33 | 120 | – | 0.64 |
| 12 | 32 | 18.03 | 18.52 | 22.22 | 25.93 | 81 | 0.49 | – |
| 13 | 32 | 18.03 | 18.52 | 22.22 | 25.93 | 81 | 0.49 | – |
| 14 | 32 | 18.41 | 18.52 | 22.22 | 25.93 | 83 | 0.11 | – |
| 15 | 32 | 18.49 | 18.52 | 22.22 | 25.93 | 83 | 0.03 | – |
| 16 | 32 | 18.55 | 18.52 | 22.22 | 25.93 | 83 | – | – |
| 17 | 32 | 19.05 | 18.52 | 22.22 | 25.93 | 86 | – | – |
| 18 | 32 | 19.44 | 18.52 | 22.22 | 25.93 | 87 | – | – |
| 19 | 32 | 19.53 | 18.52 | 22.22 | 25.93 | 88 | – | – |
| 20 | 32 | 19.84 | 18.52 | 22.22 | 25.93 | 89 | – | – |
| 21 | 32 | 20.00 | 18.52 | 22.22 | 25.93 | 90 | – | – |
| 22 | 32 | 20.02 | 18.52 | 22.22 | 25.93 | 90 | – | – |
| 23 | 32 | 20.04 | 18.52 | 22.22 | 25.93 | 90 | – | – |
| 24 | 32 | 20.05 | 18.52 | 22.22 | 25.93 | 90 | – | – |
| 25 | 32 | 20.30 | 18.52 | 22.22 | 25.93 | 91 | – | – |
| 26 | 32 | 20.52 | 18.52 | 22.22 | 25.93 | 92 | – | – |
| 27 | 32 | 22.55 | 18.52 | 22.22 | 25.93 | 101 | – | – |
| 28 | 32 | 23.34 | 18.52 | 22.22 | 25.93 | 105 | – | – |
| 29 | 32 | 23.40 | 18.52 | 22.22 | 25.93 | 105 | – | – |
| 30 | 32 | 24.19 | 18.52 | 22.22 | 25.93 | 109 | – | – |
| 31 | 33 | 20.76 | 21.08 | 25.29 | 29.51 | 82 | 0.32 | – |
| 32 | 33 | 23.38 | 21.08 | 25.29 | 29.51 | 92 | – | – |
| 33 | 33 | 24.21 | 21.08 | 25.29 | 29.51 | 96 | – | – |
| 34 | 33 | 24.77 | 21.08 | 25.29 | 29.51 | 98 | – | – |
| 35 | 33 | 24.85 | 21.08 | 25.29 | 29.51 | 98 | – | – |
| 36 | 33 | 25.90 | 21.08 | 25.29 | 29.51 | 102 | – | – |
| 37 | 33 | 27.46 | 21.08 | 25.29 | 29.51 | 109 | – | – |
| 38 | 33 | 30.19 | 21.08 | 25.29 | 29.51 | 119 | – | 0.68 |
| 39 | 34 | 23.85 | 23.64 | 28.37 | 33.10 | 84 | – | – |
| 40 | 34 | 24.19 | 23.64 | 28.37 | 33.10 | 85 | – | – |
| 41 | 34 | 25.32 | 23.64 | 28.37 | 33.10 | 89 | – | – |
| 42 | 34 | 25.75 | 23.64 | 28.37 | 33.10 | 91 | – | – |
| 43 | 34 | 27.09 | 23.64 | 28.37 | 33.10 | 95 | – | – |
| 44 | 34 | 27.39 | 23.64 | 28.37 | 33.10 | 97 | – | – |
| 45 | 34 | 27.61 | 23.64 | 28.37 | 33.10 | 97 | – | – |
| 46 | 34 | 28.21 | 23.64 | 28.37 | 33.10 | 99 | – | – |
| 47 | 34 | 29.01 | 23.64 | 28.37 | 33.10 | 102 | – | – |
| 48 | 34 | 31.09 | 23.64 | 28.37 | 33.10 | 110 | – | – |

**TABLE S.35**    **MARKET ANALYSIS**

Hourly Manufacturing Market Analysis: Comparison of Employee Pay with the Market-Based Pay Structure

| Grade | No. of Employees | Total Pay Annual Dollars | Total Market Midpoint Annual Dollars | Difference | % from Mkt Mid | % to Meet Mkt Mid | No. of Employees Under Min | No. of Employees Over Max | Dollars Under Min | Dollars Over Max |
|---|---|---|---|---|---|---|---|---|---|---|
| 31 | 11 | 430,425 | 439,186 | 8,761 | −2.0 | 2.0 | 2 | 1 | 2,962 | 1,335 |
| 32 | 19 | 800,565 | 880,667 | 80,102 | −9.1 | 10.0 | 4 | 0 | 2,336 | 0 |
| 33 | 8 | 420,371 | 422,040 | 1,669 | −0.4 | 0.4 | 1 | 1 | 668 | 1,418 |
| 34 | 10 | 562,198 | 591,798 | 29,600 | −5.0 | 5.3 | 0 | 0 | 0 | 0 |
| **Total** | **48** | **2,213,559** | **2,333,692** | **120,133** | **−5.1** | **5.4** | **7** | **2** | **5,966** | **2,754** |

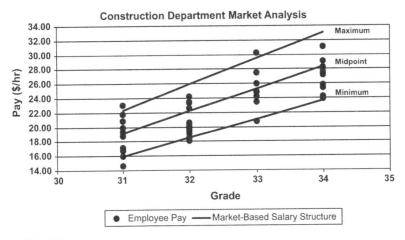

**Construction Department Market Analysis**

**FIGURE S.28    EMPLOYEE PAY AND MARKET-BASED SALARY STRUCTURE**

**TABLE S.36    MARKET-BASED SALARY INCREASE BUDGET RECOMMENDATION**

| | |
|---|---|
| Catch-up (or fall back) | 5.4% |
| Anticipated market movement (4%)(0.5) | 2.0% |
| Pay policy | 0.0% |
| Market-based salary increase budget recommendation | 7.4% |

12.9    See Table S.36.

12.10    Lower the market–based salary structure by 0.6%.

# CHAPTER 13 INTEGRATED MARKET MODEL: EXPONENTIAL

13.1    Survey A will be aged 1.8%. Survey B will be aged 3.0%. Survey C will be aged 6.0%.

13.2    See Table S.37.

13.3    The type of relationship is positive, exponential, and not perfect. See Figure S.29. Note that the relationship is almost linear, but since you are going to develop a salary structure with a constant percent midpoint progression, you will use an exponential model.

13.4    Predicted average log market pay $= 2.716 + (0.0497)(\text{grade})$.

$$r^2 = 85.0\%$$
$$\text{SEE} = 0.059$$

13.5    See Figure S.30 and Table S.38.

**TABLE S.37** **AGED SURVEY DATA FOR MINING JOBS**

| Job | Company Job Title | Grade | Survey | Market Weighted Average | Aging Factor | Aged Market Weighted Average |
|-----|-------------------|-------|--------|-------------------------|--------------|------------------------------|
| 1 | Mining Engineer 1 | 41 | A | 44,399 | 1.018 | 45,198 |
| 2 | Production Scheduler | 41 | A | 48,586 | 1.018 | 49,461 |
| 3 | Accountant 2 | 42 | B | 38,678 | 1.030 | 39,838 |
| 4 | Chemist 2 | 42 | C | 42,142 | 1.060 | 44,671 |
| 5 | Geologist 2 | 42 | A | 50,011 | 1.018 | 50,911 |
| 6 | Cost Accountant 2 | 42 | B | 52,469 | 1.030 | 54,043 |
| 7 | Mining Engineer 2 | 42 | A | 61,699 | 1.018 | 62,810 |
| 8 | Accountant 3 | 43 | B | 46,411 | 1.030 | 47,803 |
| 9 | Geologist 3 | 43 | A | 54,842 | 1.018 | 55,829 |
| 10 | Network Engineer 2 | 43 | C | 57,194 | 1.060 | 60,626 |
| 11 | HR Generalist 3 | 43 | C | 57,331 | 1.060 | 60,771 |
| 12 | Mining Engineer 3 | 43 | A | 65,315 | 1.018 | 66,491 |
| 13 | Billing Supervisor | 44 | B | 51,762 | 1.030 | 53,315 |
| 14 | Accountant 4 | 44 | B | 60,367 | 1.030 | 62,178 |
| 15 | Geologist 4 | 44 | A | 69,293 | 1.018 | 70,540 |
| 16 | Accounting Supervisor | 45 | B | 58,597 | 1.030 | 60,355 |
| 17 | Mining Engineer 4 | 45 | A | 86,716 | 1.018 | 88,277 |
| 18 | Lead Mining Engineer | 46 | A | 61,211 | 1.018 | 62,313 |
| 19 | Mine Foreman | 46 | A | 75,912 | 1.018 | 77,278 |
| 20 | Operations Supervisor | 46 | A | 84,648 | 1.018 | 86,172 |
| 21 | Project Engineer Manager | 47 | A | 81,160 | 1.018 | 82,621 |
| 22 | Health and Safety Manager | 47 | A | 92,537 | 1.018 | 94,203 |
| 23 | Accounting Manager | 48 | B | 110,577 | 1.030 | 113,894 |
| 24 | Engineering Supervisor | 49 | A | 94,642 | 1.018 | 96,346 |
| 25 | Sr. Engineering Manager | 50 | A | 138,444 | 1.018 | 140,936 |
| 26 | Mine Manager | 51 | A | 145,205 | 1.018 | 147,819 |

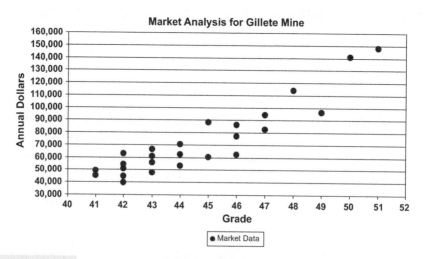

**FIGURE S.29** **AGED SALARIES AND GRADE**

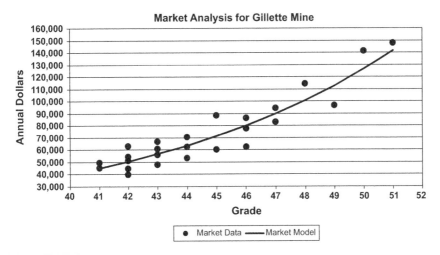

**FIGURE S.30    PREDICTED AVERAGE SALARY AND GRADE**

**TABLE S.38    EVALUATION OF EXPONENTIAL MODEL**

| Term or Evaluation Criteria | Symbol | Problem Interpretation or Evaluation |
|---|---|---|
| Appearance | | Goes through middle of data and describes trend nicely |
| Progression from one grade to the next | | The progression from one grade to the next is 12.1% |
| Coefficient of determination | $r^2$ | 85% of the variability in log market pay can be attributed to the relationship with grade. If you were to calculate the "real" $r^2$, it would be 89%, meaning 89% of the variability in market pay can be attributed to the relationship with grade |
| Standard error of estimate | SEE | On average, the log market pay varies from the predicted average log market pay by about 0.059. On average, the market pay varies from the predicted average market pay by about 14.6% |
| Common sense | | Makes sense with respect to pay levels and increases with grade |

13.6   See Table S.39 and Figure S.31.

13.7   See Table S.40 and Figure S.32.

13.8   See Table S.41.

| TABLE S.39 | MARKET-BASED SALARY STRUCTURE | | |
|---|---|---|---|
| **Market-Based Salary Structure** | | | |
| Grade | Minimum | Midpoint | Maximum |
| 41 | 36,123 | 45,154 | 54,185 |
| 42 | 40,503 | 50,629 | 60,755 |
| 43 | 45,414 | 56,768 | 68,121 |
| 44 | 50,920 | 63,650 | 76,380 |
| 45 | 57,094 | 71,367 | 85,641 |
| 46 | 64,016 | 80,020 | 96,024 |
| 47 | 71,778 | 89,722 | 107,667 |
| 48 | 80,480 | 100,600 | 120,720 |
| 49 | 90,238 | 112,798 | 135,357 |
| 50 | 101,179 | 126,474 | 151,769 |
| 51 | 113,446 | 141,808 | 170,169 |

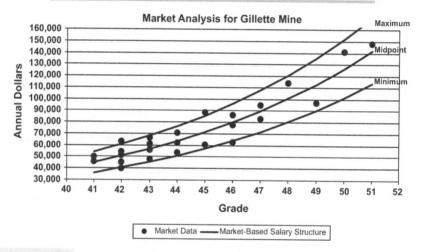

FIGURE S.31    MARKET-BASED SALARY STRUCTURE

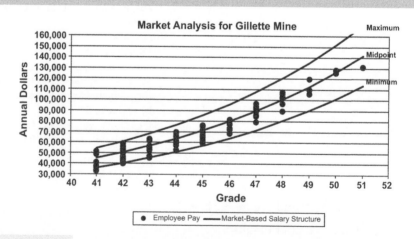

FIGURE S.32    EMPLOYEE PAY AND MARKET-BASED SALARY STRUCTURE

**TABLE S.40  EMPLOYEE PAY AND MARKET-BASED SALARY STRUCTURE**

| Emp. No. | Grade | Pay | Minimum | Midpoint | Maximum | Pay as % of Mid | Dollars Under Min | Dollars Over Max |
|---|---|---|---|---|---|---|---|---|
| 1 | 41 | 49,589 | 36,123 | 45,154 | 54,185 | 110 | – | – |
| 2 | 41 | 32,966 | 36,123 | 45,154 | 54,185 | 73 | 3,157 | – |
| 3 | 41 | 50,926 | 36,123 | 45,154 | 54,185 | 113 | – | – |
| 4 | 41 | 35,737 | 36,123 | 45,154 | 54,185 | 79 | 386 | – |
| 5 | 41 | 48,079 | 36,123 | 45,154 | 54,185 | 106 | – | – |
| 6 | 41 | 33,131 | 36,123 | 45,154 | 54,185 | 73 | 2,992 | – |
| 7 | 41 | 40,944 | 36,123 | 45,154 | 54,185 | 91 | – | – |
| 8 | 41 | 37,412 | 36,123 | 45,154 | 54,185 | 83 | – | – |
| 9 | 41 | 50,785 | 36,123 | 45,154 | 54,185 | 112 | – | – |
| 10 | 42 | 53,315 | 40,503 | 50,629 | 60,755 | 105 | – | – |
| 11 | 42 | 44,145 | 40,503 | 50,629 | 60,755 | 87 | – | – |
| 12 | 42 | 43,293 | 40,503 | 50,629 | 60,755 | 86 | – | – |
| 13 | 42 | 41,579 | 40,503 | 50,629 | 60,755 | 82 | – | – |
| 14 | 42 | 55,839 | 40,503 | 50,629 | 60,755 | 110 | – | – |
| 15 | 42 | 51,452 | 40,503 | 50,629 | 60,755 | 102 | – | – |
| 16 | 42 | 52,710 | 40,503 | 50,629 | 60,755 | 104 | – | – |
| 17 | 42 | 46,133 | 40,503 | 50,629 | 60,755 | 91 | – | – |
| 18 | 42 | 40,377 | 40,503 | 50,629 | 60,755 | 80 | 126 | – |
| 19 | 42 | 41,164 | 40,503 | 50,629 | 60,755 | 81 | – | – |
| 20 | 42 | 55,621 | 40,503 | 50,629 | 60,755 | 110 | – | – |
| 21 | 42 | 49,554 | 40,503 | 50,629 | 60,755 | 98 | – | – |
| 22 | 42 | 44,230 | 40,503 | 50,629 | 60,755 | 87 | – | – |
| 23 | 42 | 58,548 | 40,503 | 50,629 | 60,755 | 116 | – | – |
| 24 | 42 | 39,699 | 40,503 | 50,629 | 60,755 | 78 | 804 | – |
| 25 | 43 | 45,137 | 45,414 | 56,768 | 68,121 | 80 | 277 | – |
| 26 | 43 | 45,294 | 45,414 | 56,768 | 68,121 | 80 | 120 | – |
| 27 | 43 | 63,014 | 45,414 | 56,768 | 68,121 | 111 | – | – |
| 28 | 43 | 61,366 | 45,414 | 56,768 | 68,121 | 108 | – | – |
| 29 | 43 | 46,962 | 45,414 | 56,768 | 68,121 | 83 | – | – |
| 30 | 43 | 56,621 | 45,414 | 56,768 | 68,121 | 100 | – | – |
| 31 | 43 | 52,560 | 45,414 | 56,768 | 68,121 | 93 | – | – |
| 32 | 43 | 51,940 | 45,414 | 56,768 | 68,121 | 91 | – | – |
| 33 | 43 | 62,143 | 45,414 | 56,768 | 68,121 | 109 | – | – |
| 34 | 43 | 62,517 | 45,414 | 56,768 | 68,121 | 110 | – | – |
| 35 | 43 | 54,481 | 45,414 | 56,768 | 68,121 | 96 | – | – |
| 36 | 44 | 69,606 | 50,920 | 63,650 | 76,380 | 109 | – | – |
| 37 | 44 | 69,485 | 50,920 | 63,650 | 76,380 | 109 | – | – |
| 38 | 44 | 65,384 | 50,920 | 63,650 | 76,380 | 103 | – | – |
| 39 | 44 | 58,981 | 50,920 | 63,650 | 76,380 | 93 | – | – |
| 40 | 44 | 61,168 | 50,920 | 63,650 | 76,380 | 96 | – | – |
| 41 | 44 | 59,550 | 50,920 | 63,650 | 76,380 | 94 | – | – |
| 42 | 44 | 68,277 | 50,920 | 63,650 | 76,380 | 107 | – | – |
| 43 | 44 | 52,525 | 50,920 | 63,650 | 76,380 | 83 | – | – |
| 44 | 44 | 57,091 | 50,920 | 63,650 | 76,380 | 90 | – | – |
| 45 | 44 | 65,198 | 50,920 | 63,650 | 76,380 | 102 | – | – |
| 46 | 45 | 61,503 | 57,094 | 71,367 | 85,641 | 86 | – | – |
| 47 | 45 | 69,445 | 57,094 | 71,367 | 85,641 | 97 | – | – |

*(continued)*

**TABLE S.40**   (*CONTINUED*)

| Emp. No. | Grade | Pay | Minimum | Midpoint | Maximum | Pay as % of Mid | Dollars Under Min | Dollars Over Max |
|---|---|---|---|---|---|---|---|---|
| 48 | 45 | 64,933 | 57,094 | 71,367 | 85,641 | 91 | – | – |
| 49 | 45 | 62,706 | 57,094 | 71,367 | 85,641 | 88 | – | – |
| 50 | 45 | 74,966 | 57,094 | 71,367 | 85,641 | 105 | – | – |
| 51 | 45 | 71,783 | 57,094 | 71,367 | 85,641 | 101 | – | – |
| 52 | 45 | 66,652 | 57,094 | 71,367 | 85,641 | 93 | – | – |
| 53 | 45 | 69,482 | 57,094 | 71,367 | 85,641 | 97 | – | – |
| 54 | 45 | 59,057 | 57,094 | 71,367 | 85,641 | 83 | – | – |
| 55 | 45 | 75,953 | 57,094 | 71,367 | 85,641 | 106 | – | – |
| 56 | 46 | 81,003 | 64,016 | 80,020 | 96,024 | 101 | – | – |
| 57 | 46 | 80,001 | 64,016 | 80,020 | 96,024 | 100 | – | – |
| 58 | 46 | 81,396 | 64,016 | 80,020 | 96,024 | 102 | – | – |
| 59 | 46 | 77,193 | 64,016 | 80,020 | 96,024 | 96 | – | – |
| 60 | 46 | 73,153 | 64,016 | 80,020 | 96,024 | 91 | – | – |
| 61 | 46 | 68,169 | 64,016 | 80,020 | 96,024 | 85 | – | – |
| 62 | 46 | 72,018 | 64,016 | 80,020 | 96,024 | 90 | – | – |
| 63 | 46 | 72,993 | 64,016 | 80,020 | 96,024 | 91 | – | – |
| 64 | 47 | 88,756 | 71,778 | 89,722 | 107,667 | 99 | – | – |
| 65 | 47 | 96,272 | 71,778 | 89,722 | 107,667 | 107 | – | – |
| 66 | 47 | 96,463 | 71,778 | 89,722 | 107,667 | 108 | – | – |
| 67 | 47 | 85,399 | 71,778 | 89,722 | 107,667 | 95 | – | – |
| 68 | 47 | 79,192 | 71,778 | 89,722 | 107,667 | 88 | – | – |
| 69 | 47 | 93,717 | 71,778 | 89,722 | 107,667 | 104 | – | – |
| 70 | 47 | 84,452 | 71,778 | 89,722 | 107,667 | 94 | – | – |
| 71 | 47 | 91,343 | 71,778 | 89,722 | 107,667 | 102 | – | – |
| 72 | 47 | 92,344 | 71,778 | 89,722 | 107,667 | 103 | – | – |
| 73 | 48 | 101,977 | 80,480 | 100,600 | 120,720 | 101 | – | – |
| 74 | 48 | 102,711 | 80,480 | 100,600 | 120,720 | 102 | – | – |
| 75 | 48 | 105,316 | 80,480 | 100,600 | 120,720 | 105 | – | – |
| 76 | 48 | 107,232 | 80,480 | 100,600 | 120,720 | 107 | – | – |
| 77 | 48 | 96,502 | 80,480 | 100,600 | 120,720 | 96 | – | – |
| 78 | 48 | 89,574 | 80,480 | 100,600 | 120,720 | 89 | – | – |
| 79 | 48 | 96,360 | 80,480 | 100,600 | 120,720 | 96 | – | – |
| 80 | 49 | 105,383 | 90,238 | 112,798 | 135,357 | 93 | – | – |
| 81 | 49 | 109,437 | 90,238 | 112,798 | 135,357 | 97 | – | – |
| 82 | 49 | 119,611 | 90,238 | 112,798 | 135,357 | 106 | – | – |
| 83 | 50 | 127,978 | 101,179 | 126,474 | 151,769 | 101 | – | – |
| 84 | 50 | 125,578 | 101,179 | 126,474 | 151,769 | 99 | – | – |
| 85 | 51 | 130,733 | 113,446 | 141,808 | 170,169 | 92 | – | – |

**TABLE S.41    MARKET ANALYSIS**

Market Analysis for Gillette Mine: Comparison of Employee Pay with the Market-Based Salary Structure

| Grade | No. of Employees | Total Pay | Total Market Midpoint | Difference | % from Mkt Mid | % to Meet Mkt Mid | No. of Employees Under Min | No. of Employees Over Max | Dollars Under Min | Dollars Over Max |
|---|---|---|---|---|---|---|---|---|---|---|
| 41 | 9 | 379,569 | 406,386 | 26,817 | −6.6 | 7.1 | 3 | 0 | 6,534 | 0 |
| 42 | 15 | 717,661 | 759,435 | 41,774 | −5.5 | 5.8 | 2 | 0 | 929 | 0 |
| 43 | 11 | 602,035 | 624,448 | 22,413 | −3.6 | 3.7 | 2 | 0 | 397 | 0 |
| 44 | 10 | 627,264 | 636,500 | 9,236 | −1.5 | 1.5 | 0 | 0 | 0 | 0 |
| 45 | 10 | 676,480 | 713,670 | 37,190 | −5.2 | 5.5 | 0 | 0 | 0 | 0 |
| 46 | 8 | 605,927 | 640,160 | 34,233 | −5.3 | 5.6 | 0 | 0 | 0 | 0 |
| 47 | 9 | 807,939 | 807,498 | −441 | 0.1 | −0.1 | 0 | 0 | 0 | 0 |
| 48 | 7 | 699,672 | 704,200 | 4,528 | −0.6 | 0.6 | 0 | 0 | 0 | 0 |
| 49 | 3 | 334,430 | 338,394 | 3,964 | −1.2 | 1.2 | 0 | 0 | 0 | 0 |
| 50 | 2 | 253,556 | 252,948 | −608 | 0.2 | −0.2 | 0 | 0 | 0 | 0 |
| 51 | 1 | 130,733 | 141,808 | 11,075 | −7.8 | 8.5 | 0 | 0 | 0 | 0 |
| **Total** | **85** | **5,835,266** | **6,025,447** | **190,181** | **−3.2** | **3.3** | **7** | **0** | **7,861** | **0** |

415

| TABLE S.42 | MARKET-BASED SALARY INCREASE BUDGET RECOMMENDATION |
|---|---|
| Catch-up (or fall back) | 3.3% |
| Anticipated market movement | 4.0% |
| Pay policy— | −3.0% |
| Market-based salary increase budget recommendation | 4.3% |

13.9 See Table S.42.

13.10 Raise the market-based salary structure by 0.3%.

# CHAPTER 14 INTEGRATED MARKET MODEL: MATURITY CURVE

14.1 You age the data 4.0%.

14.2 See Table S.43.

14.3 The type of relationship is positive, maturity curve, and not perfect. See Figure S.33. You will fit a cubic polynomial to the data to see if that is appropriate.

14.4 Predicted average pay $= 41,979 + (10,741.1)(\text{YSBS}) - (348.72)(\text{YSBS}^2) + (3.7880)(\text{YSBS}^3)$

$$r^2 = 60.6\%$$
$$\text{SEE} = 13,961$$

14.5 See Figure S.34 and Table S.44.

14.6 See Table S.45 and Figure S.35.

14.7 See Table S.46 and Figure S.36.

14.8 See Table S.47.

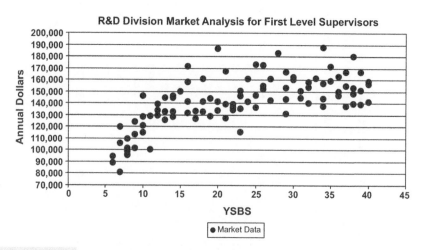

FIGURE S.33    AGED MARKET PAY AND YSBS

**TABLE S.43**    AGED MARKET PAY AND YSBS

| Inc. No. | YSBS | Market Pay | Aged Market Pay | Inc. No. | YSBS | Market Pay | Aged Market Pay | Inc. No. | YSBS | Market Pay | Aged Market Pay |
|---|---|---|---|---|---|---|---|---|---|---|---|
| 1 | 6 | 85,600 | 89,024 | 34 | 16 | 164,900 | 171,496 | 67 | 29 | 138,100 | 143,624 |
| 2 | 6 | 90,600 | 94,224 | 35 | 17 | 121,900 | 126,776 | 68 | 29 | 147,300 | 153,192 |
| 3 | 7 | 77,700 | 80,808 | 36 | 17 | 127,900 | 133,016 | 69 | 29 | 160,000 | 166,400 |
| 4 | 7 | 101,300 | 105,352 | 37 | 18 | 127,500 | 132,600 | 70 | 30 | 153,700 | 159,848 |
| 5 | 7 | 115,000 | 119,600 | 38 | 18 | 135,800 | 141,232 | 71 | 30 | 155,800 | 162,032 |
| 6 | 8 | 91,900 | 95,576 | 39 | 18 | 154,200 | 160,368 | 72 | 31 | 138,700 | 144,248 |
| 7 | 8 | 92,300 | 95,992 | 40 | 19 | 124,000 | 128,960 | 73 | 31 | 144,600 | 150,384 |
| 8 | 8 | 95,000 | 98,800 | 41 | 19 | 138,300 | 143,832 | 74 | 32 | 134,600 | 139,984 |
| 9 | 8 | 97,300 | 101,192 | 42 | 20 | 128,500 | 133,640 | 75 | 32 | 147,700 | 153,608 |
| 10 | 8 | 105,200 | 109,408 | 43 | 20 | 135,800 | 141,232 | 76 | 32 | 151,700 | 157,768 |
| 11 | 9 | 97,500 | 101,400 | 44 | 20 | 179,200 | 186,368 | 77 | 33 | 155,000 | 161,200 |
| 12 | 9 | 108,700 | 113,048 | 45 | 21 | 122,300 | 127,192 | 78 | 34 | 132,100 | 137,384 |
| 13 | 9 | 119,400 | 124,176 | 46 | 21 | 133,500 | 138,840 | 79 | 34 | 138,500 | 144,040 |
| 14 | 10 | 110,200 | 114,608 | 47 | 21 | 160,600 | 167,024 | 80 | 34 | 150,600 | 156,624 |
| 15 | 10 | 116,200 | 120,848 | 48 | 22 | 129,200 | 134,368 | 81 | 34 | 180,600 | 187,824 |
| 16 | 10 | 123,100 | 128,024 | 49 | 22 | 131,300 | 136,552 | 82 | 35 | 153,100 | 159,224 |
| 17 | 10 | 140,600 | 146,224 | 50 | 22 | 133,500 | 138,840 | 83 | 35 | 164,800 | 171,392 |
| 18 | 11 | 96,600 | 100,464 | 51 | 23 | 110,800 | 115,232 | 84 | 36 | 141,000 | 146,640 |
| 19 | 11 | 123,700 | 128,648 | 52 | 23 | 130,200 | 135,408 | 85 | 36 | 144,200 | 149,968 |
| 20 | 12 | 124,200 | 129,168 | 53 | 23 | 140,400 | 146,016 | 86 | 36 | 156,500 | 162,760 |
| 21 | 12 | 128,700 | 133,848 | 54 | 23 | 144,800 | 150,592 | 87 | 37 | 131,900 | 137,176 |
| 22 | 12 | 133,700 | 139,048 | 55 | 24 | 135,600 | 141,024 | 88 | 37 | 148,500 | 154,440 |
| 23 | 13 | 120,800 | 125,632 | 56 | 24 | 154,600 | 160,784 | 89 | 37 | 160,200 | 166,608 |
| 24 | 13 | 127,700 | 132,808 | 57 | 25 | 131,200 | 136,448 | 90 | 38 | 134,000 | 139,360 |
| 25 | 13 | 138,800 | 144,352 | 58 | 25 | 141,000 | 146,640 | 91 | 38 | 142,300 | 147,992 |
| 26 | 14 | 138,800 | 144,352 | 59 | 25 | 166,500 | 173,160 | 92 | 38 | 147,300 | 153,192 |
| 27 | 14 | 123,500 | 128,440 | 60 | 26 | 146,200 | 152,048 | 93 | 38 | 172,900 | 179,816 |
| 28 | 14 | 127,500 | 132,600 | 61 | 26 | 148,800 | 154,752 | 94 | 39 | 133,500 | 138,840 |
| 29 | 14 | 140,400 | 146,016 | 62 | 26 | 165,600 | 172,224 | 95 | 39 | 145,200 | 151,008 |
| 30 | 15 | 144,000 | 149,760 | 63 | 27 | 136,900 | 142,376 | 96 | 39 | 160,200 | 166,608 |
| 31 | 16 | 126,200 | 131,248 | 64 | 27 | 154,400 | 160,576 | 97 | 40 | 135,600 | 141,024 |
| 32 | 16 | 136,000 | 141,440 | 65 | 28 | 175,800 | 182,832 | 98 | 40 | 150,200 | 156,208 |
| 33 | 16 | 151,900 | 157,976 | 66 | 29 | 126,000 | 131,040 | 99 | 40 | 152,300 | 158,392 |

**TABLE S.44**    EVALUATION OF CUBIC MODEL

| Evaluation Criteria | Symbol | Problem Interpretation or Evaluation |
|---|---|---|
| Appearance | | Goes through middle of data and describes trend nicely |
| Coefficient of determination | $r^2$ | 60.6% of variability of survey pay associated with YSBS |
| Standard error of estimate | SEE | 13,961 "average" deviation of survey pay from predicted average survey pay |
| Common sense | | Makes sense with respect to pay levels and general shape of the relationship |

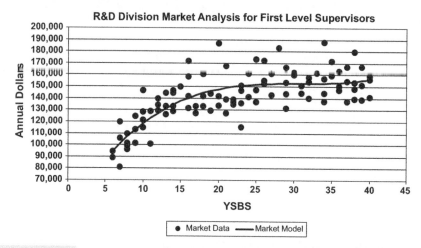

**FIGURE S.34**    PREDICTED AVERAGE MARKET PAY AND YSBS USING CUBIC MODEL

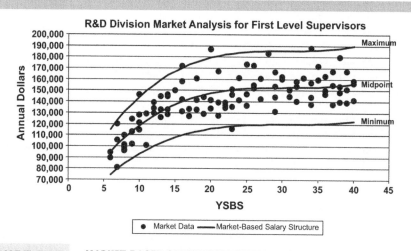

**FIGURE S.35**    MARKET-BASED SALARY STRUCTURE

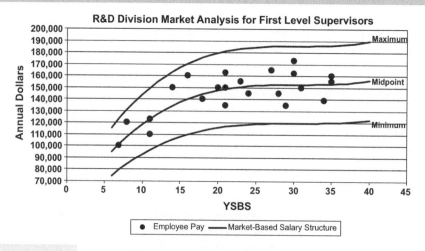

**FIGURE S.36**    EMPLOYEE PAY AND MARKET-BASED SALARY STRUCTURE

## TABLE S.45  MARKET-BASED SALARY STRUCTURE

| YSBS | Minimum | Midpoint | Maximum | YSBS | Minimum | Midpoint | Maximum |
|---|---|---|---|---|---|---|---|
| 6 | 74,267 | 94,690 | 115,114 | 23 | 118,149 | 150,640 | 183,131 |
| 7 | 79,513 | 101,379 | 123,245 | 24 | 118,642 | 151,268 | 183,895 |
| 8 | 84,336 | 107,529 | 130,721 | 25 | 119,015 | 151,744 | 184,473 |
| 9 | 88,756 | 113,164 | 137,572 | 26 | 119,287 | 152,091 | 184,895 |
| 10 | 92,789 | 118,306 | 143,823 | 27 | 119,475 | 152,331 | 185,186 |
| 11 | 96,453 | 122,978 | 149,502 | 28 | 119,598 | 152,487 | 185,377 |
| 12 | 99,766 | 127,202 | 154,637 | 29 | 119,673 | 152,583 | 185,493 |
| 13 | 102,747 | 131,002 | 159,258 | 30 | 119,718 | 152,640 | 185,563 |
| 14 | 105,412 | 134,400 | 163,389 | 31 | 119,750 | 152,681 | 185,613 |
| 15 | 107,779 | 137,418 | 167,057 | 32 | 119,788 | 152,730 | 185,671 |
| 16 | 109,867 | 140,080 | 170,294 | 33 | 119,850 | 152,809 | 185,768 |
| 17 | 111,693 | 142,408 | 173,124 | 34 | 119,953 | 152,940 | 185,927 |
| 18 | 113,275 | 144,425 | 175,576 | 35 | 120,115 | 153,146 | 186,178 |
| 19 | 114,631 | 146,154 | 177,678 | 36 | 120,353 | 153,450 | 186,547 |
| 20 | 115,778 | 147,617 | 179,456 | 38 | 121,133 | 154,444 | 187,756 |
| 21 | 116,735 | 148,837 | 180,939 | 39 | 121,709 | 155,179 | 188,649 |
| 22 | 117,519 | 149,837 | 182,154 | 40 | 122,434 | 156,103 | 189,773 |

## TABLE S.46  EMPLOYEE PAY AND MARKET-BASED SALARY STRUCTURE

| Emp. No. | YSBS | Pay | Minimum | Midpoint | Maximum | Pay as % of Mid | Dollars Under Min | Dollars Over Max |
|---|---|---|---|---|---|---|---|---|
| 1 | 7 | 100,000 | 79,513 | 101,379 | 123,245 | 99 | – | – |
| 2 | 8 | 120,000 | 84,336 | 107,529 | 130,721 | 112 | – | – |
| 3 | 11 | 110,000 | 96,453 | 122,978 | 149,502 | 89 | – | – |
| 4 | 11 | 123,000 | 96,453 | 122,978 | 149,502 | 100 | – | – |
| 5 | 14 | 150,000 | 105,412 | 134,400 | 163,389 | 112 | – | – |
| 6 | 16 | 160,000 | 109,867 | 140,080 | 170,294 | 114 | – | – |
| 7 | 18 | 140,000 | 113,275 | 144,425 | 175,576 | 97 | – | – |
| 8 | 20 | 150,000 | 115,778 | 147,617 | 179,456 | 102 | – | – |
| 9 | 21 | 135,000 | 116,735 | 148,837 | 180,939 | 91 | – | – |
| 10 | 21 | 150,000 | 116,735 | 148,837 | 180,939 | 101 | – | – |
| 11 | 21 | 163,000 | 116,735 | 148,837 | 180,939 | 110 | – | – |
| 12 | 23 | 155,000 | 118,149 | 150,640 | 183,131 | 103 | – | – |
| 13 | 24 | 145,000 | 118,642 | 151,268 | 183,895 | 96 | – | – |
| 14 | 27 | 165,000 | 119,475 | 152,331 | 185,186 | 108 | – | – |
| 15 | 28 | 145,000 | 119,598 | 152,487 | 185,377 | 95 | – | – |
| 16 | 29 | 135,000 | 119,673 | 152,583 | 185,493 | 88 | – | – |
| 17 | 30 | 162,000 | 119,718 | 152,640 | 185,563 | 106 | – | – |
| 18 | 30 | 173,000 | 119,718 | 152,640 | 185,563 | 113 | – | – |
| 19 | 31 | 150,000 | 119,750 | 152,681 | 185,613 | 98 | – | – |
| 20 | 34 | 139,000 | 119,953 | 152,940 | 185,927 | 91 | – | – |
| 21 | 35 | 155,000 | 120,115 | 153,146 | 186,178 | 101 | – | – |
| 22 | 35 | 160,000 | 120,115 | 153,146 | 186,178 | 104 | – | – |

**TABLE S.47    MARKET ANALYSIS**

Market Analysis of First-Level R&D Supervisors: Comparison of Employee Pay with the Market-Based Salary Structure

| YSBS | No. of Employees | Total Pay | Total Market Midpoint | Difference | % from Mkt Mid | % to Meet Mkt Mid | No. of Employees Under Min | No. of Employees Over Max | Dollars Under Min | Dollars Over Max |
|---|---|---|---|---|---|---|---|---|---|---|
| 7 | 1 | 100,000 | 101,379 | 1,379 | -1.4 | 1.4 | 0 | 0 | 0 | 0 |
| 8 | 1 | 120,000 | 107,529 | -12,471 | 11.6 | -10.4 | 0 | 0 | 0 | 0 |
| 11 | 2 | 233,000 | 245,956 | 12,956 | -5.3 | 5.6 | 0 | 0 | 0 | 0 |
| 14 | 1 | 150,000 | 134,400 | -15,600 | 11.6 | -10.4 | 0 | 0 | 0 | 0 |
| 16 | 1 | 160,000 | 140,080 | -19,920 | 14.2 | -12.5 | 0 | 0 | 0 | 0 |
| 18 | 1 | 140,000 | 144,425 | 4,425 | -3.1 | 3.2 | 0 | 0 | 0 | 0 |
| 20 | 1 | 150,000 | 147,617 | -2,383 | 1.6 | -1.6 | 0 | 0 | 0 | 0 |
| 21 | 3 | 448,000 | 446,511 | -1,489 | 0.3 | -0.3 | 0 | 0 | 0 | 0 |
| 23 | 1 | 155,000 | 150,640 | -4,360 | 2.9 | -2.8 | 0 | 0 | 0 | 0 |
| 24 | 1 | 145,000 | 151,268 | 6,268 | -4.1 | 4.3 | 0 | 0 | 0 | 0 |
| 27 | 1 | 165,000 | 152,331 | -12,669 | 8.3 | -7.7 | 0 | 0 | 0 | 0 |
| 28 | 1 | 145,000 | 152,487 | 7,487 | -4.9 | 5.2 | 0 | 0 | 0 | 0 |
| 29 | 1 | 135,000 | 152,583 | 17,583 | -11.5 | 13.0 | 0 | 0 | 0 | 0 |
| 30 | 2 | 335,000 | 305,280 | -29,720 | 9.7 | -8.9 | 0 | 0 | 0 | 0 |
| 31 | 1 | 150,000 | 152,681 | 2,681 | -1.8 | 1.8 | 0 | 0 | 0 | 0 |
| 34 | 1 | 139,000 | 152,940 | 13,940 | -9.1 | 10.0 | 0 | 0 | 0 | 0 |
| 35 | 2 | 315,000 | 306,292 | -8,708 | 2.8 | -2.8 | 0 | 0 | 0 | 0 |
| **Total** | **22** | **3,185,000** | **3,144,399** | **-40,601** | **1.3** | **-1.3** | **0** | **0** | **0** | **0** |

**TABLE S.48    MARKET-BASED SALARY INCREASE BUDGET RECOMMENDATION**

| | |
|---|---|
| Catch-up (or fall back) | −1.3% |
| Anticipated market movement | 3.0% |
| Pay policy | 2.0% |
| Market-based salary increase budget recommendation | 3.7% |

14.9 See Table S.48.

14.10 Raise the market–based salary structure by 5.3%.

# CHAPTER 15 JOB PRICING MARKET MODEL: GROUP OF JOBS

15.1 You age the data 4.0%.

15.2 See Table S.49.

15.3 See Figure S.37.

15.4 The relationship is linear, positive, and not perfect. It does not matter what kind of relationship there is because no regression will be done.

15.5 See Table S.50.

15.6 See Figure S.38.

15.7 With a 50% range spread, a dot above the top line (the maximum of the salary structure) means that the employee's pay is 20% higher than the market pay, as measured by the median.

15.8 See Table S.51.

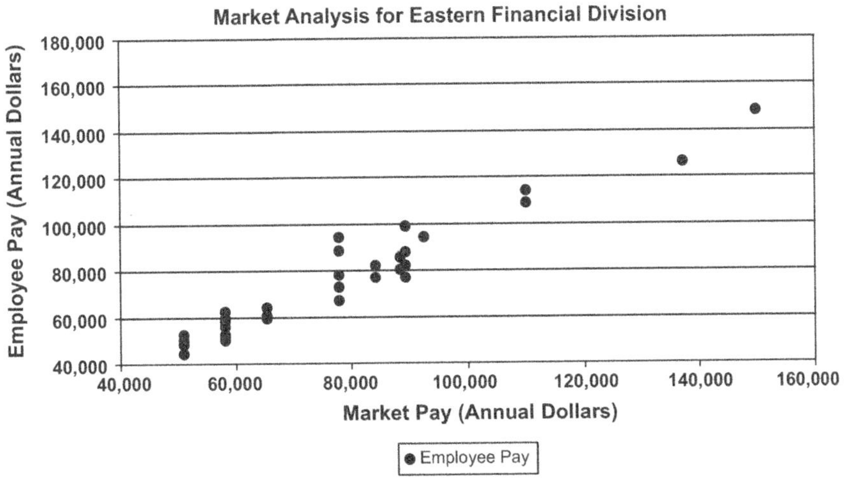

**FIGURE S.37    EMPLOYEE PAY AND MARKET PAY**

TABLE S.49    EMPLOYEE PAY AND AGED MARKET PAY

| Emp. No. | Family | Job Title | Survey Median | Aged Survey Median | Employee Pay |
|---|---|---|---|---|---|
| 1 | Accounting | Accountant | 49,000 | 50,960 | 43,800 |
| 2 | Accounting | Accountant | 49,000 | 50,960 | 47,900 |
| 3 | Accounting | Accountant | 49,000 | 50,960 | 50,100 |
| 4 | Accounting | Accountant | 49,000 | 50,960 | 52,300 |
| 5 | Accounting | Sr. Accountant | 56,000 | 58,240 | 50,100 |
| 6 | Accounting | Sr. Accountant | 56,000 | 58,240 | 51,200 |
| 7 | Accounting | Sr. Accountant | 56,000 | 58,240 | 52,300 |
| 8 | Accounting | Sr. Accountant | 56,000 | 58,240 | 55,600 |
| 9 | Accounting | Sr. Accountant | 56,000 | 58,240 | 57,800 |
| 10 | Accounting | Sr. Accountant | 56,000 | 58,240 | 58,900 |
| 11 | Accounting | Sr. Accountant | 56,000 | 58,240 | 62,000 |
| 12 | Accounting | Staff Accountant | 75,000 | 78,000 | 66,500 |
| 13 | Accounting | Staff Accountant | 75,000 | 78,000 | 72,700 |
| 14 | Accounting | Staff Accountant | 75,000 | 78,000 | 77,500 |
| 15 | Accounting | Staff Accountant | 75,000 | 78,000 | 88,500 |
| 16 | Accounting | Staff Accountant | 75,000 | 78,000 | 94,000 |
| 17 | Financial Analysis | Financial Analyst | 63,000 | 65,520 | 60,600 |
| 18 | Financial Analysis | Financial Analyst | 63,000 | 65,520 | 63,800 |
| 19 | Financial Analysis | Financial Analyst | 63,000 | 65,520 | 58,900 |
| 20 | Financial Analysis | Financial Analyst | 63,000 | 65,520 | 60,000 |
| 21 | Financial Analysis | Sr. Financial Analyst | 86,000 | 89,440 | 76,500 |
| 22 | Financial Analysis | Sr. Financial Analyst | 86,000 | 89,440 | 82,000 |
| 23 | Financial Analysis | Sr. Financial Analyst | 86,000 | 89,440 | 87,500 |
| 24 | Financial Analysis | Sr. Financial Analyst | 86,000 | 89,440 | 98,500 |
| 25 | Financial Analysis | Staff Financial Analyst | 106,000 | 110,240 | 108,400 |
| 26 | Financial Analysis | Staff Financial Analyst | 106,000 | 110,240 | 113,900 |
| 27 | Mgmt/Supv | Supv Accts Payable | 81,000 | 84,240 | 76,500 |
| 28 | Mgmt/Supv | Supv Accts Payable | 81,000 | 84,240 | 82,000 |
| 29 | Mgmt/Supv | Supv Accts Receivable | 85,000 | 88,400 | 79,800 |
| 30 | Mgmt/Supv | Supv Accts Receivable | 85,000 | 88,400 | 85,300 |
| 31 | Mgmt/Supv | Supv Payroll | 89,000 | 92,560 | 94,100 |
| 32 | Mgmt/Supv | Manager Accounting | 132,000 | 137,280 | 126,000 |
| 33 | Mgmt/Supv | Manager Financial Analysis | 144,000 | 149,760 | 148,000 |

**TABLE S.50**   EMPLOYEE PAY AND MARKET-BASED SALARY STRUCTURE

| Emp. No. | Employee Pay | Minimum | Midpoint | Maximum | Pay as % of Mid | Dollars Under Min | Dollars Above Max |
|---|---|---|---|---|---|---|---|
| 1 | 43,800 | 40,768 | 50,960 | 61,152 | 86 | | |
| 2 | 47,900 | 40,768 | 50,960 | 61,152 | 94 | | |
| 3 | 50,100 | 40,768 | 50,960 | 61,152 | 98 | | |
| 4 | 52,300 | 40,768 | 50,960 | 61,152 | 103 | | |
| 5 | 50,100 | 46,592 | 58,240 | 69,888 | 86 | | |
| 6 | 51,200 | 46,592 | 58,240 | 69,888 | 88 | | |
| 7 | 52,300 | 46,592 | 58,240 | 69,888 | 90 | | |
| 8 | 55,600 | 46,592 | 58,240 | 69,888 | 95 | | |
| 9 | 57,800 | 46,592 | 58,240 | 69,888 | 99 | | |
| 10 | 58,900 | 46,592 | 58,240 | 69,888 | 101 | | |
| 11 | 62,000 | 46,592 | 58,240 | 69,888 | 106 | | |
| 12 | 66,500 | 62,400 | 78,000 | 93,600 | 85 | | |
| 13 | 72,700 | 62,400 | 78,000 | 93,600 | 93 | | |
| 14 | 77,500 | 62,400 | 78,000 | 93,600 | 99 | | |
| 15 | 88,500 | 62,400 | 78,000 | 93,600 | 113 | | |
| 16 | 94,000 | 62,400 | 78,000 | 93,600 | 121 | | 400 |
| 17 | 60,600 | 52,416 | 65,520 | 78,624 | 92 | | |
| 18 | 63,800 | 52,416 | 65,520 | 78,624 | 97 | | |
| 19 | 58,900 | 52,416 | 65,520 | 78,624 | 90 | | |
| 20 | 60,000 | 52,416 | 65,520 | 78,624 | 92 | | |
| 21 | 76,500 | 71,552 | 89,440 | 107,328 | 86 | | |
| 22 | 82,000 | 71,552 | 89,440 | 107,328 | 92 | | |
| 23 | 87,500 | 71,552 | 89,440 | 107,328 | 98 | | |
| 24 | 98,500 | 71,552 | 89,440 | 107,328 | 110 | | |
| 25 | 108,400 | 88,192 | 110,240 | 132,288 | 98 | | |
| 26 | 113,900 | 88,192 | 110,240 | 132,288 | 103 | | |
| 27 | 76,500 | 67,392 | 84,240 | 101,088 | 91 | | |
| 28 | 82,000 | 67,392 | 84,240 | 101,088 | 97 | | |
| 29 | 79,800 | 70,720 | 88,400 | 106,080 | 90 | | |
| 30 | 85,300 | 70,720 | 88,400 | 106,080 | 96 | | |
| 31 | 94,100 | 74,048 | 92,560 | 111,072 | 102 | | |
| 32 | 126,000 | 109,824 | 137,280 | 164,736 | 92 | | |
| 33 | 148,000 | 119,808 | 149,760 | 179,712 | 99 | | |

**TABLE S.51**   MARKET ANALYSIS

Market Analysis for Eastern Financial Division: Comparison of Employee Pay with the Market-Based Salary Structure

| Job | No. of Employees | Total Employee Pay | Total Mkt Midpoint | Difference | % from Mkt Mid | % to Meet Mkt Mid | No. of Employees Under Min | No. of Employees Over Max | Total Dollars Under Min | Total Dollars Over Max |
|---|---|---|---|---|---|---|---|---|---|---|
| Acct | 4 | 194,100 | 203,840 | 9,740 | −4.8 | 5.0 | 0 | 0 | 0 | 0 |
| Sr. Acct | 7 | 387,900 | 407,680 | 19,780 | −4.9 | 5.1 | 0 | 0 | 0 | 0 |
| Staff Acct | 5 | 399,200 | 390,000 | −9,200 | 2.4 | −2.3 | 0 | 1 | 0 | 400 |
| **Acctg Total** | **16** | **981,200** | **1,001,520** | **20,320** | **−2.0** | **2.1** | **0** | **1** | **0** | **400** |
| Fin Analyst | 4 | 243,300 | 262,080 | 18,780 | −7.2 | 7.7 | 0 | 0 | 0 | 0 |
| Sr. Fin Analyst | 4 | 344,500 | 357,760 | 13,260 | −3.7 | 3.8 | 0 | 0 | 0 | 0 |
| Staff Fin Analyst | 2 | 222,300 | 220,480 | −1,820 | 0.8 | −0.8 | 0 | 0 | 0 | 0 |
| **Fin Anal Total** | **10** | **810,100** | **840,320** | **30,220** | **−3.6** | **3.7** | **0** | **0** | **0** | **0** |
| Mgr Acctg | 1 | 126,000 | 137,280 | 11,280 | −8.2 | 9.0 | 0 | 0 | 0 | 0 |
| Mgr Fin Anal | 1 | 148,000 | 149,760 | 1,760 | −1.2 | 1.2 | 0 | 0 | 0 | 0 |
| Supv AP | 2 | 158,500 | 168,480 | 9,980 | −5.9 | 6.3 | 0 | 0 | 0 | 0 |
| Supv AR | 2 | 165,100 | 176,800 | 11,700 | −6.6 | 7.1 | 0 | 0 | 0 | 0 |
| Supv Payroll | 1 | 94,100 | 92,560 | −1,540 | 1.7 | −1.6 | 0 | 0 | 0 | 0 |
| **Mgt/Supv Total** | **7** | **691,700** | **724,880** | **33,180** | **−4.6** | **4.8** | **0** | **0** | **0** | **0** |
| **Grand Total** | **33** | **2,483,000** | **2,566,720** | **83,720** | **−3.3** | **3.4** | **0** | **1** | **0** | **400** |

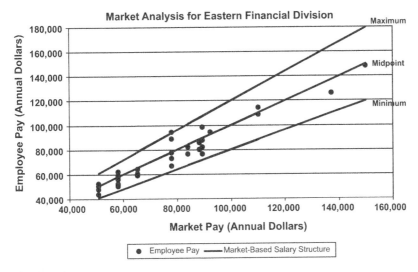

**FIGURE S.38    EMPLOYEE PAY AND MARKET-BASED SALARY STRUCTURE**

**TABLE S.52    MARKET-BASED SALARY INCREASE BUDGET RECOMMENDATION**

| | |
|---|---|
| Catch-up (or fall back) | 3.4% |
| Anticipated market movement | 3.0% |
| Pay policy | 0.0% |
| Market-based salary increase budget recommendation | 6.4% |

15.9    See Table S.52.

15.10    You adjust the market–based salary structure by 0.7%.

# CHAPTER 16 JOB PRICING MARKET MODEL: POWER MODEL

16.1    You age the data 5.0%.

16.2    See Table S.53.

16.3    See Figure S.39. The plot shows a "comet" relationship, which means that a power model may be appropriate. Also, both company revenue and president salary vary by orders of magnitude.

16.4    See Figure S.40. The relationship between the log pay and log sales is linear, positive, and not perfect.

| TABLE S.53 | AGED MARKET PAY AND COMPANY SALES | |
|---|---|---|

| Sales (€ Million) | President Pay (€ 000) | Aged President Pay (€ 000) |
|---|---|---|
| 160 | 420 | 441 |
| 180 | 240 | 252 |
| 200 | 300 | 315 |
| 220 | 650 | 683 |
| 300 | 400 | 420 |
| 500 | 800 | 840 |
| 600 | 500 | 525 |
| 800 | 1,100 | 1,155 |
| 900 | 800 | 840 |
| 1,000 | 500 | 525 |
| 1,400 | 1,800 | 1,890 |
| 2,000 | 800 | 840 |
| 3,000 | 1,500 | 1,575 |
| 5,000 | 1,000 | 1,050 |
| 7,000 | 3,800 | 3,990 |
| 9,000 | 1,200 | 1,260 |
| 10,000 | 3,300 | 3,465 |
| 17,000 | 1,400 | 1,470 |
| 20,000 | 4,500 | 4,725 |
| 35,000 | 2,100 | 2,205 |
| 50,000 | 7,000 | 7,350 |
| 87,000 | 5,000 | 5,250 |

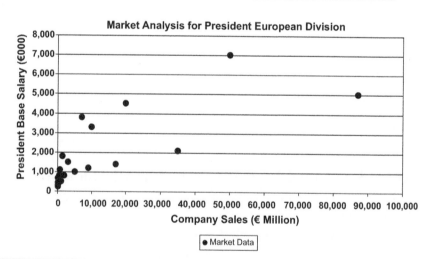

| FIGURE S.39 | AGED PRESIDENT PAY AND COMPANY SALES ON LINEAR PLOT |
|---|---|

16.5   $\log y' = 1.6345 + 0.42977 \log x$

Predicted average log president pay $= 1.6345 + 0.42977 \log$ sales

and   Coefficient of determination, $r^2 = 0.787$

Standard error of estimate (SEE) $= 0.196$

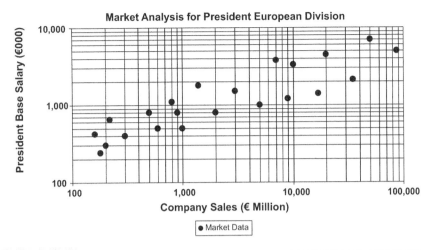

**FIGURE S.40    AGED PRESIDENT PAY AND COMPANY SALES ON LOG–LOG PLOT**

16.6   See Figure S.41 and Table S.54.

16.7   € 839,000. See Table S.55.

16.8   See Figure S.42.

16.9   The base salary is on the high side, but looking at the plot, it does not seem to be an outlier.

16.10   Other items in addition to base salary are annual bonus and long-term incentive.

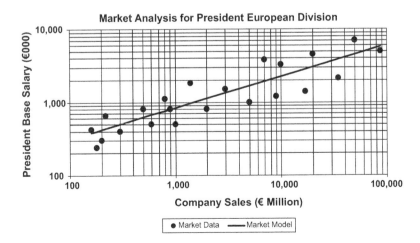

**FIGURE S.41    PREDICTED AVERAGE MARKET PAY AND COMPANY SALES**

**TABLE S.54** EVALUATION OF POWER MODEL

| Evaluation Criteria | Symbol | In Terms of Log President Pay | In Terms of President Pay |
|---|---|---|---|
| Appearance | | Goes through middle of data and describes trend nicely | |
| Coefficient of determination | $r^2$ | 78.7% of variability of log president pay is associated with log company sales | 68.3% of variability of president pay is associated with company sales |
| Standard error of estimate | SEE | 0.196 is "average" deviation of log president pay from predicted average log president pay | 57% is "average" deviation of president pay from predicted average president pay |
| Common sense | | | Makes sense based on experience |

**TABLE S.55** MARKET ANALYSIS

Market Analysis of Base Pay: President European Division

| BPD Sales (Euros) | BPD President Base Pay (Euros) | Market Market Model Predicted Average Base Pay | Difference | % BPD from Market Model | % to Meet Market Model | Average% from Market Model |
|---|---|---|---|---|---|---|
| 1 Billion | 1,500,000 | 839,000 | −661,000 | 78.8 | −44.1 | 57 |

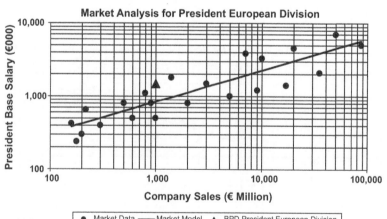

**FIGURE S.42** BPD PAY WITH PREDICTED AVERAGE MARKET PAY AND COMPANY SALES

**TABLE S.56    CORRELATION MATRIX**

|  | Pay | Grade | Time in Grade | Company Service | Attended Training |
|---|---|---|---|---|---|
| Pay | 1 |  |  |  |  |
| Grade | 0.675 | 1 |  |  |  |
| Time in Grade | 0.298 | 0.412 | 1 |  |  |
| Company Service | 0.600 | 0.816 | 0.777 | 1 |  |
| Attend Training | 0.517 | 0.142 | 0.008 | 0.109 | 1 |

# CHAPTER 17 MULTIPLE LINEAR REGRESSION

17.1   See Table S.56. You would choose grade as the first independent variable to build a model, because it has the highest correlation with pay (0.675).

17.2   There is a very high correlation of 0.816 between company service and grade, and this might cause multicollinearity problems if they are both entered into the same model. Similarly, there is a very high correlation of 0.777 between company service and time in grade, and this might cause multicollinearity problems if they are both entered into the same model.

17.3   See Figures S.43,–S.46. In each one, the relationship is positive, linear, and not perfect.

17.4   *Pay Versus Grade*
Predicted average pay $= -135{,}899 + (5{,}614)(\text{grade})$, $r^2 = 0.456$, SEE $= 4{,}068$.

The intercept has no problem interpretation because there is no grade zero. The predicted average pay increases by 5,614 for each increase in grade. Of the variability of pay, 45.6% can be attributed

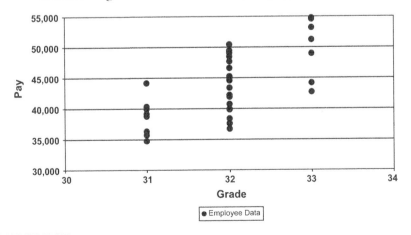

**FIGURE S.43    PAY AND GRADE**

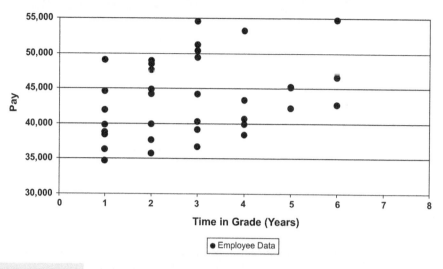

to the relationship with grade. Pay varies by about 4,068 on average from the predicted average pay.

*Pay Versus Time in Grade*

Predicted average pay $= 40{,}594 + (1{,}022)$(time in grade), $r^2 = 0.089$, SEE $= 5{,}266$.

The predicted average pay is 40,594 for no time in grade. The predicted average pay increases by 1,022 for each additional year of time in grade. Of the variability of pay, 8.9% can be attributed to the relationship with time in grade. Pay varies by about 5,266 on average from the predicted average pay.

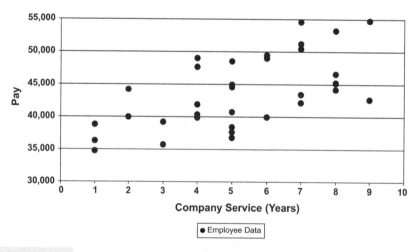

**FIGURE S.45**   **PAY AND COMPANY SERVICE**

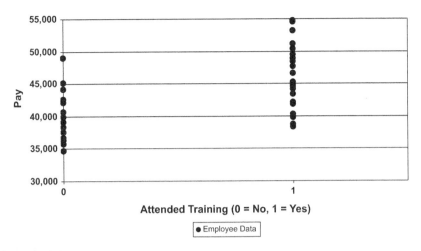

**FIGURE S.46    PAY AND ATTENDANCE AT TRAINING**

*Pay Versus Company Service*
Predicted average pay $= 35,966 + (1,440)$(company service), $r^2 = 0.360$, SEE $= 4,413$.

The predicted average pay is 35,966 for no company service. The predicted average pay increases by 1,440 for each additional year of company service. Of the variability of pay, 36.0% can be attributed to the relationship with company service. Pay varies by about 4,413 on average from the predicted average pay.

*Pay Versus Attended Training*
Predicted average pay $= 40,129 + (5,687)$(company service), $r^2 = 0.267$, SEE $= 4,721$.

The predicted average pay is 40,129 for not attended training. The predicted average pay increases by 5,687 for attended training. Of the variability of pay, 26.7% can be attributed to the relationship with attended training. Pay varies by about 4,721 on average from the predicted average pay.

17.5  Predicted average pay $= -122,530 + (5,016)$(grade) $+ (4,725)$(attended training), $r^2 = 0.637$, SEE $= 3,373$.

The intercept has no problem interpretation because there is no grade zero. The predicted average pay increases 5,016 for each increase in grade, taking into account attended training. The predicted average pay increases by 4,725 for attended training, taking into account grade. Of the variability of pay, 63.7% can be attributed to the relationship with the linear combination of grade and attended training. Pay varies by about 3,373 on average from the predicted average pay.

17.6   Yes. Keep both grade and attended training in the model. The coefficient of determination for the model with grade only is 45.6%, and increased to 63.7% with the addition of attended training, an increase of 18.1%. Similarly, the coefficient of determination with attended training only is 26.7%, increasing to 63.7% with the addition of grade, an increase of 37.0%. In either case from whichever direction you are coming from, the increase of the second variable was much more than 5.0%, and, using the decision rule in Chapter 17 , would indicate that the second variable should be kept.

17.7   Predicted average pay $= -117,469 + (4,931)(\text{grade}) + (172)(\text{time in grade}) + (4,754)(\text{attended training})$, $r^2 = 0.639$, SEE $= 3,416$.

The intercept has no problem interpretation because there is no grade zero. The predicted average pay increases 4,931 for each additional grade, taking into account time in grade and attended training. The predicted average pay increases by 172 for each additional year of time in grade, taking into account grade and attended training. The predicted average pay increases by 4,754 for attended training, taking into account grade and time in grade. Of the variability of pay, 63.9% can be attributed to the relationship with the linear combination of grade, time in grade, and attended training. Pay varies by 3,416 on average from the predicted average pay.

17.8   No, you do not keep time in grade in this new model, because the coefficient of determination increased from 63.7% in the previous model to 63.9% in this new model. According to the decision rule in Chapter 17 , since the increase is less than 1.0%, we will not keep this additional variable.

17.9   Predicted average $z$-score of pay $= (0.593)(z\text{-score of grade}) + (0.050)(z\text{-score of time in grade}) + (0.432)(z\text{-score of attended training})$.

17.10   Using the perspective of comparing beta-weights, grade has the most influence. Attended training has about 73% as much influence as grade, and time in grade has only about 8% as much influence as grade. The low influence of time in grade is supporting evidence that it should not be included in the model.

# Index

Please look in the Glossary for additional information on items listed in this index as well as for items you don't find here.

*Statistics for Compensation: A Practical Guide to Compensation Analysis,* By John H. Davis
Copyright © 2011 John Wiley & Sons Inc.